About Island Press

Island Press is the only nonprofit organization in the United States whose principal purpose is the publication of books on environmental issues and natural resource management. We provide solutions-oriented information to professionals, public officials, business and community leaders, and concerned citizens who are shaping responses to environmental problems.

In 1998, Island Press celebrates its fourteenth anniversary as the leading provider of timely and practical books that take a multidisciplinary approach to critical environmental concerns. Our growing list of titles reflects our commitment to bringing the best of an expanding body of literature to the environmental community throughout North America and the world.

Support for Island Press is provided by The Jenifer Altman Foundation, The Bullitt Foundation, The Mary Flagler Cary Charitable Trust, The Nathan Cummings Foundation, The Geraldine R. Dodge Foundation, The Charles Engelhard Foundation, The Ford Foundation, The Vira I. Heinz Endowment, The W. Alton Jones Foundation, The John D. and Catherine T. MacArthur Foundation, The Andrew W. Mellon Foundation, The Charles Stewart Mott Foundation, The Curtis and Edith Munson Foundation, The National Fish and Wildlife Foundation, The National Science Foundation, The New-Land Foundation, The David and Lucile Packard Foundation, The Surdna Foundation, The Winslow Foundation, The Pew Charitable Trusts, and individual donors.

To my parents, Beth and Donen Gleick

THE WORLD'S WATER
1998–1999

The Biennial Report on Freshwater Resources

Peter H. Gleick

Pacific Institute for Studies in Development,
Environment, and Security
Oakland, California

Island Press

Washington, D.C. • Covelo, California

Gleick, Peter H.

    The world's water 1998–1999 : the biennial report on freshwater resources / Peter H. Gleick.

      p.   cm.

    Includes bibliographical references and index.

    ISBN 1-55963-592-4 (pbk. : acid-free paper)

    1. Fresh water.   2. Water-supply.   3. Water resources development.

  4. Water resources development—International cooperation.

I. Title.

TD345.G54   1998

333.91—dc21   98-24877

                                                              CIP

Contents

SEVEN   Moving Toward a Sustainable Vision for
the Earth's Fresh Water

WATER BRIEFS

DATA SECTION

Foreword

It is no accident that Earth is often referred to as "the water planet." Earth is unique at least among planets of our solar system largely because of its abundant water; in oceans, in the atmosphere, and—most crucially—as fresh water on land. Without water, life as we know it could not exist. Try to imagine a day totally without water—water to drink, to bathe in, to wash things. It's almost impossible, unless you were stranded in the midst of an extreme desert. And that situation would be not merely miserable, but life-threatening. Of all the resources and components of our environment, water is probably the one most taken for granted, with the possible exception of the oxygen in the atmosphere. Yet it is among the most essential.

Abundant though water is on Earth, the amount of fresh water available on land surfaces is a tiny fraction of the total amount in the world. The vast majority is in the oceans; but because of the salts in ocean water, it is largely unusable for people and all terrestrial organisms. Water is continuously cycled between the planet's surface and atmosphere through evaporation and precipitation, powered by the sun. It is the fresh water that falls on land as rain or snow, or that has been accumulated and stored over eons as groundwater, that people depend on to meet most human needs: domestic, industrial, and agricultural. That supply, although replenished daily (but not evenly distributed in time or space), is both limited and vulnerable to human actions and abuse. Freshwater resources around the world have been overused, polluted, fought over, and squandered with little regard for the human and ecological consequences.

Humanity is beginning to pay the price for that abuse, and the cost is mounting steeply. Water shortages are becoming more frequent and widespread as populations grow and each person's demand for water rises as societies modernize. The number of people suffering from water-related diseases is also rising; rates of extinction of numerous organisms dependent on freshwater ecosystems are unprecedented as the water they require is taken away or contaminated; and people are increasingly in conflict over water that must be shared with neighbors. Furthermore, global climate change is rewriting everything we thought we knew about how the planet's hydrologic processes work.

As is often the case in environmental science, what we don't know about water grossly exceeds what we do know. We don't really know how much freshwater is reliably available on a renewable basis, either globally or locally. We are only beginning to understand the full impacts of actions such as damming rivers and creating temporary lakes, irrigating cropland over the long term, draining and filling wetlands, or overdrawing groundwater. Indeed, even our understanding of how all societies use water, how much they use, and what humanity's actual needs for water are is sketchy at best. But the proliferation of water-related problems, from shortages to pollution, indicates that old ways of thinking about water development are no longer appropriate to the kind of world we live in. Humanity cannot go on building huge mega-dams and appropriating more and more of our small planet's fresh water exclusively for its own use without disrupting societies, threatening their security, and doing serious damage to our life-supporting ecosystems.

Into this huge knowledge gap steps Peter Gleick with *The World's Water*. This book, planned to be a regularly updated volume, offers a comprehensive look at the crucial water problems facing humanity and the natural world. It introduces the lay reader, as well as the water expert, to the state of the world's freshwater resources, to recent water problems, and to the progress being made to solve them. Everything you ever wanted to know about water can be found here: detailed information and observations on freshwater problems that cannot be found in any other single source. Besides basic information on water, the book includes discussion of such diverse topics as water and human health, the benefits and costs of large dams, conflicting demands for water use, international conflict over water resources, new scientific theories about the origin of the planet's water, the potential impacts of climate change on water supplies, the status of human efforts to manage water, and the development of new water institutions.

The author, Peter Gleick, is a scientist with a strong background in hydrology and climatology who approaches these problems from a legal, policy, and institutional perspective as well. He pioneered studies on the implications of climate change for water resources and traced the links between water supply and international security and conflict. Known around the world as an authority on water issues, Gleick recently has estimated and defined a basic water requirement for each individual in a modern society. Armed with this information, the world's leaders may at last be able to come to grips with water-supply problems, which sadly have been so long ignored. By consistently asking the right questions and working to answer them, Gleick has performed an invaluable service to civilization and the natural systems that sustain it. But much remains to be done to assure the availability of clean fresh water for people everywhere, especially in the face of growing populations. *The World's Water* is an invaluable contribution that will greatly assist the work of development specialists and decision makers in this vital task.

ANNE H. EHRLICH AND PAUL R. EHRLICH

Acknowledgments

The world is filled with many hard-working people who have dedicated their lives to trying to solve problems related to fresh water. This book would not have been possible without their assistance, hard work, and contributions, and they all deserve to be acknowledged. Special thanks must go to Anna Steding, a research associate of the Pacific Institute whose patience, good will, and consistent efforts on behalf of this project were exceeded only by her intellectual and analytical skills. Many people thought this book would be a good idea, but Steven Nightingale kept pushing the idea until I found it irresistible, and then Todd Baldwin at Island Press found it irresistible as well. Thanks to them for their foresight and support.

Others who helped are identified in the references cited throughout the book. Some, however, were particularly useful in finding specific pieces of information, obscure or hidden data, or the latest information on the topics found within this book. Several people reviewed chapters and offered suggestions and corrections, though any remaining errors are, of course, my responsibility. I extend particular thanks to Len Abrams, Tony Broccoli, Michael Cernea, Cynthia Crossen, Alan Dupont, Paul Epstein, Larry Farwell, Beth and Donen Gleick, Betsy Gleick, Jim Gleick, Ken Gray, Richard Helmer, Anita Highsmith, José Hueb, Tom Karl, Dick Knight, L. Kuppens, Deborah Moore, Patrick McCully, Maria Neira, Bikash Pandey, Sue Partridge, Howard Perlman, Sandra Postel, Lori Pottinger, Guy Preston, Jerry Delli Priscoli, Phillipe Ranque, Paul Raskin, Tom Ross, Michel Saint Lot, Helga Schmidtke, Anders Seim, Andras Szöllösi-Nagy, Larry Stephens, George van der Merwe, Klaus Wangnick, Bob Wilkinson, and Aaron Wolf.

The Pacific Institute for Studies in Development, Environment, and Security is a wonderful home for this work. Thanks to all of the Institute staff who provided support during the past year, including Arlene Wong, Santos Gomez, Jason Morrison, Pat Brenner, Yolanda Todd, Jerry Bass, Lisa Owens-Viani, and Karin Witte. Particular thanks are due to Greg Robillard who helped make the figures and maps an important part of the book.

This book would not have been possible without the financial support of several foundations that believe in what the Institute is trying to accomplish. This first edition of *The*

World's Water is partly due to their support. Thanks to the Kendall Foundation, the Joyce Mertz-Gilmore Foundation, the Horace W. Goldsmith Foundation, the Hewlett Foundation, and the Ford Foundation.

Finally, thank you Nicki, for your patience, ear, and understanding during the long process that has produced this book.

PETER H. GLEICK

Introduction

The earth is undergoing rapid and unprecedented change. Every rotation around our axis and revolution around the sun sees new and different transformations of our geophysical, social, cultural, and economic world. Some of these changes are intentional and desired. But many are unanticipated, inadvertent, and unwanted. The richer countries are becoming more integrated, computerized, and connected, while much of the population of the planet still struggles to survive on less than what some pay to carry around a cellular phone. A revolution in communication offers the promise of more freedom, higher productivity, and less misunderstanding. But billions of people remain mired in poverty; imperiled ecosystems are failing; cultural diversity is disappearing; and large-scale environmental changes are disrupting our soils, oceans, and atmosphere. No wonder the phrase *global change* has become so current—humans are truly changing the globe.

Humans now have the power to modify the earth's basic form and behavior, to destroy whole populations of living creatures, and to alter global geophysical processes once thought invulnerable to human action. We know, for example, that humanity has invented the means of destroying civilization and has even gone so far as to build the arsenal capable of doing so. We have unintentionally created a hole in the very fabric of the atmosphere that protects us from dangerous ultraviolet radiation from the sun. We are beginning to change our climate, possibly with dangerous consequences for the hydrologic cycle and many other natural systems that support us. And humans are appropriating or contaminating an increasingly large fraction of the earth's freshwater supplies.

It is not surprising that ancient civilizations considered water one of the most basic and important elements. Water touches, and is touched by, all of our daily actions, and many serious problems involving water face current and future generations. Among the most disturbing characteristics of our current situation are the following observations:

- Per-capita water demands are increasing and per-capita water availability is declining due to population growth and trends in economic development.

- Half the world's people lack basic sanitation services. More than a billion lack potable drinking water. In much of the world these numbers are rising, not falling.

- Incidences of some water-related diseases are rising; resistance to drugs is increasing; and disease ranges are expanding.

- The amount of land irrigated per capita is falling, and competition for agricultural water from cities is growing.

- More than 700 species of fish are formally acknowledged to be threatened or endangered; the ecological disasters of the Aral Sea, Lake Victoria, and other bodies of water are now in our textbooks as examples of what we do either intentionally or unintentionally to our aquatic environment.

- Political and military conflicts over shared water resources are on the rise in some regions.

- Groundwater overdraft is accelerating. Unsustainable groundwater use occurs on every continent except Antarctica.

- The scientific community has acknowledged that human interference in global climate is now evident and that the hydrologic cycle will be seriously affected in ways we are only now beginning to study.

Welcome to the first edition of *The World's Water*—a new biennial publication of the Pacific Institute for Studies in Development, Environment, and Security. The past few years have seen a resurgence of interest and worry about the state of the world's freshwater resources among the public, media, academia, and policymakers. New research programs, publications, and conferences are all manifestations of this growing concern. There remains, however, a serious lack of good information and data about water problems and solutions. I hope *The World's Water* will play a part in filling that void.

Two principal factors have led to the increased interest in fresh water—a growing understanding of the role that water plays in maintaining human and ecological well-being and the growing worry that problems of water availability and quality are threatening that well-being. *The World's Water* addresses each of these factors by identifying the critical water issues that need to be addressed and by offering updated information and data on those issues.

This first edition of *The World's Water* reviews several new and potentially exciting institutional developments in the world of water; describes progress made in defeating endemic, and sometimes epidemic, water-related diseases; reviews concerns about regional conflicts over water and the implications of global climatic change; offers an update on the ongoing debate over large dams with details about the Three Gorges Project in China and the Lesotho Highlands Project in southern Africa; and offers insights into the changing water-development paradigm as the world slowly shifts from supply-side structural developments toward a more comprehensive and integrated look at both water supply and demand. Two chronologies of water-related conflicts are also included: one from the ancient Middle East and one from A.D. 1500 to the present. This edition ends with a positive "Vision for 2050: Sustaining Our Waters"—an essay offering a glimpse at a future worth aspiring to and moving toward. The Water Briefs section highlights potentially significant new developments, including a new theory on the very origin of water on earth, as well as the complete text of the new UN Convention on Non-Navigational

Uses of International Watercourses and the recent India-Bangladesh treaty on the Ganges. A list of water-related Internet sites, where an ever-increasing amount of good information on water can be found, is also included. The Data section provides a wide range of new and updated quantitative data.

No single water publication can adequately address all of the issues of interest to water experts, students, and the general public. The next edition of *The World's Water*, to be published in the year 2000, will look at the issues of water and food security, water quality, water and natural ecosystems, and the human use of reclaimed or recycled water. It will also provide updates on water data, treaties, laws, and changes in our scientific understanding of hydrologic processes worldwide.

Finally, the world of water is changing very rapidly. While this series offers the opportunity for regular updates on important issues and data, I also invite readers to visit *www. worldwater.org* for up-to-date water-related information and other items of interest.

The Changing Water Paradigm

Water-resources development around the world has taken many different forms and directions over past millennia. Humans have long sought ways of reducing our vulnerability to irregular river flows and variable rainfall by moving, storing, and redirecting natural waters. The earliest agricultural civilizations formed in regions with reliable rainfall and perennial rivers. The first irrigation canals led to crop production in drier and drier regions and permitted longer growing seasons. As cities grew larger, water supplies had to be brought from increasingly distant sources, leading to advances in civil engineering and the hydrologic sciences. The industrial revolution and population growth of the nineteenth and twentieth centuries led to even more dramatic and extensive modifications of the hydrologic cycle, and the last few decades have witnessed unprecedented construction of massive engineering projects for flood control, water supply, hydropower, and irrigation.

As the new millennium approaches, the way we think about managing freshwater resources and human demands for water is changing again. Traditional planning approaches and a reliance on physical solutions continue to dominate, but new methods are being developed to use existing infrastructure to meet the demands of growing populations without requiring major new construction or new large-scale water transfers from one region to another. More and more water suppliers and planning agencies are beginning to shift their focus to explore efficiency improvements, implement options for managing demand, and reallocate water among users to reduce projected gaps and meet future needs. These shifts have not come easily; they have not come without strong internal opposition; they are still not universally accepted; and they may not be permanent. Nevertheless, the apparent changes in philosophy and approach are significant in their potential for addressing unresolved water problems.

These changes represent a real shift in the paradigm of human water use, and therefore, any chapter trying to present the components of such a shift must necessarily only skim the surface. Other chapters of this book address pieces of these issues in more detail, and subsequent editions will return to this theme in the future. This chapter discusses current water-management approaches and looks at the new paths being explored for planning and managing water resources. While much of the chapter focuses on changes in the United States and other developed nations, examples are also drawn from the developing world, where similar trends can be seen. The chapter evaluates the major reasons for the change in approach and discusses the applicability of these new concepts in different parts of the world.

Twentieth-Century Water-Resources Development

There have been three major drivers to the enormous expansion of water-resources infrastructure in the past century: (1) population growth; (2) industrial development; and (3) expansion of irrigated agriculture. All three factors have increased dramatically. Between 1900 and 1995, the population of the world has grown from 1,600 million to nearly 6,000 million people. Land under irrigation increased from around 50 million hectares at the turn of the century to over 250 million hectares today. These and other factors have led to a nearly sevenfold increase in freshwater withdrawals (see Figure 1.1.) Postel et al. (1996) estimate that humans already appropriate 54 percent of accessible runoff on earth and that future population growth and economic development could lead humans to use more than 70 percent of accessible runoff by 2025.

Twentieth-century water-resources planning and development have relied on projections of future populations, per-capita water demand, agricultural production, and levels of economic productivity. Because each of these variables has always been projected to rise, water needs have also always been expected to rise. As a result, traditional water planning regularly concludes that future water demands will exceed actual water supplies. The water-management problem then becomes an exercise in coming up with ways of bridging this anticipated gap. Prior to the 1980s, these exercises led planners to focus on supply-side solutions: they assumed that projected shortfalls would be met by taming more of the natural hydrologic cycle through construction of more physical infrastructure, usually reservoirs for water storage and new aqueducts and pipelines for inter-basin transfers.

Providing this infrastructure has required an enormous economic investment. In the United States alone, total capital investment for water over the past century is estimated at $400 billion (unnormalized) (Rogers 1993), mostly for large-scale engineering projects. In the arid western United States, limited availability and growing demand for water led to the construction of the Central Arizona Project, the California Central Valley and State Water projects, and dozens of the world's most massive dams on the Columbia, Colorado, and Missouri rivers. By the mid-1990s, the United States had built over 80,000 dams and reservoirs, nearly 90,000 megawatts of hydroelectric capacity, and more than 15,000 municipal wastewater treatment plants. During the same boom in construction, more than 60 percent of the inland wetlands of the United States were lost, 50 percent of stream miles were polluted to a significant extent, and many major fish runs were decimated or destroyed (Rogers 1993).

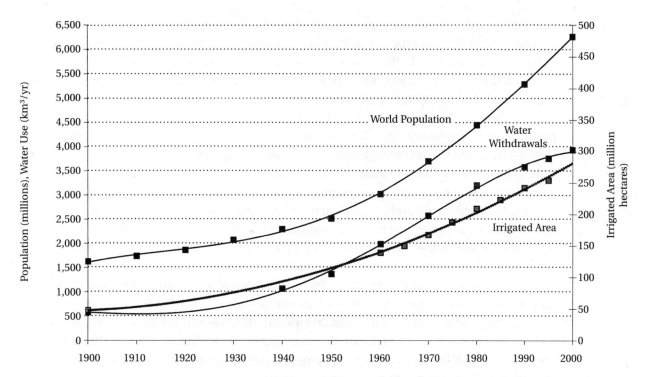

FIGURE 1.1   GLOBAL TRENDS IN WATER WITHDRAWALS, WORLD POPULATION, AND IRRIGATED AREA, 1900 TO 1995.
Shown here are global data on population (1900 to 1990), water withdrawals (1900 and 1940 to 1990), and irrigated area (1900 and 1940–1990). Estimates for the year 2000 are also included. *Sources:* Population data are from 1992 United Nations Population Reference Bureau estimates. Water use data are from Shiklomanov 1996. Irrigated area data are from United Nations Food and Agriculture Organization web site at *www.fao.org.*

Throughout the century, water developments in most countries have relied on government funds and management. In the United States, China, the former Soviet Union, and India, where huge investments have been made, only governments have had sufficient resources to build desired projects. In the United States, two government agencies, the U.S. Bureau of Reclamation (BoR) and the U.S. Army Corps of Engineers (ACoE), have shared the responsibility for implementing national water policies aimed at providing water for cities and agriculture in the arid west.

The BoR and the ACoE have been the most important actors in U.S. water policy: they built the largest water projects in the country; they control far more water than anyone else; and they sell that water at extremely low prices. The BoR alone has built more than 350 major storage reservoirs, 250 diversion dams, 25,000 kilometers of canals, 2,200 kilometers of pipeline, and 50 hydroelectric plants with an installed capacity of approximately 14,000 megawatts. Actual current value and replacement cost of these projects exceed $100 billion (Reisner and Bates 1990). BoR projects provide over 35 billion cubic meters of water every year, mostly for irrigation users. The ACoE is responsible for maintaining and improving navigation in U.S. waterways (rivers and canals) and for developing hydroelectric systems at dams on navigable rivers. Eventually, it was to play a major role in the construction of hundreds of large projects for multiple uses on the Columbia, Missouri, and other rivers. Perhaps even more important, the two agencies exported their vision of water development around the world by training foreign water engineers and advising other governments.

In addition to having deep financial pockets, there were two other reasons for early governmental involvement in U.S. water projects: the desire in the late 1800s to develop the western United States through large-scale settlement practices; and the desire of officials to bring money back to their home districts to provide jobs and industry for their constituents. In the eastern United States, most agricultural production is sustained by rainfall. As agricultural development expanded into the western United States, however, it soon became apparent that the fertile soils there would require a water supply far more reliable than the modest and inconsistent precipitation in the region could provide. Encouraging large-scale settlement of the arid west required irrigation water for agriculture, large dams to provide electricity for major industrial developments, and major aqueducts to supply cities. As western lands began to be settled, the need for major water-supply projects to deliver water to agricultural lands and urban areas led to the Federal Reclamation Act of 1902, which created the BoR (originally called the Reclamation Service). This law set in motion the system that led to government development and operation of almost all major dams and reservoirs.

The vast amounts of money required to build major water projects encouraged members of Congress to vie to bring them to their districts, leading to an atmosphere in which project after project was approved by Congress not on their merits but as quid pro quo for Congressional colleagues. This history is described in detail in Stegner (1953), Reisner (1986), Rogers (1993), and other books; popular movies (such as *Chinatown*); and documentaries (such as *Cadillac Desert/Last Oasis*) about the era.

Although the initial idea was that federal projects would pay for themselves by charging farmers and cities for the water provided, it became clear as early as the 1920s that repayment would not occur as planned. The government eased repayment obligations, institutionalizing large direct water subsidies by relieving farmers of the burden of paying any interest on capital expenditures and by extending the repayment period by many decades. It is now widely acknowledged that many of these projects will never be repaid and will represent a permanent government subsidy (U.S. Congress 1994; Myers and Kent 1997).

Twentieth-Century Water Planning: The Status Quo

In a constantly growing society with untapped and underutilized water resources, planning for growth requires planning new physical infrastructure. As a result, water agencies have always focused on engineering solutions and planned for more and more supply, just as energy planners in electric utilities did during the 1950s, 1960s, and 1970s. And just as energy planners did, water planners are suddenly coming up against physical resource constraints, economic limitations, and the accumulated environmental implications of their prior actions. Even ignoring the difficulty of projecting future populations and levels of economic activities, there are many limitations to traditional water planning. Perhaps the greatest problem is that traditional planning routinely produces scenarios with irrational conclusions, such as water demands exceeding supply and water withdrawals unconstrained by environmental, ecological, or economic limits.

California water planning is a good example. Every several years, the California Department of Water Resources (CDWR) produces an update to the official projections of California water supply and demand. These projections originated in 1957 when the first version, "The California Water Plan," was produced. At this time, the water-development philosophy was succinctly described by CDWR Director Harvey Banks, who proclaimed that "the full solution of California's water problems thus becomes essentially a financial and engineering problem" (CDWR 1957). The most recent version,[1] officially released in late 1994, is little different in the nature of its projections and proposed solutions from the plans developed over the past 35 years (CDWR 1994). According to the CDWR, which uses a variety of models and approaches for peering into the future, California water policies and problems in 2020 will be little changed from today. The state will grow the same kinds of crops on about the same amount of land. The larger urban population will improve water-use efficiency slightly, but unnecessarily large amounts of water will still go for household use and residential and municipal landscaping. Many groundwater aquifers will still be pumped faster than they are replenished. Hundreds of millions of cubic meters of treated wastewater will be dumped into the ocean rather than recycled and reused where appropriate. Water needed to maintain threatened California ecosystems and aquatic species will come and go with the rains and with human demands. And official projections of total water demands exceed available supplies by several billion cubic meters—a gap projected in every state water report since 1957. While this vision of the future may be accurate, there are more and more signs that substantial changes in direction have already begun. The assumption that the future will look like the past no longer seems appropriate or helpful as a planning tool (Gleick et al. 1995).

Major problems afflicting water planning—not just in California but in most places—include the failure to analyze the details of what water is actually used for or how much water is required to meet different types of demands. Official water-planning efforts usually make no attempt to identify common goals for water development among conflicting stakeholders or to seek agreement on principles to resolve conflicts over water. The current lack of consensus on a guiding ethic for water policy leads to fragmented decision making and incremental changes that satisfy none of the many affected parties.

The End of Twentieth-Century Water Planning

The twentieth-century water-development paradigm, driven by an ethic of growth, has now stalled as social values and political and economic conditions have changed. More people now place a high value on maintaining the integrity of water resources and the flora, fauna, and human societies that have developed around them. There are growing calls for the costs and benefits of water management and development to be distributed in a more fair and prudent manner and for unmet basic human needs to be addressed. And more and more, efforts are being made to understand and meet the diverse interests and needs of all affected stakeholders.

---

1. A new version of the California Water Plan (Bulletin 160) should be released in late 1998. The draft, released in the spring of 1998, differs little from its predecessors.

Traditional approaches to water planning, while still firmly entrenched in many water-planning institutions, *are* beginning to change. Among the factors driving these changes are high costs of construction, tight budgets, deep environmental concerns, new technological advances, and the development of innovative alternative approaches to water management. The search for new solutions is also being pushed along in some places by the changing nature of demand for water, particularly in North America and western Europe.

The Changing Nature of Demand

Throughout the first three-quarters of this century, absolute and per-capita demand for water throughout the world increased, as shown in Figure 1.2. Freshwater withdrawals increased from an estimated 580 cubic kilometers per year (km^3/yr) in 1900 to 3,580 km^3/yr in 1990 (Shiklomanov 1996). (See Box 1.1 for definitions of water use, withdrawal, and consumption.) In the United States, the world's leading industrial power, these increases were even more dramatic. In 1900, estimated water withdrawals for all purposes were 56 km^3/yr. In twenty years, this had more than doubled to over 120 km^3/yr. By 1950 total water withdrawals were 250 cubic kilometers annually. By 1970, withdrawals doubled a third time to over 500 km^3/yr. And water use in the United States peaked in 1980 at over 610 km^3/yr (U.S. Geological Survey 1993)—a tenfold increase in water withdrawals during a period when population increased by a factor of four (see Table 3 in the Data section at the end of the book and Figure 1.3). Water withdrawals were not only growing in an absolute sense, they were growing in a per-capita sense. In 1900 in the United States, annual per-capita freshwater use was less than 700 cubic meters per person (m^3/p/yr). By the middle of the century, freshwater demand had more than doubled to 1,570 m^3/p/yr. By the late 1970s and early 1980s it had increased to nearly 2,300 m^3/p/yr (see Table 3 in the Data section) (CEQ 1991; Solley et al. 1993; Perlman 1997). These increases in water demands in the United States and elsewhere, more than any other factor, drove the incredible burst of construction of water infrastructure.

Beginning in the mid-1980s and early 1990s, these trends ended in the United States and water use began to decrease despite continued increases in population and economic wealth—a radical departure from the expectations and experiences of water planners. Water withdrawals have now declined nearly 10 percent from their peak in 1980 to 554 cubic kilometers in 1995 (see Figure 1.3). The two largest components of U.S. water use, irrigation and water for thermoelectric power plant cooling, have both declined by about 10 percent. Industrial water use has dropped even more dramatically, falling nearly 40 percent from its height in 1970 as industrial water-use efficiency has improved and as the mix of U.S. industries has changed. Yet industrial output and productivity have continued to soar, clearly demonstrating the possibility of breaking the link between water use and industrial production. Urban water use, as measured by withdrawals for public supply, has continued to increase; but this use represents only 10 percent of total U.S. water withdrawals.

The decline in water use is even more dramatic when data on per-capita withdrawals are analyzed. Per-capita freshwater withdrawals peaked in 1980 and dropped more than 20 percent by 1995. Figure 1.4 shows total per-capita water use in the United States and

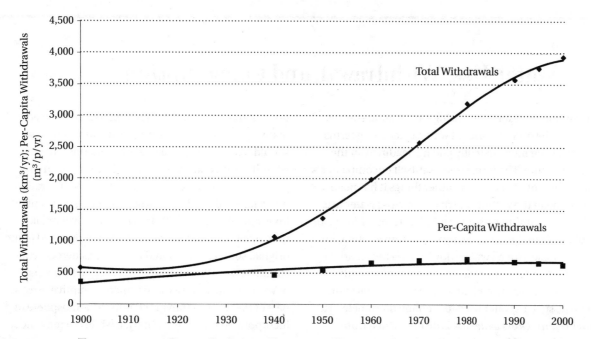

FIGURE 1.2 TOTAL AND PER-CAPITA GLOBAL WATER WITHDRAWALS, 1900 TO 2000.
Total global water withdrawals for 1900 and for various periods between 1940 and 1995 are shown here, together with per-capita withdrawals for those periods. Note the slowing of total withdrawals in recent years and the actual decline in per-capita withdrawals. *Sources:* Population data are from 1992 United Nations Population Reference Bureau estimates. Water use data are from Shiklomanov 1996.

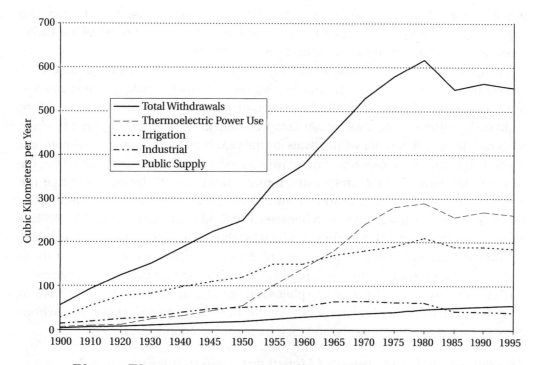

FIGURE 1.3 WATER WITHDRAWALS IN THE UNITED STATES, 1900 TO 1995.
Data on water withdrawals expressed in cubic kilometers per year are disaggregated here according to economic sector. *Sources:* Council on Environmental Quality 1991; U.S. Geological Survey 1993. Data for 1995 from H. Perlman, USGS, via FTP from 144.47.32.102.

B O X   1 . 1
Water Use, Withdrawal, and Consumption

Care should be taken when using—or reading—terms that describe different uses of water, since these terms are often used inconsistently and misleadingly in the water literature. The term *water use* itself encompasses many different ideas. Among other things it has been used to mean the withdrawal of water, gross water use, and the consumptive use of water. The term *withdrawal* should refer to the act of taking water from a source for storage or use. Not all water withdrawn is necessarily consumed. For many processes, water is often withdrawn and then returned directly to the original source after use, such as water used for cooling thermoelectric power plants. *Gross water use* is distinguished from water withdrawal by the inclusion of recirculated or reused water. Thus for many industrial processes, far more water is required than is actually withdrawn for

use, but that water may be recirculated and reused many times. Water *consumption* or *consumptive use* should refer to the use of water in a manner that prevents its immediate reuse, such as through evaporation, plant transpiration, contamination, or incorporation into a finished product. Water for cooling power plants, for example, may be withdrawn from a river or lake, used once or more than once, and then returned to the original source. This should not be considered a consumptive use. Water withdrawn for agriculture may have both consumptive and nonconsumptive uses, as part of the water is transpired into the atmosphere or incorporated into plant material, while the remainder may return to groundwater or the surface source from which it originated.

---

population growth from 1900 to 1995. Per-capita consumptive use of water has also begun to drop, falling 15 percent between 1950 and 1995. Figure 1.5 shows per-capita freshwater withdrawal and consumption from 1950 to 1995.

Though many regions continue to develop and therefore need more water, long-term projections of future global water needs have been steadily declining. Many conventional development water scenarios have been prepared over the past quarter century. Eight such projections for the year 2000 are presented in Table 1.1 and Figure 1.6, which plots estimated global water withdrawals for the year 2000 in the year the estimate was made. This figure also shows actual water withdrawals over time. As these data show, the earlier projections greatly overestimated future water demands by assuming that use would continue to grow at the historical rate (the dotted lines). Actual global withdrawals for 1995 were estimated to have been between 3,500 and 3,700 cubic kilometers (Shiklomanov 1996; Raskin et al. 1997), about half of what they were expected to be 30 years ago.

Without constantly increasing demand of 3 to 4 percent per year, the pressure to build new water infrastructure has diminished because existing supplies can be reallocated to other users. But the changing philosophy away from new development has also been driven by two other important factors: the increasing concern about the environmental impacts of water projects, and their increasing economic costs.

The Role of the Environmental Movement: 1960 to the Present

Until the late 1970s and early 1980s, water planning and management rarely took into account the environmental consequences of major water projects or the water required to maintain natural environmental resources and values. Recently, however, because of a

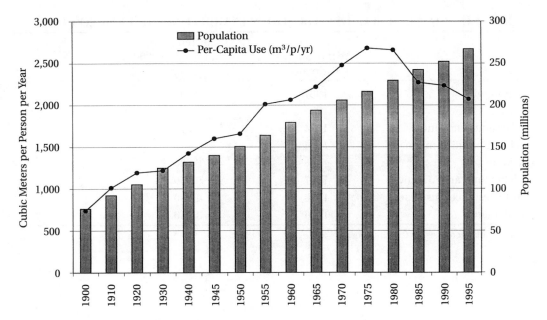

FIGURE 1.4 WATER WITHDRAWALS AND POPULATION IN THE UNITED STATES, 1900 TO 1995.
The growth of the U.S. population is shown here between 1900 and 1995 (bars), along with per-capita water withdrawals (fresh and saline) in m³/person/yr (line). Note the decline in per-capita water withdrawals beginning in 1985. *Sources:* Population data are from Population Action International, personal communication (1997). Water use data are from Council on Environmental Quality 1991; U.S. Geological Survey 1993. Data for 1995 from H. Perlman, USGS, via FTP from 144.47.32.102.

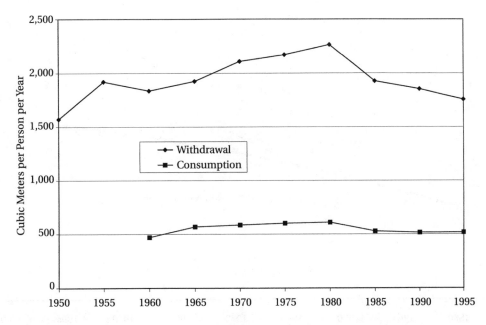

FIGURE 1.5 U.S. PER-CAPITA FRESHWATER WITHDRAWALS AND CONSUMPTION.
Per-capita freshwater withdrawals between 1950 and 1995 are compared here with per-capita water consumption in the United States between 1960 and 1995. All data are in cubic meters per person per year. Both measures show declines after 1980. *Sources:* Council on Environmental Quality 1991; U.S. Geological Survey 1993. Data for 1995 from H. Perlman, USGS, via FTP from 144.47.32.102.

TABLE 1.1   CONVENTIONAL DEVELOPMENT SCENARIOS FOR THE YEAR 2000

Study	Year of Study Estimate	Projected Water Withdrawals in 2000 (km³/yr)
Nikitopoulos	1967	6,730
L'vovich	1974	6,825
Falkenmark and Lindh	1974	6,030
Kalinin and Shiklomanov	1974	5,970
De Mare	1976	6,080
Shiklomanov and Markova	1987	5,190
World Resources Institute	1990	4,350
Shiklomanov	1996	3,940

Sources: See references for full citations.

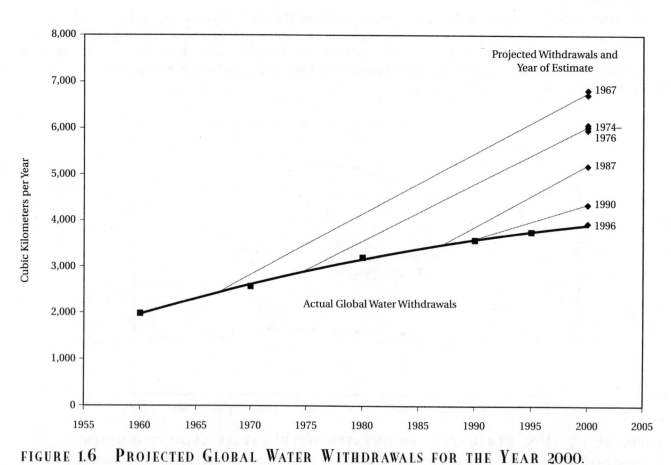

FIGURE 1.6   PROJECTED GLOBAL WATER WITHDRAWALS FOR THE YEAR 2000.

Shown here are eight estimates of global water withdrawals for the year 2000. These eight estimates were made between 1967 and 1996. The earliest estimates predicted far greater water demands than have actually materialized. See Table 1.1 for details on these estimates and their sources.

wide range of well-publicized environmental problems and changing public opinions, even people who might previously have been willing to pay the economic costs of new structures are unwilling to accept their environmental costs.

In the industrialized nations, most of the good dam sites—and many of the bad ones—have already been developed, often with major environmental sacrifices. As a result, free-flowing rivers, natural riparian systems, and many aquatic species have become increasingly rare and valued. As overall environmental awareness has increased worldwide, the desire to protect some of these remaining natural resources has grown.

Environmental concerns over rivers and watersheds in the United States began in the early 1900s with John Muir's opposition to the Hetch Hetchy dam in the mountains of California. The environmental movement in the United States was further stimulated in the 1960s by the apparent unwillingness of federal dam builders to recognize any environmental values of wild rivers and their various proposals to build several particularly large and damaging reservoirs. In one of the most astounding proposals of all, the U.S. government announced plans in the 1960s to build a series of massive dams in the Grand Canyon, one of the most important national symbols of America. These dams provoked such an enormous outcry of dismay from environmental groups—and then from the broad American public itself—that the plans were halted: the first time such a major project had been stopped. Many conservationists believe that the successful battle to stop dams, and the Grand Canyon dams in particular, led to the modern conservation movement in the United States.

In the Soviet Union in the 1970s, plans by the central government to divert the flow of Siberian rivers away from the Arctic stimulated unprecedented open opposition from the public and academic communities not normally able or willing to criticize the government. This opposition—on both environmental and economic grounds—played a major role in halting development. Even in China, where public opposition to government policies is strongly discouraged, opposition to the Three Gorges Project has been unusually vocal and persistent (see Chapter 3).

By the late 1960s and early 1970s, environmental movements in many countries were growing in strength. In the United States during this period, major laws were passed protecting water quality and major river systems, including the Clean Water Act, the Safe Drinking Water Act, the Endangered Species Act, and several others. The Federal Wild and Scenic Rivers Act of 1966 recognized the "scenic, recreational, geologic, fish and wildlife," and other natural values of pristine rivers and removes them from consideration for water projects. By 1994 nearly 18,000 kilometers of 150 rivers were protected by this law, and several other countries now have similar protections.

The environmental movement worldwide continues to develop. While there are some concerns in developing nations that "environmental" limits may simply mean constraints on economic development in those nations for the benefit of industrialized nations, there is growing grassroots opposition to big projects because of their serious local costs. Cultural and social factors have also played an important role in stimulating this opposition. In particular, many large dam projects in developing countries displace large numbers of people from areas to be flooded. Table 15 in the Data section at the back of the book lists some of the major population displacements from the construction of large dams in developing countries. In recent years, several major projects have been delayed or halted because of opposition from local groups (Cernea 1988; World Bank

1993; McCully 1996). Better communications and coordination between nongovern-mental organizations in developed and developing countries have also fueled many of these debates. Whether or not the environmental and social costs of such projects exceed the benefits to be derived from them (which is often very difficult to determine), the local opposition is very real and must be considered in any water-planning project.

Economics of Major Water Projects

Environmental concerns are not the only constraints on new water projects. Economic factors are also playing a role in changing the nature of water development. New water-supply systems have increasingly become expensive compared to nonstructural alterna-tives. When the first major dam projects were being built, it was relatively unimportant that they be economically justifiable. In the United States and elsewhere, these projects were often built as part of efforts to support a local economy, as part of broader land and agricultural settlement policies, or to bring electricity to rural communities. As these fac-tors became less urgent, new projects became harder to justify and the concept of cost/benefit analysis was introduced. The U.S. Flood Control Act of 1936 stated that pro-jects should be built only "if the benefits to whomsoever they may accrue are in excess of the estimated costs." While in principle this requirement is appropriate, no standard methods existed for doing economic assessments of large water projects. This led to a series of economic analyses done with incomplete information and questionable assumptions, which were then used to justify marginal projects. For example, all non-market environmental and social costs were simply excluded because they were unquantified or unquantifiable. Economic games were also played with stretched-out repayment periods, high discount rates, low-interest loans, and a transfer of costs to parts of water developments other than dams.

As the best and most economically viable dam sites or aqueducts have been devel-oped, each new proposed project has become less and less justifiable. The large capital costs associated with them also makes new projects harder and harder to fund. The Lesotho Highlands Project in its entirety could cost nearly US $8 billion. The Three Gorges Project in China now seems likely to cost between US $25 billion and 60 billion. The Itaipu Dam cost US $19 billion and helped Brazil dig itself into a serious financial hole (Reisner 1986). The cost of the Tehri Dam on the Bhagirathi River in India has esca-lated far beyond its original one billion dollar cost, and in late 1997 work was halted due to public protests (FTGWR 1997d). The Bakun Dam in Malaysia, one of southeast Asia's biggest and most controversial projects, has been having severe difficulties raising the necessary financing (FTGWR 1997c). The massive Nam Theun I dam in Laos would cost at least US $1.5 billion. The Sapta Koshi High Dam, a joint project between India and Nepal on the Mahakali River, could cost at least US $8 billion (FTGWR 1997b). By early 1998, the estimated cost of building the Kalabagh Dam on the Indus had soared to US $12 billion and the financing for the project is not yet arranged (FTGWR 1998). The current estimated cost of the Auburn Dam on the American River in California would be between $1 billion and 4 billion, depending on the design; it would only develop one-fiftieth of the water of the Hoover Dam, built for a fraction of the cost (Reisner and Bates 1990). These costs are simply too high even for most developed countries to support.

A second aspect of these new economic constraints is that almost all past water-infrastructure development has been subsidized or fully paid for by governments and international financial organizations. For the foreseeable future, governmental budgets in Asia and many other regions will be under increasing pressure and there are serious constraints on new money for major water projects. While this pressure has been felt in every sector of society, it plays a particularly important and direct role in changing national water policies by limiting central governmental involvement in new capital-intensive projects and devolving more responsibility to regional and local governments. Having seen large amounts of development in the past, and having borne many of the nonmonetary impacts of that development, many people are no longer willing to pay for new structures to solve water problems. Often, local people actively oppose them.

Local opposition to recently proposed major water projects has had an important secondary economic impact. In the past, international aid organizations, such as the World Bank and the Asian Development Bank, have provided a significant part of the funds necessary to build such projects. In the face of local and international opposition, however, these organizations have begun to rethink their water policies and in some cases to reduce funding for large projects. There is a strong parallel here to the situation in the United States where much of the earliest development was fully funded by the government, while current financial responsibility for new projects falls on local communities. When international aid organizations withdraw from major projects, such as the Narmada Project in India, the countries themselves must decide whether or not to go forward with funding and construction. Sometimes the costs are so enormous that developing nations are unable to proceed. The proposed Arun Himalayan project in Nepal, for example, would cost several times the entire annual budget of the country. Nepal would certainly not be able to build it on its own, and local opposition may prevent international funding.

The growing opposition to major water projects internationally has led aid organizations to consider different ways of managing water resources, including "integrated water planning." The strength of these new trends toward more integrated water planning can be seen in the latest water policy of the World Bank, which states, among other things, that

> The proposed [new World Bank] approach to managing water resources builds on the lessons of experience. At its core is the adoption of a comprehensive policy framework and the treatment of water as an economic good, combined with decentralized management and delivery structures, greater reliance on pricing, and fuller participation by stakeholders. . . . The adoption of a comprehensive framework for analyzing policies and options would help guide decisions about managing water resources in countries where significant problems exist, or are emerging, concerning the scarcity of water, the efficiency of service, the allocation of water, or environmental damage. (World Bank 1993)

As an alternative to new infrastructure, efforts are now under way to rethink international water planning and management, with some strong similarities to the changes now under way in U.S. water planning and management. In 1995, Gleick et al. proposed a definition of sustainable water use and seven principles for water managers and planners in the western United States (see Box 1.2). Similar principles are reiterated in the

---

BOX 1.2

Sustainability Criteria for Water Planning

Gleick et al. (1995) offer a working definition of sustainable water use as "the use of water that supports the ability of human society to endure and flourish into the indefinite future without undermining the integrity of the hydrological cycle or the ecological systems that depend on it." They also present seven sustainability criteria:

1. A basic water requirement will be guaranteed to all humans to maintain human health.

2. A basic water requirement will be guaranteed to restore and maintain the health of ecosystems.

3. Water quality will be maintained to meet certain minimum standards. These standards will vary depending on location and how the water is to be used.

4. Human actions will not impair the long-term renewability of freshwater stocks and flows.

5. Data on water-resources availability, use, and quality will be collected and made accessible to all parties.

6. Institutional mechanisms will be set up to prevent and resolve conflicts over water.

7. Water planning and decision making will be democratic, ensuring representation of all affected parties and fostering direct participation of affected interests.

---

new global freshwater assessment recently completed for the Commission for Sustainable Development (United Nations 1997). Many individual nations are also rethinking water policy and putting greater emphasis on development principles that reflect environmental, social, and cultural values. Among the major principles that appear to be common to all these new approaches are:

1. Environmental values must be a fundamental part of decision making and used to limit choices of sites and designs of new systems.

2. Economic principles must be applied to water use and management.

3. The use of nonstructural alternatives to meet demand must receive higher priority.

4. New supply systems, if needed, must be flexible and maximally efficient.

5. Nongovernmental organizations, individuals, independent research organizations, and other affected stakeholders must all be involved in water-management decisions.

Versions of these principles are also presented and discussed in Agenda 21 from the Earth Summit in Rio. In particular, Chapter 18 of Agenda 21 is devoted to freshwater issues (United Nations 1992). What is beginning to be seen now is a mix of large infrastructure development, nonstructural approaches, and a growing interest in exploring changes to traditional water-management and demand strategies.

Meeting Water Demands in the Next Century

Some new dams, aqueducts, and water infrastructure will certainly be built, particularly in developing countries where the basic water requirements for humans have still not been met. But even in these regions, new approaches are being developed—or old ones rediscovered—that permit water needs to be met with fewer resources, less ecological disruption, and less money. Successfully meeting human demands for water in the next century will increasingly depend on nonstructural solutions and a completely new approach to planning and management. The most important single goal of this new paradigm is to rethink water use with the objective of increasing the productive use of water. Two approaches are needed: increasing the efficiency with which current needs are met; and increasing the efficiency with which water is allocated among different uses. Major new projects will also compete with innovative small-scale approaches, including micro-dams, run-of-river hydro, land management and protection methods, and other locally managed solutions. In addition to this, nontraditional sources of supply will play an increasing role, including reclaimed or recycled water and, in some limited circumstances, desalinated brackish water or seawater.

Increasing the Efficient Use of Water

A key component of nonstructural approaches to water-resources management is a focus on using water more efficiently and reallocating water among existing uses. In the mid-1970s, arguments against developing new supplies of energy began to gain favor, driven in part by concerns over nuclear power's high costs and the potential for catastrophic accidents, and over the accumulating environmental consequences of fossil-fuel combustion. During this period, a number of analysts argued that more efficient use of energy could significantly reduce future demand and delay or avoid the need for new construction (e.g., Lovins 1977). The enormous differences in energy productivity among industrialized nations were among the factors cited in this argument over whether the apparent link between increasing energy use and economic wealth (in particular GNP) could be broken. These arguments have turned out to be largely true, and the proper incentives have led to tremendous drops in energy demand, while economic well-being has continued to improve.

The same arguments are now beginning to be heard over water: the need to develop new sources of water supply can largely be avoided by implementing intelligent water conservation and demand management programs, installing new efficient equipment, and applying appropriate economic and institutional incentives to shift water among users. Improvements in water-use efficiency will come about through changes in technology, economics, and institutions (Vickers 1991).

Vast improvements in water-use efficiency are possible in almost all sectors. In developing countries, large losses occur in distribution systems, faulty or old equipment, and poorly designed or maintained irrigation systems. In Jordan, at least 30 percent of the domestic water supply never reaches users because of flaws and inadequacies in the water-supply network, and the losses reach 50 percent in Jordan's capital, Amman (Salameh and Bannayan 1993). It has been estimated that the amount of water lost in Mexico City's supply system is equal to the amount needed to supply a city the size of Rome (Falkenmark and Lindh 1993). Great improvements are possible in the industrialized world as well. In California, for example, where water use has been subject to close scrutiny for years, great potential still exists for reducing water use without sacrificing economic productivity or personal welfare. A recent assessment of California's water policy and planning concluded that future water needs could be met for the next quarter century through better application of existing technology, well-understood economic and pricing policies, and restructuring of existing institutions and water allocations

(Gleick et al. 1995). The potential for similar improvements in other regions of the world has been described by others (Morrison et al. 1996; Postel 1997; United Nations 1997). Some of these changes are already happening.

Industrial Water Use

Technological innovation will play an important role in every water-using sector from producing goods and services to growing food. In all economic activities, water demands depend on two factors: what is being produced and the efficiency with which it is produced. Total industrial water use thus depends on the mix of goods and services demanded by society and on the processes chosen to generate those goods and services. Making a ton of steel in the 1930s consumed 60 to 100 tons of water. Today that same steel can be produced with less than six tons of water. Yet producing a ton of aluminum today only requires one and a half tons of water (Gleick 1993). Replacing old steel-making technology with new can reduce water needs. Replacing steel with aluminum, as has been happening for many years in the automobile industry for other reasons, can also reduce water needs.

Where water is plentiful and inexpensive, there may be few incentives for more water-efficient technology to be installed and used. But even in many water-short or water-scarce regions, attention is increasingly being paid to the water requirements of different technological choices and significant changes in water use are already occurring. A few examples will help highlight the large variety of options available for almost any human activity that uses water and their potential to reduce demands for new water supplies when high water productivity is an important goal.

Industrial output in Japan has steadily risen since the 1970s, while total industrial water use there has dropped more than 25 percent. In 1965, Japan used nearly 48,000 cubic meters of water to produce a million dollars of industrial output; by 1989, this had dropped to 13,000 cubic meters of water per million dollars of output (in real terms)—a tripling of industrial water productivity (Postel 1997). Similar changes have occurred in California, where total industrial water use dropped 30 percent between 1980 and 1990 without any formal or intentional efforts because of natural economic and technological changes that occurred during the decade. Over the same period, total gross industrial production rose 30 percent in real terms (CDOF 1994). In 1979, on an industry-wide level, it took an average of 13,500 cubic meters of water to produce a million dollars of industrial output. By 1990, this figure had dropped to under 7,400 cubic meters. The improvement in industrial water-use efficiency was the result of two important trends: an improvement in the efficiency with which water is used in many industrial processes, and a shift in the California economy away from water-intensive industries. These changes were driven by new water-quality standards, the cost of providing and treating water, technological improvements, and a naturally changing economic structure (Gleick et al. 1995).

Between 1985 and 1990 seven major industrial groups in California (fruits and vegetables, beverages, paperboard and boxes, refining, concrete, communications, and motor vehicles) showed positive annual economic growth rates and absolute declines in annual water use. Six of these groups improved water-use efficiency more than 35 percent (see Table 1.2). Five other major industries increased their economic output at rates substantially higher than the rates at which water use increased (meat, bakery, and

TABLE 1.2 IMPROVEMENTS IN CALIFORNIA INDUSTRIAL WATER-USE EFFICIENCY, 1985 TO 1989

Standard Industrial Classification Code (SIC)	Industry Group	1989 Water Use Index (1985 = 100)
285	Paint	46
357	Computers	50
371	Vehicles	57
367	Electronic Components	56
203	Fruits and Vegetables	61
372	Aircraft	63

Source: Wade et al. 1991.

foods; metal cans; computers; computer components; and missiles/space) (Gleick et al. 1995).

The potential still exists to reduce industrial water use through cost-effectiveness means by another 20 to 30 percent or more (Gleick et al. 1995; Sweeten and Chaput 1997), and similar savings are available in the commercial and institutional sectors in California. A recent survey of over 900 commercial, industrial, and institutional users by the Metropolitan Water District of Southern California identified overall cost-effective water-savings potential averaging 29 percent (Sweeten and Chaput 1997).

Comparable industrial water savings are often possible in developing countries as well, though detailed information is often less available. Gupta et al. (1989) describe how increased water prices and government restrictions on wastewater discharges encouraged the Zuari Agro-Chemical fertilizer plant in Goa, India, to reduce total daily water use 50 percent between 1982 and 1988. Similar interventions in Tianjin, China, reduced industrial water use per unit of industrial output there by about 60 percent, and economic incentives led to improvements in industrial water-use efficiency by between 42 and 62 percent in three industrial plants in São Paulo, Brazil, in the early 1980s (Bhatia and Falkenmark 1992).

Residential Water Use

Technological changes have the potential to reduce residential water use substantially without altering lifestyle. In the United States, most toilets used about 23 liters per flush (6 gallons) prior to the early 1990s. Beginning in the late 1980s, toilets using 13.5 liters per flush (3.4 gallons) began to appear on the market, but initially there were few requirements that they be installed. Then, laws requiring installation began to be passed, starting in 1988 with Massachusetts and soon followed by 15 other states. In 1992, Congress passed the National Energy Policy Act (NEPA). A section of NEPA essentially mandated that all new residential toilets sold after January 1994 be high-efficiency, low-flow toilets requiring only 6 liters (1.6 gallons) per flush. Thus, with a single change in technology and no change in lifestyle, the water required to flush toilets decreased more than 70 percent. This law also required changes in other bathroom and kitchen fixtures. NEPA is an example of a modest national regulatory change that led to significant water savings as well as standardizing and simplifying different and often contradictory state and local requirements (Vickers 1993). And, despite a few widely reported, sometimes humorous, and usually grossly exaggerated exceptions, the new toilets and fixtures perform as well or better than those they are replacing. (See the extensive discussion of these issues at www.waterwiser.org.)

It will take time to replace millions of old toilets with more efficient ones, and there is, as yet, no federal law requiring such retrofit. A number of cities in the United States are finding the water savings of the new toilets to be so significant, and the cost of saving that water so low, that they have instituted programs to speed up the natural replacement of existing toilets with low-flow models. Many cities have rebate programs that offer residential customers money for every high-flow toilet that they replace with low-flow models. San Diego estimates that an average single-family home that installs one low-flow toilet will save about $48 per year on water and sewer bills (City of San Diego 1995a, 1995b). The city of Austin, Texas, has a toilet-rebate program that pays $40 per toilet replaced. They also offer credits for installing drought-resistant plants in gardens and rebates for upgrading an existing residential underground irrigation system with new efficient garden systems (City of Austin 1995). The Metropolitan Water District of Southern California (MWD) estimated that each low-flow toilet installed in its district saves over 170 liters per day. Together with community groups like the Mothers of East Los Angeles Santa Ines, MWD has actively worked to retrofit old toilets. In just the first four years of their programs, over 750,000 new toilets were installed (Chesnutt et al. 1994). Local community programs employ local residents, door-to-door marketing efforts, and cash rebates to households replacing inefficient toilets. Revenues generated by such community partnerships are reinvested in job creation, school programs, community activities, and promotion of inner-city businesses (Gomez and Owens-Viani 1998).

Completely replacing existing toilets in California with low-flow models over the next 25 years would reduce annual urban water use by over 1 billion cubic meters (Gleick et al. 1995). This one change saves nearly 25 percent of the total residential indoor water demand projected in 2020 by state water planners (CDWR 1994). Even in the developing world, technologies like efficient toilets have a role to play. Because of the growing difficulty of finding new sources of water to supply Mexico City, city officials launched a water conservation program. As part of this program, 350,000 toilet replacements have already saved enough water to meet the needs of 250,000 residents (Postel 1997).

Technological change is not static; it is a dynamic and ongoing process, even for a technology as mundane (and important) as toilets. A recent political dispute over the supply of water from Malaysia to Singapore raised serious worries in Singapore about vulnerability to intentional water-supply disruptions in the event of serious political disputes with their neighbors (see Chapter 4). Singapore immediately launched an intensive campaign to improve water-use efficiency, which included replacing what the United States now considers "ultra-low flow toilets" with even higher efficiency models. Pressure-assisted toilets that use under 2 liters per flush and are even more effective than current toilets, for example, are available on the market. These kinds of changes have already led to further water and economic savings in Singapore (Zachary 1997). And there are no technological reasons why toilets have to use any water at all. Some experts argue that in regions of the world where water is scarce, it makes little sense to use water for disposing of human wastes when other satisfactory alternatives exist (Kalbermatten et al. 1982; Rogers 1997).

Additional savings can result from water-efficiency improvements in other home appliances using existing technology. Eighty-six percent of all U.S. homes had automatic washing machines in 1990 (AHAM 1993). The vast majority of these machines are top-loaders, which are far less efficient than front-loading machines more commonly used in Europe and Japan. By the late 1990s, appliance manufacturers in the United States began to offer more efficient front-loading machines, touting their energy, water, and detergent savings. A recent study suggests that replacing existing top-loading machines with efficient front-loading models has the potential to reduce water used for washing clothes by between 33 and 39 percent. Overall domestic water consumption in three western U.S. cities studied could be reduced by between 5 and 7 percent (Steding et al. 1996). In a field test in Bern, Iowa, in 1997, some 100 households received a free front-loading washing machine manufactured by the Maytag Corporation. The U.S. Department of Energy and Oak Ridge National Laboratory then monitored water use and wash loads for five months and estimated that the new machines used 39 percent less water and 58 percent less energy than the machines they replaced (*U.S. Water News* 1997). While these machines currently have a higher initial cost, their life-cycle costs are lower than those of conventional washers, and it is likely that the prices will drop as competition increases among the manufacturers.

Water productivity can also be improved in outdoor gardens, municipal lawns, golf courses, and other urban landscapes. In some parts of the United States, as much as half of all residential or institutional water demand goes to water gardens and lawns. Improvements in watering efficiency could reduce that demand substantially, as could changes in the composition of these gardens. "Xeriscaping"—the use of drought-resistant plants—is being pursued in almost all major western U.S. cities. Innovative garden designs, combined with new computer controllers, moisture sensors, and water technology, can reduce outdoor water use in homes by 25 to 50 percent or more depending on homeowner preferences, the price of water, and the cost of alternatives (Gleick et al. 1995). In some regions, outdoor municipal and institutional landscape irrigation is being done with reclaimed water, completely eliminating the use of potable water for this purpose. This use of reclaimed water is described in more detail later.

Agricultural Water Use

The largest single use of water is in agriculture, and this water use is largely inefficient—water is lost as it moves through leaky pipes and unlined aqueducts, as it is distributed to farmers, and as it is applied to grow crops. Some analysts estimate that the overall efficiency of agricultural water use worldwide is only 40 percent (Postel 1997), meaning that more than half of all water diverted for agriculture never produces food. Much of this water returns to rivers or aquifers for possible reuse, but there is little doubt that vast improvements in irrigation efficiency are possible.

In water-short areas, new techniques and new technologies are already changing the face of irrigation. New sprinkler designs, such as low-energy precision application (LEPA) can increase sprinkler efficiencies from 60 to 70 percent to as high as 95 percent (Postel 1997). Drip irrigation, invented in Israel to deal with both water scarcity and salinity problems, has expanded worldwide. Over half of Israel's irrigated land is served by drip irrigation. In California, more than 400,000 hectares of crops are watered using drip systems and more and more crops are being covered by such methods. Where high-valued crops are grown in relatively permanent settings such as orchards and vineyards, drip irrigation is now the dominant irrigation method. But even for row crops, such as strawberries, asparagus, peppers, melons, tomatoes, cotton, and sugar cane, drip systems are becoming more common.

Simple improvements in delivery systems can have enormous impacts on irrigation water needs. In most developing countries, surface canal systems are grossly inefficient. Much water is either wasted or inequitably distributed, so that crop yields of users near the end of distribution systems are far below optimal levels (Postel 1997). Identifying technical and institutional ways of improving the efficiency of these systems will go a long way toward increasing agricultural production without having to develop new supplies of water.

Economics and Water Pricing

Major barriers to the more efficient use of water include inappropriate pricing policies and economic subsidies that encourage profligate use. There are growing efforts, however, to treat water as an economic good—one of the four principles adopted at the Dublin conference in 1992 (see Box 6.3, in Chapter 6). A variety of new economic and pricing approaches are now contributing to the shift in water-resources development approaches. In the past, widespread subsidies have encouraged rapid development of supply systems and hindered water-efficiency efforts. These subsidies have been very effective at accomplishing their goals, in both urban and agricultural settings. Urban centers in the western United States, Mexico City, Singapore, Beijing, and many other big cities now support very large populations where there would otherwise have been inadequate water supplies. Semi-arid deserts now produce vast amounts of food where rainfall alone would be insufficient.

But these subsidies have also been responsible for unplanned and undesired side effects. Subsidized cotton production in central Asia expanded so much that the inflows of water to the Aral Sea were cut off by irrigation demands, leading to a shrinking of the sea, the extinction of endemic species, and adverse impacts on human health. Fossil groundwater in Saudi Arabia has been overpumped to grow subsidized wheat. One estimate is that between 1980 and 1995 Saudi Arabia consumed more than 75 percent of the proven reserves of water in its major aquifers (FTGWR 1997a). Groundwater overdraft in India, encouraged by subsidized energy costs for pumping and a lack of groundwater regulation, threatens that country's agricultural self-sufficiency.

The agricultural sector has particularly benefited from water subsidies. In California, as in much of the rest of the world, about 75 percent of all water consumed goes to agriculture, yet the contribution of agriculture to the state's gross domestic product is less than 5 percent (Gleick et al. 1995). The extremely low cost of water from federal irrigation projects in the western United States encourages the production of a range of crops that are both low-valued and highly water-intensive. In California, four crops—rice, cotton, alfalfa, and irrigated pasture—consume 57 percent of agricultural water while producing only 17 percent of agricultural revenue (Gleick et al. 1995). Many of these crops receive highly subsidized water.

Because these crops use such a high volume of water, urban and environmental interests are beginning to question the continuation of these subsidies, particularly since these interests are willing and capable of paying higher prices for the same water. In the current fiscal environment, calls for a decrease in water subsidies and a reallocation of supplies among users are increasingly being heard. Even modest changes in agricultural practices would free up substantial amounts of water for urban and environmental water uses while still permitting vibrant agricultural production. In regions where maintaining agricultural production is considered important, permitting reallocation of water among different growers and crops can also lead to increases in the efficiency of use in both a technical and economic sense. Such increases in efficiency will permit either more agricultural production or the production of higher valued crops, with the same amount of water (or less).

Ultimately, however, there will be growing pressures to take water from agriculture and put it to use in other economic sectors. One million cubic meters of water in Southern California can produce 13,000 jobs in high-tech industries; the same amount of water used to grow grass for livestock creates only six jobs. Reisner (1986, 1993) estimated that enough water to meet the entire needs of the city of Los Angeles, with a population of 13 million people, was being used in California to grow irrigated pasture for livestock. In the mid-1980s, this pasture was worth $100 million, while the economy of Southern California was worth $300 *billion*. Similar figures apply in many different countries where agriculture consumes the vast majority of water but urban water use produces far more income and wealth.

Under purely free-market economic conditions, agricultural water use would decline as cities purchased inexpensive water from farmers. Purely free-market conditions, however, never exist. In reality the legal constraints on water transfers, the complex system of water rights, subsidies for water, and other barriers act to keep much of the water used by agriculture in that sector. There are many reasons to maintain that water use—growing populations around the world must have food, and in water-rich areas, agriculture is a logical and necessary product. Countries want to encourage rural employment. Some cultural histories and values can only be maintained through subsidization of agricultural communities. But the past system of blind, unexamined subsidies of irrigated agriculture must now give way to a more reasoned and sophisticated approach.

Some changes are already beginning to occur. The growth of cities and their far greater economic productivity are now beginning to demand more water. In Israel, urban development combined with constraints on new supply options have led some to speculate that within a couple of decades all water will go first to cities and industries and the only reliable agricultural production will come from the use of treated wastewater for irrigation (Shuval 1996).

Economic factors and pricing decisions also lead to inefficient water use in the urban sector. In many cities water use is not measured or "metered," which leads to overuse of water and provides no incentives for efficient use. Even in regions with water metering, the inappropriate design of rate structures can lead to misuse of water. As a result, there is growing interest in the use of so-called "conservation" rates, such as increasing block rates, where increasing amounts of water are charged at higher and higher rates. More and more water utilities are implementing such rate structures. Different kinds of rate structures are shown in Box 1.3. In Beijing, China, a new pricing system links the cost of water to the amount of water used, encouraging conservation. A similar pricing system decreased average monthly residential water use by nearly 30 percent in Bogor, Indonesia (Postel 1992, 1997). Regional water providers in South Africa have been able to delay the construction of new regional water supply systems by imposing higher rates,

BOX 1.3

Common Municipal Pricing Options

Many possible rate structures exist. The three most common options and their significance to individuals and communities are discussed here.

Average cost pricing estimates the cost of providing the water service and selects a rate that averages the cost across the entire customer base. Average cost pricing can be applied as a single flat fee charged regardless of volume of water used, or as a uniform per unit charge. Whatever method is used, average cost pricing generally does not reflect the full economic cost to customers. That is, if the costs to service new customers are rising, by averaging them across all customers the new customers will not pay the full costs and their use will, in essence, be subsidized by existing customers.

Increasing block (tier) pricing sets different prices for different quantities of water use. Such pricing schemes are increasingly common and are used by water agencies facing water shortages or limited supply capacity. Tier rates are used to encourage water conservation and more efficient water use. While tier pricing is not truly marginal cost pricing, it attempts to reflect the higher cost of producing each additional unit of water. Given the high cost of building infrastructure and expanding supplies, it is likely that the marginal cost of water is rising. Because customers pay more per unit as their volume of water use increases, agencies hope to use price to discourage inefficient use.

Seasonal or peak-load pricing is another application of marginal cost pricing. Seasonal pricing in the water industry recognizes that there are seasonal patterns of water use—for example, higher water usage during hot months due to higher temperatures and greater outdoor water use. Water providers build their system capacity to meet these peak demands, despite its not being used during average or normal periods. For price to truly reflect the cost of providing this extra capacity during peak demand periods, water rates should include the additional cost of capacity that makes that consumption level possible. It can be applied to tiered pricing or to other pricing structures.

Source: Gomez and Wong 1997.

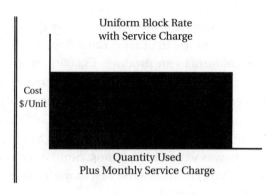

Uniform Block Rate
with Service Charge

Cost $/Unit

Quantity Used
Plus Monthly Service Charge

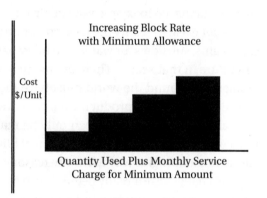

Increasing Block Rate
with Minimum Allowance

Cost $/Unit

Quantity Used Plus Monthly Service
Charge for Minimum Amount

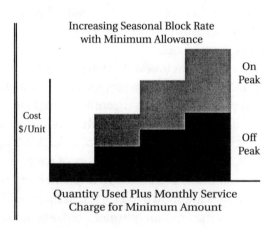

Increasing Seasonal Block Rate
with Minimum Allowance

Cost $/Unit

On Peak

Off Peak

Quantity Used Plus Monthly Service
Charge for Minimum Amount

distributing water-conservation equipment, and educating the public (Rand Water 1996). In Hermanus, South Africa, a major conservation effort includes a new 11-step rate structure that encourages technological improvements and careful attention to wasteful practices. In just the first year of operation of the program, water use in the city dropped more than 30 percent (see Chapter 6).

From an economic point of view, an efficient and sound means for solving water shortages is to permit water marketing. Large users of water could improve the efficiency of water use and sell or lease the saved water to another user, encouraging higher valued uses. In some marketing proposals, part or all of the water involved in a trade can be reserved to meet basic environmental water needs.

Large barriers to this approach are the complex systems of water laws and water rights used in different nations. Under traditional water law in many countries, a farmer who installs new efficient irrigation equipment would lose the right to the water saved, and so has no incentive to save it. The recognition of this problem has recently led to the proposal of new laws and new institutions to try to remedy these kinds of distortions. (See the discussion of South Africa's new water rights laws, for example, in Chapter 6.)

So far, only limited water transfers have occurred, but there appears to be an inexorable trend toward permitting them. A few states in the United States have begun to pass laws permitting such transfers, and the federal government has made changes in the way water-supply projects are operated that will further encourage certain kinds of water transfers. One of the most significant changes was the result of a federal law, the Central Valley Project Improvement Act of 1992 (CVPIA), which for the first time reallocates water back to fish and wildlife and allows farmers in California who receive water from the federal Central Valley Project (CVP) to sell their water to buyers outside the project's boundaries (NRLC 1997).

Under the law, farmers who agree to sell water would have to pay "full" costs for their supply, instead of the much lower subsidized federal rates. Urban areas, in particular, consider this bill to be extremely important in their search for new water supplies for growing populations. An official of the Metropolitan Water District of Southern California, the largest urban water agency in California, stated, "It's tantamount to creating a reservoir overnight" (Hundley 1992). While many questions remain about the implementation of this and related policy changes, the broad intent is to create an economic market in which water would go to the highest bidder and presumably to higher valued uses.

One note of caution: while the new emphasis on treating water as an economic good has the potential to both eliminate wasteful practices and encourage increased efficiency and conservation, a purely economic or market approach cannot adequately protect the natural ecosystems that also depend on water. We depend on nature's services to keep us alive, but these services are not "purchased," rarely quantified, and routinely excluded from official economic accounts (Daily 1997). In the drive toward economic rationality, care must be taken to preserve and protect those services that may fall outside of traditional economic measures.

Water transfers are increasingly receiving legislative, regulatory, and administrative support, but in the rush to encourage such transfers, some problems may arise. Rural communities may lose water necessary for employment or community well-being. Environmental water uses may be further disrupted rather than enhanced. And transfers may simply permit continued inefficient uses in urban areas. Gomez and Loh (1996) explore these issues and recommend that standards for water transfers be developed to ensure that they are fair, do not adversely impact ecosystems, and include the voices of rural communities that may be adversely affected.

Alternative Supplies

Alternative kinds of water supplies will increasingly be used to meet certain demands for water, rather than seeking new pristine sources from far away. Providing high-quality potable drinking water is expensive, and using that water to meet all needs is unnecessary where water is scarce. Meeting different needs with the appropriate quality of water may prove to be economically beneficial and at the same time reduce the need for new supplies at a higher and higher marginal cost. "Reclaimed" water, graywater, recycled water, brackish water, salt water, and desalinated water may all be considered usable for some needs and, in fact, may have environmental, economic, or political advantages. Reclaimed water, in particular, has some remarkable advantages, including a high reliability of supply, a known quality, and often, a centralized source near urban demand centers. A brief introduction to the use of reclaimed water is provided below and later editions of *The World's Water* will explore this issue in more detail.

Reclaimed Wastewater

The vast majority of urban water ends up flowing down the drain after being used once. This water goes either to wastewater-treatment plants, where it is treated and dumped into a river or the ocean, or ends up in local septic systems, where it sits before percolating to groundwater. Beginning in the 1950s and 1960s, interest has grown in capturing and treating wastewater to reduce human and environmental health problems associated with human sewage and industrial waste. More recently, attention has focused on treating this water and using it as a resource rather than considering it extraneous waste. Drought conditions limiting supply, environmental problems with sewage disposal, and growing water demands have all made water reclamation more appealing.

Reclaimed water, which is wastewater treated to a level suitable for use in a variety of applications, is widely available but only rarely used. Considerable research has been done on the applicability of reclaimed water of different qualities for different uses (Kirkpatrick and Asano 1986; Asano et al. 1992; Asano 1994). Reclaimed water can be used to recharge groundwater aquifers, supply industrial processes, irrigate certain crops, or augment potable supplies. This resource is receiving new attention in several regions of the world. In the Middle East, parts of Africa, and the western United States there has been a significant increase in the availability and use of treated wastewater for a wide range of industrial, commercial, and institutional needs. Some agricultural water needs are now being met with treated wastewater. In Windhoek, Namibia, reclaimed water has been used to augment the potable water supply since 1968, and in drought years up to 30 percent of the city's drinking water supply is treated wastewater (van der Merwe and Menge 1996). The U.S. National Academy of Sciences is completing a study on appropriate uses of highly treated wastewater for indirect augmentation of drinking water supplies (National Research Council 1998).

Israel has extensive wastewater reclamation programs. Seventy percent of Israeli wastewater is treated and used for agricultural irrigation, and Shuval (1996) estimates that 80 percent recycling is likely in the next few decades. This policy of providing reclaimed water for agriculture reduces the problem of disposing of large volumes of wastewater and helps guarantee a reliable supply to the Israeli agricultural sector, which faces growing competition from urban water users. Efforts to capture, treat, and reuse more wastewater are also being made in neighboring Jordan, where overall water supplies are also highly constrained (Ahmad 1989; Salameh and Bannayan 1993).

In California, by the mid-1990s, over 600 million cubic meters of reclaimed water were being used each year for different purposes. The California Department of Water Resources official projections for the year 2020 are for more than 1 billion cubic meters to be used (CDWR 1994). Gleick et al. (1995) estimated that more than 2 billion cubic meters of wastewater could be reclaimed and reused by 2020, while an industry source has estimated that a major effort could produce as much as three billion cubic meters of usable reclaimed water (Sheikh, personal communication, June 1995). Meeting any of these goals would require extensive capital investment and changes in institutional organizations, but there are strong indications that these resources will increasingly be tapped.

In northern California, the East Bay Municipal Utilities District (EBMUD) found it to its economic advantage to pay a major refinery to replumb its water system to replace the use of potable water for plant operations with reclaimed water of suitable quality. This had the advantage of providing a reliable source of water for the refinery that would be available even during droughts, while reducing the need of the water utility to find a new source of potable water for growing urban residential demand (EBMUD 1994).

California agriculture also uses recycled water with both secondary- and tertiary-treated water. Under California's Health and Safety Code, edible food crops, parks and playgrounds, and landscaped areas where humans come in contact with water must be irrigated with tertiary disinfected water. Secondary undisinfected water can be used on orchards and vineyards where recycled water does not touch edible crops, on fiber, fodder, and pasture for animals not producing milk for human consumption, and on food crops that undergo commercial processing to destroy pathogens. The city of Bakersfield has used secondary-treated water for 45 years to grow cotton, alfalfa, wheat, and corn. The city of Visalia grows walnuts with their secondary-treated water (Gleick 1998). Almost every municipal golf course in the state either already uses or will soon use reclaimed water to irrigate its greens, and more and more municipal demands will eventually be met with this source. Encouraging wider use of reclaimed water will also require rethinking water pricing, particularly how to provide economic incentives to encourage its use, how to allocate the existing costs of wastewater treatment, and how to identify appropriate uses for different qualities of supply.

Desalination

For decades, some water analysts and observers have held out desalination as the ultimate solution to the world's water woes. More than 97 percent of the water on the planet is too salty to drink or grow food. In theory, therefore, desalination offers a limitless supply of fresh water, freeing humans from the vagaries and inconsistencies of natural freshwater supplies. Yet like the unfulfilled promise of cheap nuclear power in the 1960s and

1970s, desalination remains a minor contributor to water supply, providing only a fraction of a percent of total human needs.

The desalination of salt water or brackish water is technologically well developed but remains hindered by high economic costs in part because of the large amounts of energy required to strip salt ions from water. While technological optimists continue to predict declining costs with improving technology, desalination is only an option at present for extremely water-short countries with substantial energy or economic resources—such as in the Persian Gulf, where seven of the top ten desalinating countries are located. In addition, the high cost of moving water from one place to another further constrains desalination developments to areas within a limited distance from the coasts.

By early 1997, total global desalination capacity reached 18 million cubic meters per day. Figure 1.7 shows desalination capacity worldwide from 1950 to 1996. Figure 1.8 breaks down the desalination capacity of the ten countries with the largest investment in desalination. Most of these are in the Middle East/Persian Gulf region, where water is scarce but money is not. Figure 1.9 breaks down the increments in capacity by the different processes used to remove salt from water. The two most common approaches are multi-stage flash distillation and reverse osmosis. Waters of different quality, including seawater, brackish water, or impure industrial wastewater, can be desalinated, but seawater or brackish water are the most common inputs.

Desalination cannot yet be considered a reasonable solution to domestic water shortages in most regions, even wealthy ones. Recently, a major reverse-osmosis desalination plant (more than nine million cubic meters per year capacity) was built on the California coast near the city of Santa Barbara. The city—badly hit by shortages during a recent

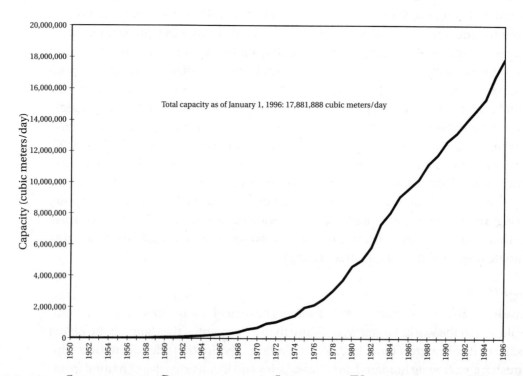

FIGURE 1.7   CUMULATIVE DESALINATION CAPACITY WORLDWIDE, 1950 TO 1996.
Global desalination capacity, measured in cubic meters per day, is shown here for the years 1950 to 1996. Land-based desalting plants with a capacity rated at more than 500 cubic meters per day and in operation as of January 1, 1996, are included. *Source:* Wangnick 1996, with permission.

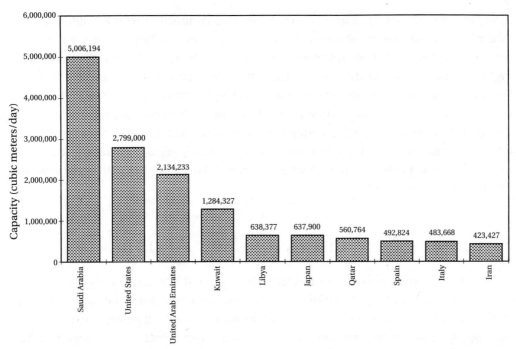

FIGURE 1.8 DESALINATION CAPACITY FOR TEN COUNTRIES WITH LARGEST CAPACITY, JANUARY 1, 1996.

The desalination capacities, measured in cubic meters per day, for the ten countries with the largest capacity are shown here. Included are land-based desalting plants with a capacity rated at more than 500 cubic meters per day and delivered or under construction as of January 1, 1996. *Source:* Wangnick 1996, with permission.

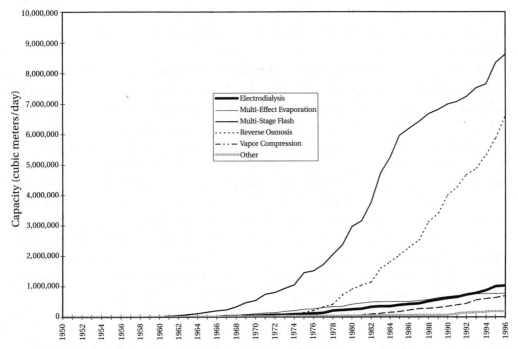

FIGURE 1.9 CUMULATIVE DESALINATION CAPACITY WORLDWIDE BY TYPE OF PROCESS, JANUARY 1, 1996.

Cumulative desalting capacity in cubic meters per day is given here by type of process. The data include land-based desalting plants rated at more than 500 m³/day and delivered or under construction as of January 1, 1996. *Source:* Wangnick 1996, with permission.

drought—also agreed to finance a major pipeline connection to the California State Water Project. Because of the high capital, operating, and maintenance costs associated with both projects, the cost of water to the citizens of Santa Barbara is now among the highest in the entire country. The average cost of water supply there has increased from $0.47 per cubic meter before the projects to over $1.55 per cubic meter in 1996. In large part because of these costs and investments in water conservation, total demand has dropped to 61 percent of the pre-drought level, eliminating the need for the desalination plant (Loaiciga and Renehan 1997). As a result, it is not in regular operation and will be used only to meet demands during severe droughts. This kind of option is not available for most regions of the world.

Summary: New Thinking, New Actions

Basic concepts and philosophies of water development are undergoing fundamental changes. Whether water is plentiful or scarce, environmental, financial, and social constraints are slowing the construction of large projects and leading planners to extend limited supplies through reallocation and efficiency improvements. Even in regions where basic human needs for water have not been met, water conservation programs are becoming an integral part of practical solutions because they permit overall water needs to be met with fewer resources, less disruption of ecosystems, and lower costs.

In many parts of the world there will continue to be strong pressures to provide major new water systems, especially if those systems are dedicated to meeting some of the basic needs remaining in developing countries. Where providing basic drinking water supply or baseline hydroelectric power is still possible without enormous social dislocations or economic and environmental costs, large new projects may still be appropriate and necessary.

But large-scale projects can no longer be expected to provide the answer to most water problems. In particular, in arid and semi-arid regions that cover about 30 percent of the earth's land area, large-scale irrigation schemes and dams are increasingly out of favor. Only 4 percent of sub-Saharan Africa's cropland is currently irrigated, yet few good dam sites remain and the costs of large irrigation projects (estimated to be $10,000 to 20,000 per hectare) are extremely high. The World Bank noted, "Favorable conditions, such as good high-yielding aquifers, rivers with sustained year-round flows, and large tracts of irrigable lands are unfortunately not available to justify the type of massive investment that has gone into the Nile Basin, the Middle East, and Asia" (Postel 1992).

Major new projects are going to compete with new opportunities for innovative smaller scale, locally managed technical, institutional, and economic solutions to water quality and quantity problems. These include micro-dams, run-of-river hydro systems, shallow wells, low-cost pumps, water-conserving land management methods, and "rainwater harvesting" approaches. Such methods are often more cost-effective and less disruptive to local communities, in part because of traditional experiences of these communities. There is already evidence that new applications of traditional methods can catalyze farmers into improving management techniques, stimulate local development, and meet local water needs in many places where large-scale irrigation projects have failed (Clarke 1991). One estimate is that 100 million people in Africa alone could benefit from the adoption of small-scale, low-cost traditional methods, but that lack of knowledge and technology continue to hinder their widespread adoption (Postel 1997). The cancellation of the massive Arun III project in Nepal recently led to a focus on providing the same benefits at lower economic, environmental, and social costs with a series of smaller scale run-of-river systems (Gyawali, personal communication, 1998).

Sandra Postel, in her book *Last Oasis,* summarizes much of this point when she says,

> A new water era has begun. In contrast to earlier decades of unfettered damming, drilling, and diverting to gain ever greater control over water, the next generation will be marked by limits and constraints—political, economic, and ecological. Yet numerous opportunities arise as well. Exploiting the market potential of new water saving technologies is an obvious one. And in many cases, achieving better water management will require decentralizing control over water, and moving from top-down decisionmaking to greater people's participation—a shift necessary for better human and economic development overall.

More rapid changes in water policy worldwide have not occurred because most economic and institutional structures still encourage inefficient use of water. Barriers to better water planning and use include national water subsidies, subsidies for growing water-intensive crops in arid regions, hydropower subsidies for particular regions or industries, inappropriate economic signals to consumers, and a range of legal constraints. Many of these "nonmarket" conditions lead to inadequate investment in new capital for efficiency equipment.

Part of the problem, however, also lies in the prevalence of old thinking among water planners and managers. An ethic of sustainability will require fundamental changes in how we think about water, and such changes come about slowly. Rather than endlessly trying to find the water to meet some projection of future desires, it is time to plan for meeting present and future human needs with the water that is available, to determine what desires can be satisfied within the limits of our resources, and to ensure that we preserve the natural ecological cycles that are so integral to human well-being (Gleick 1996). Water-resource planning in a democratic society must involve more than simply deciding what big project to build next or evaluating which scheme is the most cost-effective from a narrow economic perspective. Planning must provide information that helps people to make judgments about which "needs" and "wants" can and should be satisfied. Water is a common good and community resource, but it is also used as a private good or economic commodity; it is not only a recreational resource but also a basic necessity of life; it is imbued with cultural values and plays a part in the social fabric of our communities. Applying new principles of sustainability and equity can help bridge the gap between such diverse and competing interests.

REFERENCES

Ahmad, A.A. 1989. *Jordan Environmental Profile: Status and Abatement.* USAID/Government of Jordan, Amman, Jordan.

Asano, T. 1994. "Irrigation with treated sewage effluents," in K.K. Tanji and B. Yaron (eds.), *Advanced Series in Agricultural Sciences,* Vol. 22, pp. 199–228.

Asano, T., D. Richard, R.W. Crites, and G. Tchobanoglous. 1992. "Evolution of tertiary treatment requirements for wastewater reuse in California." *J. Water Environ. Technol.,* Vol. 4., No. 2, pp. 36–41.

Association of Home Appliance Manufacturers (AHAM). 1993. *Major Home Appliance Industry Factbook.* Chicago, Illinois.

Bhatia, R., and M. Falkenmark. 1992. "Water resource policies and the urban poor: Innovative approaches and policy imperatives." A background paper for the working group on "Water and Sustainable Urban Development," *International Conference on Water and the Environment*, 26–31 January, Dublin, Ireland.

California Department of Finance (CDOF). 1994. *California Statistical Abstract.* Compilation of U.S. Department of Commerce and Bureau of Economic Analysis data, Sacramento, California.

California Department of Water Resources (CDWR). 1957. *The California Water Plan.* DWR Bulletin No. 3. Sacramento, California.

California Department of Water Resources (CDWR). 1994. *The California Water Plan Update.* Bulletin 160-93. Resources Agency, Sacramento, California.

Cernea, M. 1988. "Involuntary resettlement in development projects." World Bank Technical Paper No. 80, The World Bank, Washington, D.C.

Chesnutt, T.W., C.N. McSpadden, and A. Bamezai. 1994. "Ultra low flush toilet programs: Evaluation of program outcomes and water savings." A report to the Metropolitan Water District of Southern California, A&N Technical Services, Inc. Santa Monica, California.

City of Austin. 1995. "Utility bill credits for ultra low flush toilets." Environmental and Conservation Services Department/Water and Wastewater Utility, City of Austin, Texas.

City of San Diego. 1995a. "Municipal Code Section 93.0208 Relating to Water-Conserving Plumbing Standards." City of San Diego Water Utilities Department, San Diego, California.

City of San Diego. 1995b. "Ultra-low flush toilet rebate program." City of San Diego brochure. San Diego, California.

Clarke, R. 1991. *Water: The International Crisis.* Earthscan Publications/MIT Press, Cambridge, Massachusetts.

Council on Environmental Quality (CEQ). 1991. *Environmental Quality: 21st Annual Report.* Washington, D.C.

Daily, G.C. (ed.). 1997. *Nature's Services: Societal Dependence on Natural Ecosystems.* Island Press, Washington, D.C.

De Mare, L. 1976. "Resources—Needs—Problems: An assessment of the world water situation by 2000." Institute of Technology/ University of Lund, Sweden.

East Bay Municipal Utility District (EBMUD). 1994. "Water Conservation Master Plan." EBMUD, Oakland, California.

Falkenmark, M., and G. Lindh. 1974. "How can we cope with the water resources situation by the year 2050?" *Ambio*, Vol. 3, No. 3-4, pp. 114–122.

Falkenmark, M., and G. Lindh. 1993. "Water and economic development," in P.H. Gleick (ed.), *Water in Crisis: A Guide to the World's Fresh Water Resources.* Oxford University Press, New York, pp. 80–91.

Financial Times Global Water Report. 1997a. "Do the right thing." *Financial Times Global Water Report*, Issue 20 (April 10), p. 16.

Financial Times Global Water Report. 1997b. "Mahakali River development." *Financial Times Global Water Report*, Issue 16 (February 12), pp. 9–10.

Financial Times Global Water Report. 1997c. "Financing crisis for Bakun Dam." *Financial Times Global Water Report*, Issue 27 (July 17), p. 11.

Financial Times Global Water Report. 1997d. "Protesters halt dam construction." *Financial Times Global Water Report*, Issue 36 (December 4), pp. 10–11.

Financial Times Global Water Report. 1998. "Cost of phantom dam soars to $12bn." *Financial Times Global Water Report*, Issue 38 (January 8), pp. 8–9.

Gleick, P.H. (ed.). 1993. *Water in Crisis: A Guide to the World's Fresh Water Resources.* Oxford University Press, New York.

Gleick, P.H. 1996. "Basic water requirements for human activities: Meeting basic needs." *Water International,* Vol. 21, pp. 83–92.

Gleick, P.H. (ed.). 1998 (in press). *Sustainable Use of Water: California Success Stories.* Pacific Institute for Studies in Development, Environment, and Security, Oakland, California.

Gleick, P.H., P. Loh, S. Gomez, and J. Morrison. 1995. *California Water 2020: A Sustainable Vision.* Pacific Institute for Studies in Development, Environment, and Security, Oakland, California.

Gomez, S., and P. Loh. 1996. "Communities and water markets: A review of the Model Water Transfer Act." *Hastings West-Northwest Journal of Environmental Law and Policy,* Vol. 4, No. 0 (Fall), San Francisco, California.

Gomez, S., and L. Owens-Viani. 1998. "Community/Water Agency partnerships to save water and revitalize communities," in P.H. Gleick (ed.), *Sustainable Use of Water: California Success Stories.* Pacific Institute for Studies in Development, Environment, and Security, Oakland, California (in press).

Gomez, S., and A. Wong. 1997. *Our Water, Our Future: The Need for New Voices in California Water Policy.* Pacific Institute for Studies in Development, Environment, and Security, Oakland, California.

Gupta, D.B., M.N. Murty, and R. Pandey. 1989. "Water conservation and pollution abatement in Indian industry." National Institute of Public Finance and Policy, New Delhi, India (April).

Hundley, Jr., N. 1992. *The Great Thirst: Californians and Water, 1770s–1990s.* University of California Press, Los Angeles.

Kalbermatten, J.M., D.S. Julius, C.G. Gunnerson, and D.D. Mara. 1982. "Appropriate sanitation alternatives: A technical and economic appraisal," and "A planning and design manual." World Bank Studies in Water Supply and Sanitation I and II. The Johns Hopkins University Press, Baltimore, Maryland.

Kalinin, G.P., and I.A. Shiklomanov. 1974. "USSR: World water balance and water resources of the Earth." USSR National Committee for IHD, Leningrad (in Russian).

Kirkpatrick, W.R., and T. Asano. 1986. "Evaluation of tertiary treatment systems for wastewater reclamation and reuse." *Water Science Technology,* Vol. 19, No. 10, pp. 83–95.

Loaiciga, H.A., and S. Renehan. 1997. "Municipal water use and water rates driven by severe drought: A case study." *Journal of the American Water Resources Association,* Vol. 33, No. 6, pp. 1313–1326.

Lovins, A.B. 1977. *Soft Energy Paths: Toward a Durable Peace.* Friends of the Earth International, Ballinger Publishing Company, San Francisco and New York.

L'vovich, M.I. 1974. *World Water Resources and Their Future.* Mysl' Publishing, Moscow. English translation (1979) by R.L. Nace. American Geophysical Union, Washington, D.C.

McCully, P. 1996. *Silenced Rivers: The Ecology and Politics of Large Dams.* Zed Books, London.

Morrison, J., S. Postel, and P. Gleick. 1996. "The sustainable use of water in the Lower Colorado River Basin." Pacific Institute for Studies in Development, Environment, and Security, Oakland, California.

Myers, N., and J. Kent. 1997. *Perverse Subsidies: Their Nature, Scale, and Impacts.* A report to the MacArthur Foundation. United Kingdom.

National Research Council (NRC). 1998 (in press). *Issues in Potable Reuse: The Viability of Augmenting Drinking Water Supplies with Reclaimed Water.* Report of the National Research Council, National Academy of Sciences Press, Washington, D.C.

Natural Resources Law Center (NRLC). 1997. "Restoring the waters." University of Colorado School of Law, Boulder, Colorado (May).

Nikitopoulos, B. 1967. "The world water problem—Water sources and water needs." RR-ACE: 106 and 113 (COF). Athens Center for Ekistics, Athens, Greece.

Perlman, H. 1997. Data for 1995 from the United States Geological Survey (USGS), via FTP from 144.47.32.102.

Postel, S. 1992, 1997. *Last Oasis: Facing Water Scarcity.* Worldwatch Institute, W.W. Norton, New York.

Postel, S.L., G.C. Daily, and P.R. Ehrlich. 1996. "Human appropriation of renewable fresh water." *Science,* Vol. 271, pp. 785–788 (February 9).

Rand Water. 1996. "Chief Executive's Review." *Annual Report of Rand Water,* Johannesburg, South Africa, pp. 6–7.

Raskin, P., P. Gleick, P. Kirshen, G. Pontius, and K. Strzepek. 1997. "Water futures: Assessment of long-range patterns and problems." Background Document of the Comprehensive Assessment of the Freshwater Resources of the World. Stockholm Enviornment Institute, Stockholm, Sweden.

Reisner, M. 1986 and 1993 revision. *Cadillac Desert. The American West and Its Disappearing Water.* Penguin Books, New York.

Reisner, M., and S. Bates. 1990. *Overtapped Oasis: Reform or Revolution for Western Water.* Island Press, Washington, D.C.

Rogers, P. 1993. *America's Water: Federal Roles and Responsibilities.* Twentieth Century Fund, MIT Press, Cambridge, Massachusetts.

Rogers, P. 1997. "Water for big cities: Big problems easy solutions?" Draft paper, Harvard University, Cambridge, Massachusetts.

Salameh, E., and H. Bannayan. 1993. *Water Resources of Jordan: Present Status and Future Potentials.* Friedrich Ebert Stiftung, Amman, Jordan.

Shiklomanov, I.A. 1996. "Assessment of water resources and water availability in the world." Report for the Comprehensive Global Freshwater Assessment of the United Nations. State Hydrological Institute, St. Petersburg, Russia (February draft).

Shiklomanov, I.A., and O.A. Markova. 1987. *Specific Water Availability and River Runoff Transfers in the World.* Gidrometeoizdat, Leningrad (in Russian). (An English version of the data can be found in Gleick, 1993.)

Shuval, H. 1996. "Sustainable water resources versus concepts of food security, water security, and water stress for arid countries." Background paper prepared for the Workshop on Chapter 4 of the Comprehensive Assessment of the Freshwater Resources of the World. 18–19, May 1996, New York.

Solley, W. B., R. R. Pierce, and H. A. Perlman. 1993. "Estimated Use of Water in the United States in 1990," U.S. Geological Survey Circular 1081. U.S. Department of the Interior, Washington, D.C.

Steding, A., J. Morrison, and P. Gleick. 1996. "Analysis of the hydrologic impact of replacing vertical-axis clothes washers with the Frigidaire Alliance horizontal-axis model in three western US metropolitan areas." Report of the Pacific Institute for Studies in Development, Environment, and Security, Oakland, California (June 19).

Stegner, W. 1953. *Beyond the Hundredth Meridian.* Penguin Books, New York.

Sweeten, J., and B. Chaput. 1997. "Identifying the conservation opportunities in the commercial, industrial, and institutional sector." Paper presented at the annual American Water Works Association meeting, June 1997.

United Nations. 1992. *Agenda 21: Chapter 18 (Freshwater).* United Nations, New York.

United Nations. 1997. *Comprehensive Assessment of the Freshwater Resources of the World.* Commission for Sustainable Development, Stockholm Environment Institute, Stockholm, Sweden.

U.S. Congress. 1994. "Taking from the taxpayer: Public subsidies for natural resource development." Majority staff report of the Subcommittee on Oversight and Investigations of the Committee on Natural Resources of the U.S. House of Representatives, 103rd Congress, 2nd Session. U.S. Government Printing Office, Washington, D.C.

U.S. Geological Survey. 1993. *Estimated Use of Water in the United States in 1990.* U.S. Geological Survey Circular 1081. U.S. Department of the Interior, Washington, D.C.

U.S. Water News. 1997. "Clothes washer test termed successful." *U.S. Water News,* Vol. 14, No. 12, p. 15 (December).

Van der Merwe, B., and J. Menge. 1996. "Water reclamation for potable reuse in Windhoek, Namibia." Department of the City Engineer, City of Windhoek, Namibia.

Vickers, A. 1991. "The emerging demand-side era in water management." *American Water Works Association,* Vol. 83, No. 10, pp. 38–43.

Vickers, A. 1993. "The National Energy Policy Act: Assessing its impacts on utilities." *Journal of the American Water Works Association,* Vol. 85, No. 8, pp. 56–62 (August).

Wade, W.W., J.A. Hewitt, and M.T. Nussbaum. 1991. Report to the California Urban Water Agencies, "Cost of Industrial Water Shortages." Spectrum Economics, Inc., San Francisco, California.

Wangnick, K. 1996. *1996 IDA Worldwide Desalting Plants Inventory, Report No. 14.* Wangnick Consulting, Gnarrenburg, Germany.

World Bank. 1993. "Water Resources Management." A World Bank Policy Paper. The World Bank, Washington, D.C.

World Resources Institute. 1990. *World Resources 1990–91.* Oxford University Press, New York.

Zachary, G.P. 1997. "Water pressure: Nations scramble to defuse fights over supplies." *The Wall Street Journal* (December 4), p. A17.

Water and Human Health

Throughout the talk of a new millennium and dreams of technological, communications, and social advances, an ugly reality seldom noted is that billions of people around the globe lack access to the most fundamental foundation of a decent civilized world: basic sanitation services and clean drinking water. As Akhtar Hameed Khan, director of the Orangi Pilot Project in Pakistan said: "Access to safe water and adequate sanitation is the foundation of development. For when you have a medieval level of sanitation, you have a medieval level of disease, and no country can advance without a healthy population"(UNICEF 1997). For nearly three billion people, access to a sanitation system comparable to that of ancient Rome would be a significant improvement in their quality of life.

The failure to provide basic sanitation services and clean water to so many people takes a serious toll on human health. In many developing countries, cholera, dysentery, and other water-related diseases are on the upswing. Nearly 250 million cases are reported every year, with between 5 and 10 million deaths. Diarrheal diseases leave millions of children underweight, mentally and physically handicapped, and vulnerable to other diseases. Yet we are falling further and further behind in our efforts to provide these basic services. Between 1990 and 1997, some 300 million more people were added to 2,600 million already without adequate sanitation services, a clear indication that the world community is failing to meet the most basic needs.

Ironically, the world community of water experts knows what needs to be done and how to do it, and sometimes moves to take definitive action. A major and successful effort is under way to eradicate one of the most dreaded water-related diseases: dracunculiasis or "guinea worm," the "fiery serpent" afflicting people in Africa and parts of Asia. Guinea worm cases have fallen from over three million in the mid-1980s to 150,000 in 1996, and there are hopes that it can be eradicated entirely by the year 2000. This success shows that progress can be made if the right political, economic, and educational tools are applied.

Water Supply and Sanitation: Falling Behind

The provision of basic water services has proven to be a key element in advancing economic and social development and eliminating a host of debilitating and costly diseases.

In Asia, excavations at Harappa and Mohenjo Daro in the Indus Valley have revealed ceramic pipes for water supply and brick conduits under the streets for drainage that are thought to have been in operation around 3000 B.C. (Rouse and Ince 1957). In the second millennium B.C., the Minoan civilization on Crete used running water and flushed latrines. The towns of Athens, Pompeii, and Morgantina, like most Greco-Roman towns of their time, maintained elaborate systems for the supply, use, and drainage of water and sewage (Crouch 1993). The first of the Roman aqueducts was completed around 312 B.C., and by the height of the Roman empire, nine major systems supplied the occupants of Rome with as much water per capita as is provided in many parts of the industrialized world today. This water was distributed through an extensive system of lead pipes in the streets. Well-built sewers drained the city (Drower 1956; Rouse and Ince 1957). Even in Thomas More's fictional *Utopia*, written in 1516, the ideal cities of the perfect republic are all provided with clean, protected water sources as part of their basic design.

Our ancient civilizations are gone, leaving only traces of their water systems to be studied and analyzed by archeologists and historians. What these traces say, however, is that while access to clean drinking water and sanitation services cannot guarantee the survival of a civilization from the many things that may threaten it, civilization cannot succeed without them. Dora Crouch, in her groundbreaking book on water management in ancient Greek cities (1993) notes, "Attention to water supply and drainage is the *sine qua non* for . . . that human condition we call civilization. In fact, development of water supply, waste removal, and drainage made dense settlement possible." It is ironic, therefore, to realize that more than half the population of our "civilized" world suffers today with water services inferior to those of the ancient Greeks and Romans.

The seriousness of this problem has long been recognized. Over 20 years ago in 1977, at the Mar del Plata conference on water organized by the United Nations, a commitment was made to focus efforts on providing access to safe drinking water and sanitation services during the 1980s, the International Drinking Water Supply and Sanitation Decade. While considerable progress was made during this period to address unmet needs, vast numbers of people are still unserved. The United Nations estimated that 1,300 million people without access to an adequate water supply at the beginning of the decade received that access by 1990, while the population with sanitation services increased by 750 million. This still left an estimated 1,200 million people without safe drinking water and 1,700 million without sanitation services (Christmas and de Rooy 1991). The true picture, however, was even worse. Due to underreporting, poor data, and new definitions of "access," the actual number of people lacking these basic services was reassessed upward in 1996 by the United Nations. In a new assessment, the World Health Organization (WHO) (1996) estimated that the population without access to sanitation in 1990 was closer to 2,600 million—nearly a billion more than their estimate just five years earlier. The 1990 population without clean drinking water was estimated to be 1,300 million. Maps 2.1 and 2.2 show the most recent estimates of populations without access to basic water services.

The failure of the decade to completely satisfy basic human needs for water and water services was the result of rapid population growth, underinvestment, growing urbanization, and misdirected priorities. The extent of the problem means many governments, organizations, and agencies must be involved in planning and implementing programs. Unfortunately, other social problems are often given higher priority and rapid population

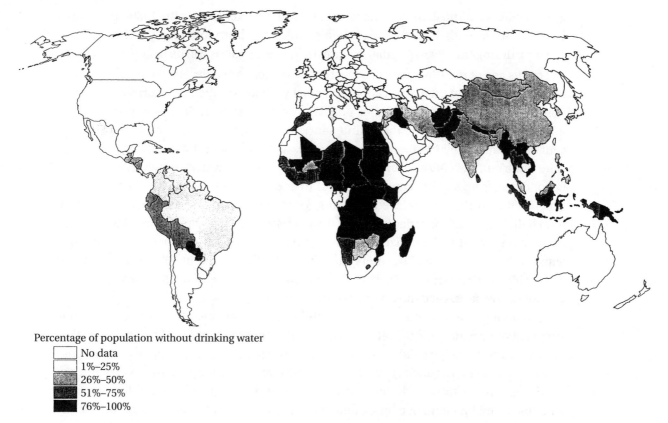

Percentage of population without drinking water

☐ No data
☐ 1%–25%
▨ 26%–50%
▨ 51%–75%
■ 76%–100%

MAP 2.1　POPULATIONS WITHOUT ACCESS TO SAFE DRINKING WATER

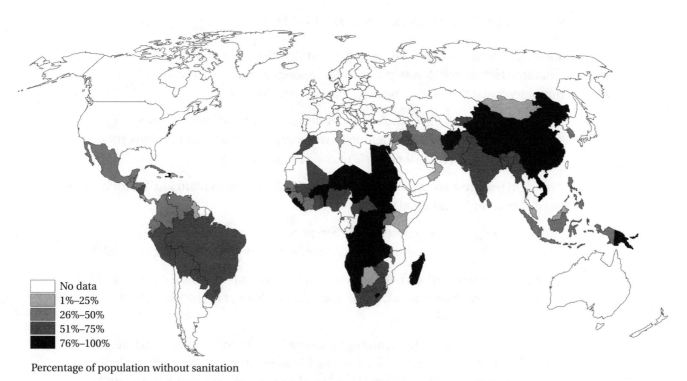

☐ No data
▨ 1%–25%
▨ 26%–50%
▨ 51%–75%
■ 76%–100%

Percentage of population without sanitation

MAP 2.2　POPULATIONS WITHOUT ACCESS TO ADEQUATE SANITATION

growth makes it difficult to catch up with basic water needs. During the 1980s, the United Nations estimates that nearly 750 million people were born into regions without these services; during the 1990s, population growth in these regions will be nearly 900 million.

The decade of the 1980s was also a period when economic development was derailed for many developing nations. Global financial policies resulted in enormous international loans to developing countries unable to both repay their debts and provide basic infrastructure development and maintenance. In Africa, for example, more than 20 percent of all exports by 1990 went to repay debt, with the result that public expenditures were often cut. In Nairobi, Kenya, capital expenditures for water and sewerage fell by a factor of ten, from $27.78 per person in 1981 to under $2.50 in 1987. In Zimbabwe, 25 percent of village water pumps failed when the government cut maintenance funds by more than half. In several countries in Africa, the percentage of the population with access to basic water services fell dramatically during this period. Even now, many of these problems continue to worsen. Dysentery rates in Kinshasha soared in 1995 when funds for water chlorination ran out (Khan 1997), and cholera cases and deaths throughout the continent now far exceed those reported in the 1970s.

Eliminating water-related diseases requires more than merely constructing infrastructure or providing clean water. It also requires maintenance and operation of that infrastructure, teaching children about adequate hygiene habits, identifying other transmission routes such as unclean handling of food, and controlling disease vectors. But the continued failure to provide basic clean water and sanitation services for so many remains the major element of one of the most significant health disasters of the twentieth century.

Basic Human Needs for Water

More than 20 years have passed since the Mar del Plata conference, one of the earliest international efforts to address global water problems. At that meeting, the world water community raised the issue of meeting "basic needs" for water:

> . . . all peoples, whatever their stage of development and their social and economic conditions, have the right to have access to drinking water in quantities and of a quality equal to their basic needs. (United Nations 1977)

This right was strongly reaffirmed during the 1992 Earth Summit in Rio de Janeiro and expanded to include ecological water needs:

> In developing and using water resources, priority has to be given to the satisfaction of basic needs and the safeguarding of ecosystems. (United Nations 1992)

In 1997, the United Nations once again reaffirmed the importance of these concepts in the Comprehensive Assessment of the Freshwater Resources of the World, prepared for the UN General Assembly:

> . . . it is essential for water planning to secure basic human and environmental needs for water [and] . . . develop sustainable water strategies that address basic human needs, as well as the preservation of ecosystems. (United Nations 1997)

Implicit in the concept of basic human needs for water is the idea of minimum resource requirements for certain human and ecological needs, and the allocation of sufficient resources to meet those needs. Different sectors of society use water for different purposes: drinking, removing or diluting wastes, producing manufactured goods, growing food, producing and using energy, and so on. The water required for each of these activities varies with climatic conditions, lifestyle, culture, tradition, diet, technology, and wealth, as shown over 25 years ago in the groundbreaking work of White et al. (1972).

The absolute minimum amount of water is that required solely for human survival. To maintain the water balance in a living human, the amount of water lost through normal activities must be regularly restored. While the amount of water required to keep a person alive depends on surrounding environmental conditions and personal physiological characteristics, the overall variability of needs is quite small. Minimum water requirements for fluid replacement have been estimated at about three liters per day under average temperate climate conditions. In a hot climate, a 70-kilogram human will lose between four and six liters per day (Gleick 1996). A further fundamental requirement not usually noted in the physiological literature is that this water should be of sufficient quality to prevent water-related diseases.

Basic water requirements for humans should also include any water necessary for disposing of human wastes. Extensive research has shown the distinct health advantages of access to adequate sanitation facilities as well as of protecting drinking water from pathogenic bacteria and viral and protozoan agents of disease. Effective disposal of human wastes controls the spread of infectious agents and interrupts the transmission of water-related diseases. In reviews of epidemiological studies related to water and sanitation, the availability of adequate sanitation services was the most direct determinant of child health after the provision of basic drinking water and water for hygiene (Esrey and Habicht 1986; Esrey et al. 1991).

In regions where absolute water quantity is a major problem, waste-disposal options that require no water are available. In most cases, however, developing economic societies have tended to prefer alternatives that use at least some water, and some societies use enormous amounts of fresh water to dispose of wastes. The choice of sanitation technology will ultimately depend on the developmental goals of a country or region, the water available, the economic cost of the alternatives, and powerful regulatory, cultural, and social factors.

Given these variables, can a recommended basic water requirement for sanitation be identified? While it is technically feasible to set a minimum at zero, additional health benefits are identifiable when up to 20 liters per capita per day (l/p/d) of clean water are provided (Esrey and Habicht 1986). Access to some water for sanitation together with education about water use decreases the incidence of diseases, increases the frequency of hygienic food preparation and washing, and reduces the consumption of contaminated food products. Accordingly, while effective disposal of human wastes can be accomplished with little or no water when necessary, a minimum of 20 l/p/d offers the maximum benefits of combining waste disposal and related hygiene and meets cultural and societal preferences for water-based disposal. A wide range of technological choices are available that can provide this service for 20 l/p/d or less (Gleick 1996).

On top of these direct sanitation requirements, additional domestic water is used for basic hygiene (washing, showering, and bathing) and for food preparation. Combining information on water to meet these needs adds another 25 l/p/d as a part of the water required for basic human needs. There are many other human uses of water, including industrial and commercial use, power plant cooling and electrical generation, and production of food, but these are not basic human needs that require local water resources. Water requirements to meet these demands depend on what precisely is being produced, on the technology used, and on a host of other characteristics. Moreover, that water can be provided from sources remote from the point of demand, unlike water for basic domestic needs.

A Recommendation for a Guaranteed Basic Water Requirement to Meet Basic Human Needs

Based on the work done in this area recently, a recommendation for a "basic water requirement" (BWR) can be made. Recommended levels are usually based on fundamental health considerations and on assumptions about technological choices made at modest levels of economic development. Considering drinking water and sanitation needs only suggests that a BWR of 25 l/p/d per day of clean water for drinking and sanitation be provided by water agencies, governments, or community organizations. This amount is just above the lower end of the 20 to 40 l/p/d target set by the U.S. Agency for International Development, the World Bank, and the World Health Organization, and it is also in line with the recommended standards of the United Nations International Drinking Water Supply and Sanitation Decade and Agenda 21 of the Earth Summit. Adding minimum levels of water for bathing and cooking raises the total overall BWR to 50 l/p/d as a new standard for meeting the four domestic basic needs—drinking, sanitation, bathing, and cooking—independent of climate, technology, and culture. While billions of people lack this standard today, it is a desirable goal from a health perspective and from a broader goal of meeting a minimum quality of life (Gleick 1996).

Providing water sufficient to meet basic needs should be an obligation of governments, water management institutions, or local communities. While in some regions governmental intervention may be necessary to provide for basic water needs, many areas will be able to use traditional water providers, municipal systems, or private purveyors within the context of market approaches. Unfortunately, there are many reasons why governments or water providers may be unable to provide this amount of water, including rapid population growth or migration, the economic cost of water-supply infrastructure in regions where capital is scarce, inadequate human resources and training, and even simple political incompetence. Nevertheless, failure to provide this basic need is a major human tragedy. Preventing that tragedy should be a major priority for local, national, and international groups.

Vast regions of the world and hundreds of millions of people lack the water required to meet the basic human needs described above. Using the BWR as a benchmark, Table 2.1 lists those countries whose reported domestic per-capita water withdrawals fail to provide 50 l/p/d. According to these data, in 1990 fifty-five countries with an aggregate population of nearly one billion fell below this level. There are actually eight countries whose total reported water use in *all* sectors falls below the recommended BWR for just basic human needs.

There are strong reasons to believe that the actual number of people failing to receive the recommended BWR is far above that reported. The data in Table 2.1 are country aver-

TABLE 2.1   COUNTRIES WITH TOTAL DOMESTIC WATER USE BELOW 50 L/P/D

Country	1990 Population (million people)	Total Domestic Water Use (liters/person/day)	Total Domestic Use as a Percentage of the BWR of 50 liters per person per day
Gambia	0.86	4.5	9
Mali	9.21	8	16
Somalia	7.5	8.9	18
Mozambique	15.66	9.3	19
Uganda	18.79	9.3	19
Cambodia	8.25	9.5	19
Tanzania	27.32	10.1	20
Central African Republic	3.04	13.2	26
Ethiopia	49.24	13.3	27
Rwanda	7.24	13.6	27
Chad	5.68	13.9	28
Bhutan	1.52	14.8	30
Albania	3.25	15.5	31
Zaire	35.57	16.7	33
Nepal	19.14	17	34
Lesotho	1.77	17	34
Sierra Leone	4.15	17.1	34
Bangladesh	115.59	17.3	35
Burundi	5.47	18	36
Angola	10.02	18.3	37
Djibouti	0.41	18.7	37
Ghana	15.03	19.1	38
Benin	4.63	19.5	39
Solomon Islands	0.32	19.7	39
Myanmar	41.68	19.8	40
Papua New Guinea	3.87	19.9	40
Cape Verde	0.37	20	40
Fiji	0.76	20.3	41
Burkina Faso	9	22.2	44
Senegal	7.33	25.4	51
Oman	1.5	26.7	53
Sri Lanka	17.22	27.6	55
Niger	7.73	28.4	57
Nigeria	108.54	28.4	57
Guinea-Bissau	0.96	28.5	57
Vietnam	66.69	28.8	58
Malawi	8.75	29.7	59
Congo	2.27	29.9	60
Jamaica	2.46	30.1	60
Haiti	6.51	30.2	60
Indonesia	184.28	34.2	68
Guatemala	9.2	34.3	69
Guinea	5.76	35.2	70
Côte d'Ivoire	12	35.6	71
Swaziland	0.79	36.4	73
Madagascar	12	37.2	74
Liberia	2.58	37.3	75
Afghanistan	16.56	39.3	79
Uruguay	3.09	39.6	79
Cameroon	11.83	42.6	85
Togo	3.53	43.5	87
Paraguay	4.28	45.6	91
Kenya	24.03	46	92
El Salvador	5.25	46.2	92
Zimbabwe	9.71	48.2	96

Sources: Data on domestic water use come from Gleick 1993, FAO 1995, and WRI 1994.

ages, and several large countries, such as India and China, report that their average domestic water use slightly exceeds 50 l/p/d. We know, however, that average national water-use data hide significant regional variations, with large segments of populations usually falling below the average, while wealthier portions of the population tend to use far more per capita. In addition, national water-use data are known to be inadequate.

In contrast to these figures, domestic water use in all industrialized countries far exceeds the BWR, though the quality of this water varies widely. In the countries of western Europe, the recommended BWR is typically less than 25 percent of total domestic use. In the United States and Canada, a BWR of 50 l/p/d is less than 10 percent of total current domestic use.

The Social and Economic Cost of Failing to Meet Basic Needs

Failing to meet basic water requirements is the direct cause of most water-related diseases, resulting in high costs to both communities and governments. One early estimate is that water-related diseases cost over $125 billion per year (in late 1970 dollars) in just direct medical expenses and lost work time (Pearce and Warford 1993). As high as this (out-of-date) estimate is, it fails to include any amounts for the social costs to communities of these diseases, the lost education and opportunities for families, lost economic productivity of sick workers, or any of a number of other hidden and poorly quantified costs.

Research and field work consistently show that the poor often already pay far more for private, poor-quality services than the wealthy do for structured, piped water systems, which are often subsidized by governments. And they often pay again by suffering from water-related diseases that proper water services could prevent. People are willing to pay for adequate water. In Brazil, for example, residents were asked how much they would be willing to pay to receive water and sewer services. They responded with numbers two to four times the actual costs that would be incurred. Similarly, Bhatia and Falkenmark (1993) reported that even in the poorest strata of society, urban residents are willing to pay water vendors four to five times the typical price of municipally supplied water. Residents of Karachi living without sanitation or hygiene education spend six times more on medical bills than do people with these basic services. In Kumasi, Ghana, the poorest pay to use neighborhood latrines and spend more on sanitation services annually than residents with in-home toilet facilities (Khan 1997).

When governments provide the water, costs are less. In Port-au-Prince, Haiti, a comprehensive survey showed that households connected to the water system typically paid around $1.00 per cubic meter, while unconnected consumers forced to purchase water from mobile vendors paid from a low of $5.50 to a staggering high of $16.50 per cubic meter (Fass 1993). Current urban residents of the United States typically pay only $0.40 to $0.80 per cubic meter for municipal water of excellent quality. Residents in Jakarta, Indonesia, purchase water for between $0.09 and $0.50 per cubic meter from the municipal water company, $1.80 per cubic meter from tanker trucks, and $1.50 to $5.20 per cubic meter from private vendors, as much as 50 times more per unit water than residents connected to the city system (Lovei and Whittington 1993). In Lima, Peru, a poor family on the edge of the city pays a vendor roughly $3.00 per cubic meter, 20 times the price paid by a family connected to the city system.

Estimates of the total cost of providing water and sanitation services vary widely and depend on assumptions of the level and quality of service. UNICEF (1997) estimates governments in Asia, Africa, and Latin America have invested about $2.1 billion per year in basic water and sanitation services for unserved urban and rural populations so far this decade, and that universal coverage would require only $6.8 billion per year (1994 dollars) for another decade. This includes $300 million annually for hygiene education, which has been shown to be a vital component of sanitation programs. The total of $68 billion over ten years comes to only 1 percent of anticipated governmental expenditures for military activities. Rogers (1997) estimates that completely meeting just basic water supply needs up to the year 2020 will require a total capital cost on the order of $24 billion per year, but that using unconventional but proven systems could reduce this cost by as much as one-third. Christmas and de Rooy (1991) estimate that the cost of meeting unmet needs at a higher level of service, including more advanced wastewater treatment, would be about $50 billion a year. This is higher than the UNESCO estimate, but still far below the actual societal costs of the failure to provide these services. The World Bank estimates that spending for water and sanitation in developing countries is approximately $26 billion per year (Rogers 1993).

Additional treatment of drinking water is a fundamental requirement of water management in most places and offers the greatest protection from many water-related diseases. In the United States and Europe, water filtration was introduced in the late 1800s, removing most microbiologic contaminants and controlling the transmission of cholera and typhoid. Chlorination of public water supplies was introduced in the United States in 1908 and rapidly adopted in most large towns and cities (Nash 1993). These two approaches remain the primary components of drinking water treatment.

Water-Related Diseases

Water-related diseases are a major concern in most of the developing world, although they have largely been eliminated as a serious health problem in industrialized countries. While data are incomplete and unreliable, the World Health Organization has estimated that there are on the order of 250 million cases of water-related diseases annually and roughly 5 to 10 million deaths (Nash 1993). The true extent of these diseases is unknown and even WHO data suggest there may be many more cases of disease annually. Most illnesses are undiagnosed and unreported, and comprehensive epidemiological studies are rare. Table 2.2 lists current estimates of the prevalence of water-related diseases worldwide.

Water-related diseases can be placed in four classes: waterborne, water-washed, water-based, and water-related insect vectors. Table 2.3 lists the diseases associated with each class.

Waterborne diseases include those for which water is the agent of transmission, particularly those pathogens transmitted from excreta to water to humans. These include most of the enteric and diarrheal diseases caused by bacteria, parasites, and viruses, such as cholera, giardia, and rotaviruses. Drinking water contaminated by human or animal excreta is the main source of water-related diseases. The first such diseases identified

TABLE 2.2   ESTIMATES OF GLOBAL MORBIDITY AND MORTALITY OF WATER-
RELATED DISEASES (EARLY 1990S)

Disease	Morbidity (episodes/year or people infected)	Mortality (deaths/year)
Diarrheal diseases	1,000,000,000	3,300,000
Intestinal helminths	1,500,000,000 (people infected)	100,000
Schistosomiasis	200,000,000 (people infected)	200,000
Dracunculiasis	150,000 (in 1996)	—
Trachoma	150,000,000 (active cases)	—
Malaria	400,000,000	1,500,000
Dengue fever	1,750,000	20,000
Poliomyelitis	114,000	—
Trypanosomiasis	275,000	130,000
Bancroftian filariasis	72,800,000 (people infected)	—
Onchocerciasis	17,700,000 (people infected; 270,000 blind)	40,000 (mortality caused by blindness)

Source: WHO 1995b.

were typhoid and cholera, both of which remain serious problems in many regions of the world. Also in this group are over 30 species of parasites that infect the human intestines, including helminths, viruses, and protozoa. While many of these are local problems, seven are distributed globally or cause serious illness: amoebiasis, giardiasis, *Taenia solium* taeniasis, ascariasis, hookworm, trichuriasis, and strongyloidiasis (WHO 1986). The best known of these is giardiasis, which is the most common animal parasite of humans and is widespread and resistant to chlorine used to treat drinking water. A billion people are thought to be infected with *Ascaris*, 500 million with the parasite responsible for amoebiasis, and 500 million with the helminth that causes trichuriasis (WHO 1986; Nash 1993).

Improvements in water quality directly reduce the incidence of waterborne diseases. In the early 1800s, Europe and North America dramatically improved public health by protecting and treating drinking water supplies, effectively eliminating cholera and typhoid. As a result, the diseases currently threatening public health are less virulent forms, such as *Giardia lamblia* and *Cryptosporidium*. Table 13 in the Data section at the back of the book shows the incidence of water-related disease outbreaks in the United States since 1970. In the United States, the incidence of waterborne diseases has dropped from roughly eight cases per 100,000 person-years between 1920 and 1940 to under four between 1970 and 1990, with the exception of a severe outbreak in Minneapolis in 1993 when over 400,000 cases caused by *Cryptosporidium* occurred, resulting in over 100 deaths.

Water-washed diseases result from inadequate sanitation or contact with contaminated water and can be prevented by washing with clean water. These include trachoma and typhus, which are eye and louse-borne diseases, and diarrheal diseases that can be passed directly from person to person.

Water-based diseases come from hosts that either live in water or require water for part of their life cycle. These diseases are passed to humans when they are ingested or come into contact with skin. The two most widespread examples in this category are schistosomiasis, which results from contact with snails that serve as hosts, and dracunculiasis

TABLE 2.3   CLASSIFICATIONS OF WATER-RELATED DISEASES

Category Agent	Infection	Pathogenic
Waterborne (fecal-oral)		
Diarrheas and dysentries	Amoebiasis	Protozoa
	Camphylobacter interitis	Bacterium
	Cholera	Bacterium
	E. coli diarrhea	Bacterium
	Giardiasis	Protozoa
	Rotavirus diarrhea	Virus
	Salmonellosis	Bacterium
	Shigellosis (bacillary dysentry)	Bacterium
Enteric fevers	Typhoid	Bacterium
	Paratyphoid	Bacterium
	Poliomyelitis	Virus
	Ascariasis (giant roundworm)	Helminth
	Trichuriasis (whipworm)	Helminth
	Strongyloidiasis	Helminth
	Taenia solium taeniasis (pork tapeworm)	Helminth
Water-washed	Infectious skin diseases	Miscellaneous
	Infectious eye diseases	Miscellaneous
	Louse-borne typhus	Rickettsia
	Louse-borne relapsing fever	Spirochete
Water-based	Schistosomiasis	Helminth
	Dracunculiasis	Helminth
	Clonorchiasis	Helminth
	Others	Helminth
Water-related insect vector	Trypanosomiasis (sleeping sickness)	Protozoa
	Filariasis	Helminth
	Malaria	Protozoa
	Onchocerciasis (river blindness)	Helminth
	Yellow fever	Virus
	Dengue fever	Virus
	Others	Virus

Source: Adapted from R.G. Feachem 1984.

(guinea worm), which results from ingesting contaminated host zooplankton. Schistosomiasis currently infects 200 million people in 70 countries, and the disease continues to spread where irrigation projects produce habitat that favors the host snails. Major outbreaks of schistosomiasis often follow the construction of large dams. In the Sudan, the construction of Sennâr Dam led to the infection of nearly the entire nearby population. Following the construction of the Aswan High Dam on the Nile, infection rates rose from close to zero to almost 100 percent. Infection rates in children near the Akosombo Dam in West Africa rose from less than 10 percent to more than 90 percent within a year of the reservoir's filling (Nash 1993).

The final category—diseases associated with *water-related insect vectors*—includes those spread by insects that breed or feed near contaminated water, such as malaria, onchocerciasis, trypanosomiasis, dengue fever, and yellow fever. While these diseases are not directly attributable to water quality, they are often spread by large-scale development of water systems that create conditions favorable for their hosts. Malaria, which is transmitted by mosquitoes, is the most serious, with more than two billion people in over 100 countries at risk of infection. Three to four hundred million people carry the parasite, and malaria accounts for 20 to 30 percent of all childhood deaths in these countries. Furthermore, the area at risk of malaria is expanding rapidly and strains of malaria with strong drug resistance are increasingly common.

Until clean drinking water and sanitation services are universally available, millions of people will die annually from preventable water-related illnesses. What might universal provision of clean drinking water and sanitation services achieve? Many studies have been done on the expected reduction in morbidity from improved water services (White et al. 1972; Cembrowicz 1984; Esrey et al. 1990). Analyzing their own field studies and the collected work of several hundred others, these authors offer estimates of how different water-related diseases would be affected. Using this work and data on current prevalence of these diseases produces rough estimates. Some diseases, like dracunculiasis (described in detail below) or typhoid may be completely eliminated. Tens or even hundreds of millions of cases of diarrheal diseases and perhaps a million or more deaths can be prevented annually. Urinary schistosomiasis can be almost completely eliminated and the incidence of intestinal schistosomiasis cut in half, saving over 150,000 lives each year and reducing the number of people infected by 100 million. The actual numbers will depend on population size, the organized response of the health community, and other factors hard to predict, but there is no doubt that vast improvements in human health and well-being will result.

Two studies—one on dracunculiasis (guinea worm), the other on cholera—presented below show what the world can achieve when the right political, economic, and medical efforts are applied, and what human suffering and misery result when we fail to make those efforts. Both diseases are closely related to poor drinking water quality and one, cholera, is related to lack of adequate sanitation as well. Cholera has undergone a resurgence in the past few years and has become epidemic in Latin America for the first time this century. At the same time, it has expanded in Africa and a new epidemic form has appeared in Asia. While treatment is inexpensive and simple, death rates in Africa remain high, and no inexpensive effective vaccine has yet been developed.

During the same period, however, the prevalence of dracunculiasis has been greatly reduced through a comprehensive preventative health program sponsored by the World Health Organization, the Centers for Disease Control and Prevention, national governments, and nongovernmental organizations such as Global 2000. The goal is complete eradication, and there is a chance that goal may be achieved within the next five years.

Update on Dracunculiasis (Guinea Worm)

The Disease

Dracunculiasis is a parasitic disease that causes dreadful suffering and disability among the world's poorest populations. It is the only disease solely associated with unhealthy drinking water. Water is often scarce in countries with endemic dracunculiasis and the parasite is passed along through contamination of local drinking water supplies, often when a single stagnant source supplies the domestic requirements for an entire community.

Dracunculiasis was first documented over 3,400 years ago, in the fifteenth century B.C. The first known mention of the disease is found in an ancient document, the "Turin Papyrus." Physical evidence was found when a pathological examination of an Egyptian mummy turned up the remains of a calcified *Dracunculus medinensis* worm (WHO 1997a). In the eleventh century A.D., Abou Ali ibn Sina gave detailed descriptions of the evolution, transmission, and treatment of dracunculiasis, which was prevalent in Persia during this period. The celebrated Swedish naturalist Linnaeus first suggested that the disease was caused by worms, and the life cycle of the parasite was described in 1870 by Alexei P. Fedchenko.

This round worm is the largest of the tissue parasites affecting humans. The adult female, which carries from 1 to 3 million embryos, can measure up to 1 meter (m) in length and 2 millimeters (mm) across. The parasite migrates through the victim's body causing severe pain, especially in the areas around the joints. The worm eventually emerges from the skin causing intense pain, a blister, and then an ulcer, accompanied by fever, nausea, and vomiting. Partial or total disability can last several months. Besides the burning pain associated with the worm's emergence, more than half of the blister sites develop secondary infections resulting in sepsis, abscesses, arthritis, or potentially fatal tetanus. A small fraction of victims may become permanently crippled.

When someone infected with the parasite steps into water, the terrible burning sensation caused by the emerging worm is relieved. The cool water also induces a contraction of the female worm at the base of the ulcer, causing the sudden expulsion of hundreds of thousands of embryos and thus contaminating the water source. If the now totally empty worm is not carefully extracted from the patient's body, it dies and becomes a source of bacterial infection.

When embryos of the guinea worm are released in stagnant water they are swallowed by zooplankton (cyclops) that serve as an intermediate host. The embryonic parasite punctures the digestive tract of the cyclops and makes its way to its abdomen where, over a period of approximately two weeks, it is transformed into an infectious larva.

The loop is closed and dracunculiasis is returned to the human hosts when water containing the infected cyclops is drunk. The cyclops are destroyed when they come into contact with the human digestive system and the infectious larvae are liberated. The larvae puncture the intestine of the drinker and enter the tissues. Three months later when they reach maturity, the young male and female worms mate. As long as the worms remain confined to the deep tissue surrounding the lymph glands, they do not present a problem to the carrier. One year after the water containing the infected zooplankton host has been consumed, the fully grown female worms start to migrate through the infected person's body, giving rise to the characteristic symptoms of dracunculiasis.

Given its transmission through unclean drinking water, the disease generally affects the poorest and most remote rural populations in the developing world, which have limited or no access to safe sources of drinking water. By affecting the most productive members (farmers, mothers, and schoolchildren) of a society, the disease severely hinders household functioning and community development. When women, who play a central role in household activities, are infected, there have been observed declines in nutritional levels, sanitary levels, and child health care. School absenteeism exceeding 60 percent has been observed either because children themselves are sick or because they must carry out chores of sick family members. When farmers and agricultural workers are affected, agricultural productivity and output decline (UNICEF 1997). Because of the ability to interrupt the life cycle of the parasite, the World Health Organization (WHO) in the 1980s, in collaboration with United Nations organizations and national governments, set as a goal the eradication of dracunculiasis, and tremendous progress toward this goal has been made.

The Eradication Program

Guinea worm disease can be eradicated for several reasons: (1) there is no human carrier beyond the one-year incubation period; (2) there is no known animal reservoir; (3) detection of infections (i.e., worms protruding from skin lesions) is an easy way to identify the presence of the disease in communities; (4) protrusion of the worm is required for transmission of the disease; (5) transmission is highly seasonal, facilitating monitoring, control, and containment; (6) the methods for controlling transmission are simple; and (7) the disease is well recognized by the local population in areas where it occurs.

By the end of the nineteenth century, the scientific community understood how the disease was transmitted and had started to advocate suitable protective measures. Between 1926 and 1931, dracunculiasis was totally eradicated from Uzbekistan following a series of effective health education, water purification, and carrier control programs. No recurrence of the disease has been recorded in this region since 1932. In the 1970s, dracunculiasis was eradicated in Iran (and officially certified free of the disease by WHO in 1997). In 1986, the World Health Assembly adopted a resolution making guinea worm the first disease to be targeted by WHO for eradication after smallpox in 1977. Other resolutions by African ministers of health in 1988, by the Executive Board of UNICEF in April 1989, and by the UNICEF-sponsored World Summit for Children in September 1990 included the goal of eradication.

In 1991, the World Health Assembly adopted resolution WHA44.5 requesting that dracunculiasis be eradicated by the end of the 1990s. In 1992, in order to speed up guinea worm eradication, the WHO and UNICEF set up the WHO/UNICEF Interagency Technical Team to assist the Programme for the Eradication of Dracunculiasis in Africa. Substantial attention was given to local health education and surveillance and the mobilization of women in rural areas.

A consortium of agencies, institutions, and organizations—including the Centers for Disease Control and Prevention (CDC), the World Health Organization, UNICEF, the United Nations Development Programme (UNDP), the World Bank, Global 2000, government and bilateral aid organizations in Japan, Saudi Arabia, the United Arab Emirates, Sweden, Denmark, France, Canada, the Netherlands, Norway, and the United States—provide assistance to affected countries. Private corporations (including DuPont Company, Precision Fabrics Group, and American Cyanamid Corporation) are also involved, donating insecticide to kill the disease hosts and cloth filters to strain drinking water.

National governments of all the endemic countries have established eradication programs and appointed national program coordinators. The head of state of Ghana visited 21 villages affected by the disease in 1988 as part of a national education campaign. In 1989, the Nigerian government committed US $1 million to dracunculiasis eradication and announced it would use the presence of the disease as the primary criterion for identifying rural water-supply improvement projects. The president of Burkina Faso, Capt. Blaise Compaore, presided at the opening ceremony of the Fifth African Regional Conference on Dracunculiasis Eradication in Ouagadougou in March 1994. In Mali, the former head-of-state, General A.T. Touré, leads the effort as chairman of the Guinea Worm Eradication Intersectorial Committee and has been instrumental in mobilizing the support of other African leaders for dracunculiasis eradication.

The overall strategy of national eradication programs consists of three phases: (1) conducting baseline surveys to identify villages where the disease is endemic; (2) training village-based health workers to use case registries for monthly surveillance and to implement control measures in affected villages; and (3) containing cases where they occur by treating lesions to prevent transmission and protecting water supply from contamination (Ruiz-Tiben et al. 1995).

The thrust of control strategies is to educate people about the origin of the disease and about the measures they and their communities can take to prevent it. Affected households are often provided with cloth filters and taught how to use them to remove the infected hosts from water. Household members are also taught how to prevent worms from entering sources of drinking water. Other control interventions include providing safe drinking water to affected villages and selectively using insecticide to reduce populations of the host.

The Current Situation

Although these actions have been effective and good progress has been made toward complete eradication, dracunculiasis is still prevalent in 17 developing countries, including Yemen, India, and 15 African nations. Map 2.3 shows the band of countries where dracunculiasis was found in 1996. Pakistan is the first country to have completely eradicated the disease during the new global program for dracunculiasis eradication, having reported zero cases every month since October 1993, and this in the face of ongoing disease surveillance strengthened by the offering of rewards throughout the country. As of December 31, 1996, five of the countries with the disease reported fewer than 100 cases for the whole year, including Kenya, which reported zero cases for the first time since its active containment program was implemented. In India, dracunculiasis was eliminated from Tamil Nadu in 1984, Gujarat in 1989, and Maharashtra in 1991, and only nine cases were reported for the whole country in 1996. Table 12 in the Data section at the back of the book shows all cases for the years 1972 to 1996.

Globally, the number of cases has dropped 97 percent during the past decade, from an estimated 3.5 million cases in 1986 to about 150,000 cases worldwide during 1996 (CDC 1995a; WHO 1997a). Of the remaining cases nearly 80 percent occurred in the Sudan, which has been racked by armed conflict that hampers prevention efforts.

As part of the eradication effort, a list has been drawn up of countries and regions with a past record of human dracunculiasis. The list covers four groups of countries:

- 18 countries in which transmission was endemic during the 1980s;

- 9 countries in which transmission was endemic between 1940 and 1980;

- 40 countries in which the disease was likely to have been prevalent before 1940; and

- Korea and Japan—where only two indigenous cases without any endemic transmission have been reported during the last 60 years or so.

A three-year monitoring period has been implemented for the 27 countries of the first two groups to ensure that the disease does not reappear among the native population

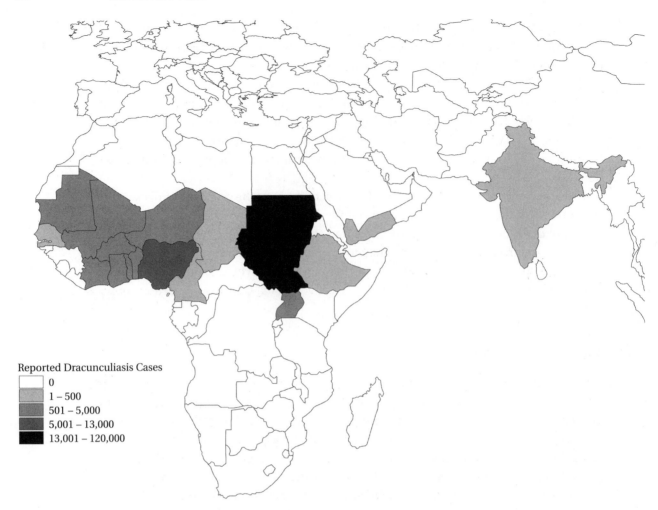

MAP 2.3   REPORTED DRACUNCULIASIS CASES, 1996

after eradication. The 40 countries in the third group will be monitored to ensure that dracunculiasis is not reintroduced from adjacent countries. This calls for the isolation of patients and meticulous checking of imported cases to ensure that they do not become the source of new outbreaks. All affected countries use a "case-containment" approach for eradication, where the goal is to detect every single case of dracunculiasis in even the most remote village no later than 24 hours after the worm has begun emerging through the skin. Village health workers then work to prevent or contain any further transmission of the disease by a combination of individual and community-wide actions. Progress in several endemic countries is described below.

Ghana conducted its first national-scale screening campaigns in 1989 with active surveillance. In Ghana, the number of cases dropped from 179,556 in 1989 to 8,432 cases in 1993 to fewer than 5,000 in 1996—a reduction of 97 percent. Ethnic violence during 1994 in the northern region of the country, which has higher rates of the disease, temporarily side-tracked efforts at eradication. Concerted efforts to reinstitute eradication efforts rapidly when the violence subsided resulted in quick detection and full containment of 97 percent of the cases by late 1995 (WHO 1994a; WHO 1995a).

Nigeria's first national screening campaign was undertaken in 1988, and its eradication efforts have reduced the annual incidence from 653,492 cases in 1988 to 35,749 cases in 1994 to 12,282 cases in 1996 (CDC 1994; WHO 1997b).

Niger and Mali, countries in West Africa, ranked fifth and seventh in the number of reported cases in 1996. In 1990, prior to active searches for the disease, Mali reported 884 cases of dracunculiasis to WHO (WHO 1993). During that year, health officials in Mali initiated a pilot project to control dracunculiasis by training village-based health workers to conduct health education, initiating active surveillance, and distributing nylon cloth to families for filtering drinking water (Hopkins and Ruiz-Tiben 1991). From December 1991 through March 1992, national village-by-village searches detected 16,060 cases of dracunculiasis in 1,264 villages in five of seven regions of the country. In March 1993, Global 2000, Inc., and the WHO Collaborating Center for Research, Training, and Eradication of Dracunculiasis at CDC began providing direct assistance for the eradication of dracunculiasis by assigning a resident public health advisor.

By December 1993, Mali's Guinea Worm Eradication Program (GWEP) had trained one village-based health worker in each of 1,100 villages with endemic dracunculiasis and had begun monthly reporting of cases from 433 such villages. By 1996 Mali reported only 2,402 cases—a drop of 85 percent. In 1989, prior to active case surveillance, Niger reported 288 cases of dracunculiasis to the World Health Organization. From October through November 1991, national village-by-village searches detected 32,829 cases of dracunculiasis in 1,690 villages and an eradication program at the village level was introduced. In 1996, even with more complete monitoring, Niger reported only 2,956 cases nationwide—a drop of 90 percent.

India launched its eradication program in 1984 and seems on the verge of success. The number of recorded cases has dropped successively from 39,792 in 1984 to 4,800 cases in 1990 to only 9 cases in 1996. Cameroon is close to having totally eradicated the disease. The number of cases has been reduced from a high of 871 in 1989 to 17 in 1996. Benin, Burkina Faso, Chad, Côte d'Ivoire, Ethiopia, Kenya, Mauritania, Senegal, Sudan, Togo, Uganda, and Yemen also all have active, local dracunculiasis eradication programs. On a world scale, the estimated cost to the World Health Organization of certifying dracunculiasis eradication up to the year 2000 is estimated at a meager US $1.5 million per year.

During eradication efforts, an active community member provides notification of all new cases, teaches community health, distributes filters, and cares for patients. In Mauritania, this role has regularly been assigned to a woman, which has proven to be particularly effective (Carter Center 1997). As countries get closer to eradicating the disease, the cost per case prevented substantially increases and the interest in dracunculiasis as an important public health issue diminishes. Effective strategies aimed at identifying and containing isolated cases would ensure the transition from low prevalence to actual eradication.

As the eradication program has progressed, villages have begun to demand safe sources of drinking water from their governments and to push water authorities to address the issue of maintenance of local systems. The obvious benefits of these actions led CDC, WHO, UNICEF, and Global 2000 to sign a joint agreement in 1993 to use the guinea worm eradication program's surveillance system to test a more integrated system to monitor for other diseases as well. Ultimately, guinea worm will be eradicated most effectively by providing protected clean drinking water in all endemic regions. This can be done by drilling protected wells, by providing basic water treatment through filtering or chemical disinfection, and by educating rural populations about the need to protect water supplies.

Update on Cholera

The Disease

Cholera is an acute, diarrheal illness caused by infection of the intestine with the bacterium *Vibrio cholerae*. Although cholera can be life-threatening, it is easily prevented and treated. Despite this, cholera is epidemic in many developing countries because of the failure to provide adequate sanitation and clean drinking water. Cholera was prevalent in the United States and many other now industrialized countries in the 1800s, but it was virtually eliminated by modern sewage and water-treatment systems. As a result, cholera has been very rare in industrialized nations for the past 100 years. Almost all cases of cholera in recent years in these countries occur in travelers returning from areas where epidemic cholera exists, including the Indian subcontinent, sub-Saharan Africa, and Latin America.

While eating contaminated food can transmit cholera, a significant majority of cases are related to waterborne transmission. In detailed epidemiological assessments of the ongoing Latin America outbreak, waterborne transmission was identified in seven of every eight cases, and health scientists concluded that the first stage in prevention was to provide safe drinking water. They declared that "the longstanding deficits in basic urban infrastructure and the need for new efforts to correct them have never been more apparent" (Tauxe et al. 1995).

A person may get cholera by drinking water or eating food contaminated with the cholera bacterium, *Vibrio cholerae*. Only rarely is cholera transmitted by direct person-to-person contact. Sudden large outbreaks are usually caused by a contaminated water supply and inadequate treatment of sewage. The bacterium can survive in fresh water for long periods. In highly endemic areas it is mainly a disease of young children, although breastfeeding infants are rarely affected. Cholera bacteria may also live in the environment in brackish rivers and coastal waters and marine shellfish and plankton serve as the main reservoirs.

Cholera has a short incubation period, from less than one day to five days. Most persons infected with *V. cholerae* do not become ill, although they play an important role in carrying *V. cholerae* from place to place, causing epidemics to spread. Approximately 1 in 20 people infected with cholera experiences severe disease characterized by profuse watery diarrhea, sometimes vomiting, and leg cramps. In these people, rapid loss of fluids leads to dehydration and shock. While treatment for cholera is straightforward, death can occur within hours without treatment (CDC 1996).

Until the 1800s cholera was found almost entirely in the Ganges/Brahmaputra River basins in Asia (Epstein 1992). From 1817 to 1923, cholera spread in six major pandemics through Europe, England, and into the United States along international trading routes. By the end of 1832, cholera was found in most major urban centers in the United States and severe cholera epidemics were prevalent in many parts of the world, including England (Epstein 1992). In the late 1840s, a London physician, John Snow, proposed that cholera was a contagious disease caused by a poison that reproduces in the human body and is found in the vomitus and stools of cholera patients. He believed that the principal means of transmission was contaminated water.

Snow was able to prove his theory in 1854, when another severe epidemic occurred. By documenting and mapping cholera cases among subscribers to London's two water companies, he showed that cholera occurred much more frequently in customers of a water company that drew its water from the lower Thames, where it had become contaminated with London sewage. The other company obtained cleaner water from the upper Thames.

Even more compelling evidence soon appeared. In one neighborhood, the concentration of cholera cases was so great that the number of deaths reached over 500 in 10 days. Snow carefully mapped the cases and observed that they centered around a particular water pump. He advised officials to remove the pump handle, successfully halting the epidemic. Snow's fame has led to the creation of a John Snow Society, whose membership consists of people who have visited the John Snow Pub, which claims ownership of the "original" pump handle (Snow 1936; Rosenberg 1962).

Treatment and Prevention

Most cases of cholera caused by *V. cholerae* can be easily and successfully treated by oral rehydration therapy: immediate replacement of the fluid and salts lost through diarrhea. Standard prepackaged mixtures of sugar and salts are available to be mixed with clean water. This solution is used throughout the world to treat diarrhea. Severe cholera cases may require intravenous fluid replacement.

When cholera occurs in an unprepared community, fatality rates may be as high as 50 percent—usually because there are no facilities for treatment, or because treatment is given too late. Up until the 1970s, fatality rates in Asia and Africa were often this high. In contrast, a well-organized response in a country with a well-established diarrheal disease control program can limit the case-fatality rate to less than 1 percent. In the recent epidemic in Latin America, effective response typically limited the death rate to about 1 percent. In Africa, where institutional resources and health care are often more limited and problematic, death rates of 10 percent have been common. In the past several years, the average death rate in all of Africa has dropped to between 3 and 6 percent. Figure 2.1 shows cholera deaths as a fraction of total reported cases, by region from 1970 to 1997. Complete data on cholera can be found in Table 9 in the Data section at the back of the book.

In severe cases, an effective antibiotic can reduce the volume and duration of diarrhea and the period of vibrio excretion. Tetracycline is the usual antibiotic of choice, but resistance to it is increasing. Routine treatment of a community with antibiotics has no effect on the spread of cholera, nor does restricting travel and trade between countries or between different regions of a country. Setting up health barriers at borders to try to control the spread of cholera uses personnel and resources that are more effectively devoted to control measures.

Oral vaccines that provide limited protection for several months against one form of cholera, *V. cholerae O1*, have recently become available in a few countries, but these vaccines do not appear to protect against all forms of epidemic cholera and the Center for Disease Control in Atlanta does not recommended the vaccine for travelers (CDC 1988; MacPherson and Tonkin 1992). New oral cholera vaccines are being developed and may

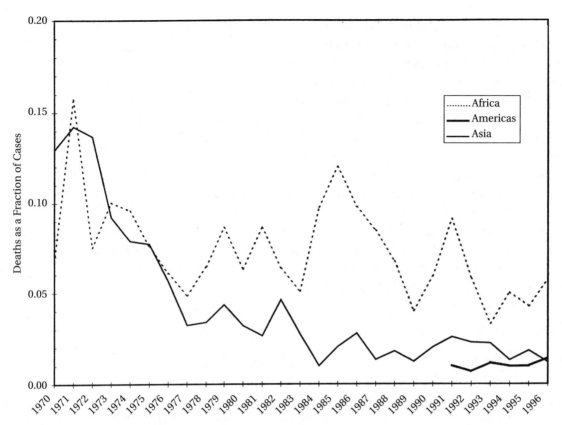

FIGURE 2.1   REPORTED CHOLERA DEATHS AS A FRACTION OF CASES BY REGION, 1970 TO 1996.

The trend in reported cholera deaths as a fraction of total reported cases from Table 9 in the Data section is plotted here for Africa, the Americas, and Asia. This fraction exhibits a downward trend, most notably in Asia, which could reflect fewer deaths from better health care, or better reporting of cases. Data for Europe and Oceania are not included in the figure because of the low numbers of cases and deaths in those regions.

Source: September 1997, April 1998, Division of Emerging and Other Communicable Diseases Surveillance and Control, Global Task Force on Cholera, World Health Organization, Geneva, personal communication.

provide more reliable protection, although at a high cost per case averted. No vaccines have the combination of high efficacy, long duration of protection, simplicity of administration, and low cost necessary to make mass vaccination feasible in cholera-affected countries.

In the absence of a low-cost, effective vaccine, treatment and prevention require three things: hygienic disposal of human wastes, an adequate supply of safe drinking water, and proper food hygiene. Effective food hygiene measures include cooking food thoroughly; preventing cooked foods from being contaminated by raw foods, contaminated surfaces, or flies; and avoiding unpeeled raw fruits or vegetables.

Cholera in the Late 1990s

The vibrio responsible for the seventh cholera pandemic, now in progress, is known as *V. cholerae O1*, biotype El Tor. This bacterium takes its name from the El Tor quarantine camp in Sinai, Egypt, where it was first isolated in 1906. The current pandemic began in 1961 in the Celebes (Sulawesi), Indonesia. The disease then spread rapidly to other countries of eastern Asia and reached Bangladesh in 1963, India in 1964, and the USSR, Iran, and Iraq in 1965 and 1966. In 1970 cholera reached West Africa, which had not experienced the disease for more than 100 years. It then quickly spread throughout most of the continent.

Cholera has become much more prevalent in the past decade. In part this may be due to improvements in reporting, but there has also been an expansion in geographical scope and intensity. Beginning in 1900, the total number of cases reported annually has rarely exceeded 100,000 (though systematic assessment began only around 1970). Even with the outbreak of the seventh pandemic in the 1960s, total reported cholera cases exceeded 100,000 only twice until 1991.

In January of that year cholera reached Peru. Within two weeks, over 12,000 cases were reported. Within three months, cholera had reached Ecuador, Colombia, Chile, and Brazil, and it continued to spread with explosive rapidity. Within a year cholera was epidemic in 11 countries in Latin America, which had been free of cholera for over 100 years (Levine 1991). By the end of 1991, nearly 600,000 cases had been reported worldwide, 390,000 of them in Latin America. Figure 2.2 shows total global cholera cases, separated by regions, for 1970 to 1997, with the significant outbreak in Latin America. Figure 2.3 shows reported cholera deaths during the same period for the same regions. Several reports suggest that the enormous outbreak was initiated by the dumping of contaminated

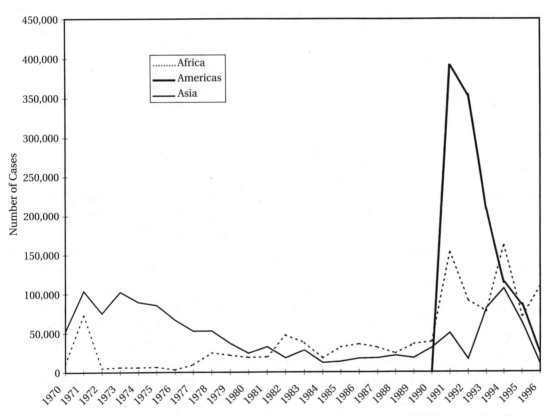

FIGURE 2.2 REPORTED CHOLERA CASES BY REGION, 1970 TO 1996.
Data on reported cholera cases from Table 9 in the Data section are plotted for Africa, the Americas, and Asia. The epidemic in the Americas is striking here, in particular both the rapid *increase* and rapid *decrease* in the number of cases. Data for Europe and Oceania are omitted because of the small number of cases in those regions. *Source:* September 1997, April 1998, Division of Emerging and Other Communicable Diseases Surveillance and Control, Global Task Force on Cholera, World Health Organization, Geneva, personal communication.

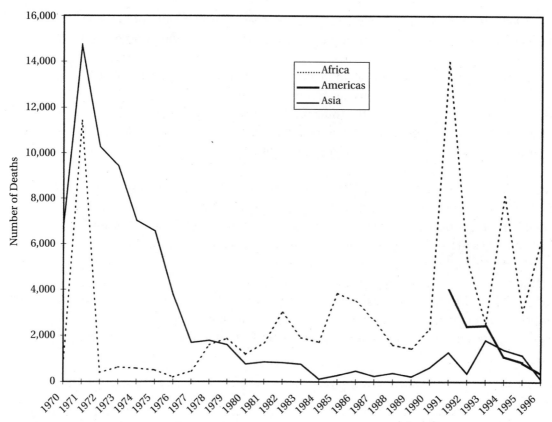

FIGURE 2.3   REPORTED CHOLERA DEATHS BY REGION, 1970 TO 1996.

Data on reported cholera deaths from Table 9 in the Data section are plotted here for Africa, the Americas, and Asia. Although the number of cases in the Americas was nearly three times that in Africa in the early 1990s, the number of deaths has been consistently much lower. Data for Europe and Oceania are omitted because of the small number of cases in those regions. *Source:* September 1997, April 1998, Division of Emerging and Other Communicable Diseases Surveillance and Control, Global Task Force on Cholera, World Health Organization, Geneva, personal communication.

bilge water from a Chinese freighter in Lima harbor (Anderson 1991) and then amplified in the plankton. The rapid spread was likely caused by the lack of urban infrastructure capable of providing clean drinking water and sanitation services in Peru and other countries of the region. Lima, Peru, has a population exceeding 4 million living in over 2,000 shantytowns without adequate water services. Some researchers also believe the outbreak in Latin America was accelerated by the spread of the cholera bacterium through coastal algal blooms associated with warmer ocean waters and heavy rains that increased nutrient runoff accompanying the El Niño (*The Economist* 1995; Patz et al. 1996). There are also reports that a cut in chlorination in the limited waste treatment plants in Lima accelerated the spread (Epstein 1992). These reports suggest that misguided fears about chlorination by-products, a growing concern in developed countries, led water managers to ignore the much larger threat associated with diseases prevented by chlorination. While this may have helped the epidemic to spread, it cannot explain the magnitude and scope of the 1991 Latin American outbreak. The cholera epidemic also had more direct economic costs. Peru alone lost over $1 billion in seafood exports and tourist revenues (Epstein and Nutter 1997).

In Latin America the *V. cholerae O1* epidemic has caused a total of 1,196,000 cases and over 11,300 deaths (overall case-fatality rate: 0.95 percent) through the end of 1997 (CDC 1993a; CDC 1995b; WHO 1996; personal communication, WHO 1998). While the number of cases has dropped from a high of 390,000 in 1991 to under 18,000 in 1997, cholera seems well entrenched. In areas with inadequate sanitation, a cholera epidemic cannot be stopped immediately, and there are no signs that the epidemic in the Americas will end soon. Major improvements in sewage and water-treatment systems are urgently needed in many of these countries before epidemic cholera can be eliminated.

While the epidemic in Latin America received the most attention, two other events occurred in the early 1990s that warrant watching. In the midst of the Latin American outbreak, the total number of cases in Africa quadrupled, and a completely new vibrio capable of causing epidemic cholera appeared in Asia. Until 1992, only the El Tor biotype caused epidemic cholera, though other types could cause sporadic cholera cases (CDC 1996). In October 1992, an epidemic of cholera-like illness began in Madras, India. In early 1993, similar epidemics began in Calcutta (with more than 13,000 cases) and in Bangladesh (with more than 10,000 cases and 500 deaths). The cholera associated with these epidemic outbreaks could not be identified as any of the 138 known types of *V. cholerae* and was given the designation of a new serogroup, O139 (nicknamed "Bengal") (Shimada et al. 1993). In early 1993, a traveler to India returned to the United States with a case of cholera subsequently identified as the Bengal biotype (CDC 1993b).

Total cholera cases in Asia exceeded 100,000 in 1994, and by 1995 the epidemic of cholera caused by *V. cholerae O139* had affected at least 11 countries in the region (CDC 1996). Specific totals for numbers of *V. cholerae O139* cases are unknown because affected countries did not report infections caused by O1 and O139 separately. A report in the British journal *Lancet,* however, suggests that more than 100,000 cases of cholera caused by *V. cholerae O139* may have occurred in just the first year of the epidemic in Asia (ICDDR 1993). This new strain largely replaced *V. cholerae O1* throughout the Bay of Bengal region in the 1994.

Surprisingly, in 1995 and early 1996, the El Tor biotype reappeared in much of the region displacing the Bengal form, leading some investigators to believe that the emergence of *V. cholerae O139* was a one-time event. Continuing surveillance, however, has now revealed the resurgence of O139 since August 1996. In the past year, the incidence of the Bengal form has exceeded that of the El Tor form (Mitra et al. 1996). Tests of 20 different strains of *V. cholerae O139* show complete resistance to ampicillin and furazolidone and significant resistance to neomycin, streptomycin, and tetracycline, antibiotics often used to treat other cholera strains (CDC 1993b; Mitra et al. 1996).

The emergence of this new strain has at least three other major public health implications (CDC 1993b). First, it expands the definition of cholera beyond the illness caused exclusively by the El Tor form. Because it appears to cause the same illness and to have similar epidemic potential, the World Health Organization has asked all nations to report illnesses caused by this strain as cholera. Second, the rapid spread of the *V. cholerae O139* epidemic in southern Asia, even among adults previously exposed to cholera, suggests that existing immunity to *V. cholerae O1,* whether the result of natural infection or cholera vaccine, offers little or no protection. Third, many laboratories use identification methods that will not detect the new strain. As a result, some laboratories may assume they have detected a nonepidemic form and fail to take appropriate preventative measures. This could contribute to preventable outbreaks of epidemic cholera.

The El Tor cholera epidemic in Africa has lasted more than 20 years and recently worsened. During the 1970s and 1980s, total reported cholera cases never exceeded 47,000 and were regularly lower. In 1991, some 21 African countries reported 153,367 cholera cases and 13,998 cholera-related deaths, representing 26 percent of all reported cases and nearly three-quarters of cholera-associated deaths for that year (WHO 1992a). Not once during the 1990s have fewer than 70,000 cases been reported in Africa. In 1996, some 108,535 cases in Africa were reported by 29 countries. In 1997, nearly 120,000 cases were reported by 25 countries, with many countries not reporting. Furthermore, data are notoriously incomplete and the number of cases and deaths may be substantially underestimated. The 1990s cholera case-fatality rates of between 3 and 9 percent reported in Africa are lower than rates of 30–50 percent reported in the 1970s but far higher than the rates found in most of the rest of the world. See Table 9 in the Data section at the back of the book.

Improving cholera surveillance and developing a coordinated response for epidemic cholera are high public health priorities in Africa. The first priority is to prevent cholera-associated deaths by providing vigorous rehydration therapy to affected persons (WHO 1992b). The proportion of childhood diarrheal episodes being treated with oral rehydration increased from an estimated 4 percent in 1984 to 40 percent in 1991 (WHO 1992c).

Determination of the routes of cholera transmission is important in developing effective prevention measures. Studies from a major outbreak in Burundi in 1992 and 1993 showed that cholera victims were significantly more likely than healthy villagers to have drunk untreated water from or bathed in Lake Tanganyika during the three days before onset of illness. Because of a shortage of potable water and an insufficient number of working water taps in Rumonge, untreated lake water was often used for domestic purposes. Access to a functioning water tap during the three days before onset of illness was highly effective in preventing the disease (CDC 1993c).

Because waterborne transmission of cholera in Africa is associated with drinking untreated water from rivers and shallow wells, one strategy for preventing cholera is the provision of disinfected drinking water to persons residing in areas at risk. Boiling water is effective but consumes scarce fuel wood and is difficult to sustain. Chlorination is the most widely used method for purifying municipal water supplies. Providing safe, treated water supplies also may prevent other waterborne diseases (e.g., typhoid fever, hepatitis, and other diarrheal illnesses in children) (CDC 1993c).

Efforts to control cholera epidemics in Africa and elsewhere by mass chemoprophylaxis, vaccination campaigns, roadblocks, and broad embargoes on commodities have been ineffective and have diverted scarce resources away from the critical activities of providing treatment and improving the safety of water and food supplies. Adequate surveillance of the marine environment and of freshwater distribution systems can guide the rational distribution of treatment and prevention supplies. Rapid and thorough investigation of outbreaks can identify unsuspected sources of the infection, can assess the adequacy of treatment, and are essential to development of future prevention efforts.

During a 1994 outbreak in Rwanda and along the Rwanda–Zaire border, a strain of *V. cholerae O1* was identified to be highly resistant to antibiotics normally recommended for the treatment, such as tetracycline (WHO 1994b). WHO has designated furazolidone as the appropriate and recommended antibiotic in this case. Box 2.1 offers some exam-

---

BOX 2.1

Examples of Cholera Outbreaks, 1996–1997

In September 1996, the Philippine government declared a cholera outbreak in Manila after 170 people were admitted to hospitals in four days. The cause was thought to be contamination of drinking water pipes by sewage (AP 1996a).

In late November 1996, Zambia's leading independent newspaper reported a cholera outbreak in the northern part of the country associated with contaminated drinking water (*The Post* 1996).

Fighting in Monrovia, Liberia, in 1996 caused the destruction of water treatment and supply facilities and led to an outbreak of cholera, including fatal cases among refugees seeking refuge in the United States embassy compound (AP 1996b).

In Mongolia, a large cholera outbreak spread to the major cities in August 1996, leading to a quarantine of 1,000 foreigners in Ulan Bator (AP 1996c).

Maputo, the capital of Mozambique, has suffered from an outbreak of cholera associated with contaminated drinking water supplies in the city starting in September 1997. The outbreak then spread into the south of the country. By November, over 6,000 cases had been recorded and 56 people were reported dead (Reuters 1997).

A renewed outbreak of cholera in Dar-es-Salaam, Tanzania, killed 25 people in ten days in November 1997. A total of 255 cases was diagnosed, suggesting either a high death rate or a low rate of reporting of illness (*USA Today* 1997).

---

ples from around the world of cholera outbreaks reported by local or international media for 1996 and 1997.

Summary

Despite repeated calls for the world community to meet basic human needs for water, too little progress has been made, in part because of a lack of effort to define the needs. Unless basic needs are met, large-scale human misery and suffering will continue and grow in the future. Diseases associated with inadequate access to clean drinking water or inadequate sanitation services remain a scourge throughout the world, despite the fact that society knows what is necessary and has the means to reduce or eliminate them. These diseases cost society hundreds of billions of dollars a year in death, illness, and lost productivity, as well as huge uncounted costs in social and cultural disruptions. Far less money is needed to meet basic human needs for water, though those paying for water services are not always those that bear the costs of water-related illness.

Some progress is being made. In particular, dracunculiasis appears on the verge of eradication. This horrific disease used to afflict millions in the endemic countries of Africa and Asia, but sustained surveillance and eradication programs since the 1980s have decreased its prevalence by over 90 percent. While the disease has a number of characteristics that make it particularly susceptible to eradication, the most important factor is provision of safe, clean drinking water supplies. With continued effort, this disease will soon follow smallpox and be eliminated from the world.

Another severe water-related disease, cholera, cannot be eradicated by the same means as dracunculiasis. Indeed, the prevalence of cholera has been growing in recent years, with new epidemics and new strains appearing. The bacterium can survive for long periods in fresh water, inhabit shellfish and plankton, survive in humans without causing adverse symptoms, and is often not recognized by the local population or health officials. Nevertheless, epidemic cholera has been largely exterminated in industrialized countries with modern operating water and waste-treatment facilities. While inexpensive and effective treatment for the disease is available, prevention is an achievable and desirable goal and requires refocusing global efforts to provide basic water services for all people.

REFERENCES

Anderson, C. 1991. "Cholera epidemic traced to risk miscalculation." *Nature,* Vol. 354, p. 255.

Associated Press (AP). 1996a. "Government declares cholera outbreak in Manila." Associated Press. September 6.

Associated Press (AP). 1996b. "Diplomats urge end to Liberian fighting; cholera at U.S. compound." B. Duff-Brown. Associated Press. April 18.

Associated Press (AP). 1996c. "More than 1,000 foreigners quarantined in cholera outbreak, Ulan Bator, Mongolia." Associated Press. August 15.

Bhatia, R., and M. Falkenmark. 1993. "Water policies and the urban poor: Innovative approaches and policy implications." *Water and Sanitation Currents.* World Bank, Washington, D.C.

Carter Center. 1997. "The Global 2000 dracunculiasis program." Overview documents. The Carter Center, Atlanta, Georgia.

Cembrowicz, R.C. 1984. "Technically, socially and economically appropriate technologies for drinking water supply in small communities," in U. Neis and A. Bittner (eds.) *Water Reuse: Selected Reports in Water Reuse in Urban and Rural Areas.* Institute of Water Resources Engineering, Karlsruhe Institute for Scientific Co-operation, Tubingen, Germany.

Centers for Disease Control (CDC). 1988. "Cholera vaccine." *Morbidity and Mortality Weekly Report (MMWR),* Vol. 37, pp. 617–624.

Centers for Disease Control (CDC). 1993a. "Update: Cholera—Western hemisphere, 1992." *Morbidity and Mortality Weekly Report (MMWR),* Vol. 42, pp. 89–91.

Centers for Disease Control (CDC). 1993b. "Imported cholera associated with a newly described toxigenic Vibrio cholerae O139 strain—California, 1993." *Morbidity and Mortality Weekly Report (MMWR),* Vol. 42, pp. 92–95.

Centers for Disease Control (CDC). 1993c. "Epidemic cholera—Burundi and Zimbabwe, 1992–1993." *Morbidity and Mortality Weekly Report (MMWR),* Vol. 42 (21), June 04, Via *http://www.cdc.gov/epo/mmwr/mmwr.html.*

Centers for Disease Control (CDC). 1994. "Update: Dracunculiasis eradication—Ghana and Nigeria, 1993." *Morbidity and Mortality Weekly Report (MMWR),* Vol. 43, pp. 293–295.

Centers for Disease Control (CDC). 1995a. "Progress toward the eradication of Dracunculiasis (guinea worm disease): 1994." *Emerging Infections Diseases (EID),* Vol. 1, No. 2, (April–June). Atlanta, Georgia.

Centers for Disease Control (CDC). 1995b. "Update: Vibrio cholerae O1—Western Hemisphere, 1991–1994, and V. cholerae O139—Asia, 1994." *Morbidity and Mortality Weekly Report (MMWR),* Vol. 44 (via the Internet).

Centers for Disease Control (CDC). 1996. *Cholera Fact Sheet,* No. 107 (March). Atlanta, Georgia.

Christmas, J., and C. de Rooy. 1991. "The water decade and beyond." *Water International,* Vol. 16, pp. 127–134.

Crouch, D. 1993. *Water Management in Ancient Greek Cities.* Oxford University Press, New York.

Drower, M.S. 1956. "Water-supply, irrigation, and agriculture in ancient Mesopotamia." *History of Technology,* Vol. 1, pp. 520–556.

The Economist. 1995. "Don't drink the plankton." *The Economist,* 23 September, 1995, p. 75.

Epstein, P.R. 1992. "Cholera and the environment: An introduction to climate change." *Physicians for Social Responsibility Quarterly,* Vol. 2, No. 3, pp. 146–160.

Epstein, P.R., and F. Nutter. 1997. "Climate change: Assessing the economic damages." Newsletter of the Center for Health and the Global Environment. Harvard Medical School. Via *http://www.med.harvard.edu/chge/.*

Esrey, S.A., and J.P. Habicht. 1986. "Epidemiological evidence for health benefits from improved water and sanitation in developing countries." *Epidemiological Reviews,* Vol. 8, pp. 117–128.

Esrey, S.A., J.B. Potash, L. Roberts, and C. Shiff. 1990. "Health benefits from improvements in water supply and sanitation: Survey and analysis of the literature on selected diseases." WASH Technical Report No. 66. Water and Sanitation for Health (WASH) Project, Office of Health, Bureau for Science and Technology. United States Agency for International Development, Arlington, Virginia.

Esrey, S.A., J.B. Potash, L. Roberts, and C. Shiff. 1991. "Effects of improved water supply and sanitation on ascariasis, diarrhoea, dracunculiasis, hookworm infection, schistosomiasis, and trachoma." *Bulletin of the World Health Organization,* Vol. 69, No. 5, pp. 609–621.

Fass, S.M. 1993. "Water and poverty: Implications for water planning." *Water Resources Research,* Vol. 29, No. 7, pp. 1975–1981.

Feachem, R.G. 1984. "Infections related to water and excreta: The health dimension of the decade," in P.G. Bourne (ed.), *Water and Sanitation.* Academic Press, Inc., Orlando, Florida, pp. 21–47.

Food and Agriculture Organization. 1995. *Irrigation in Africa in Figures, Extract from Water Report 7.* Food and Agricultural Organization of the United Nations, Rome, Italy.

Gleick, P.H. (ed.). 1993. *Water in Crisis: A Guide to the World's Fresh Water Resources.* Oxford University Press, New York.

Gleick, P.H. 1996. "Basic water requirements for human activities: Meeting basic needs." *Water International,* Vol. 21, pp. 83–92.

Hopkins, D.R., and E. Ruiz-Tiben. 1991. "Strategies for dracunculiasis eradication." *Bulletin World Health Organization,* Vol. 69, pp. 533–540.

International Center for Diarrheal Diseases Research (ICDDR). 1993. "Large epidemic of cholera-like disease in Bangladesh caused by Vibrio cholerae O139 synonym Bengal." *Lancet,* Vol. 342, pp. 387–390.

Khan, A.H. 1997. "The sanitation gap: Development's deadly menace," in *The Progress of Nations 1997.* United Nations (UNICEF), Division of Communication, New York, pp. 5–13.

Levine, M.M. 1991. "South America: The return of cholera." *Lancet,* Vol. 388, pp. 45–46.

Lovei, L., and D. Whittington. 1993. "Rent-extracting behavior by multiple agents in the provision of municipal water supply: A study of Jakarta, Indonesia." *Water Resources Research,* Vol. 29, No. 7, pp. 1965–1974.

MacPherson D., and M. Tonkin. 1992. "Cholera vaccination: A decision analysis." *Canadian Medical Association Journal,* Vol. 146, pp. 1947–1952.

Mitra, R., A. Basu, D. Dutta, G.B. Nair, and Y. Takeda. 1996. "Resurgence of Vibrio cholerae O139 Bengal with altered biogram in Calcutta, India." *Lancet,* Vol. 348 (October 26), p. 1181.

More, Thomas. 1516. *Utopia.* Paul Turner Translation 1965, Penguin Books, Middlesex, United Kingdom.

Nash, L. 1993. "Water quality and health," in P.H. Gleick (ed.), *Water in Crisis: A Guide to the World's Fresh Water Resources.* Oxford University Press, New York, pp. 25–39.

Patz, J.A., P.R. Epstein, T.A. Burke, and J.M. Balbus. 1996. "Global climate change and emerging infectious diseases." *Journal of the American Medical Association,* Vol. 275, No. 3, pp. 217–223.

Pearce, D.W., and J.J. Warford. 1993. *World Without End: Economics, Environment, and Sustainable Development.* Oxford University Press, New York.

The Post. 1996. "Cholera hits Kaputa." Zambia. No. 609 (November 27).

Reuters. 1997. "Cholera outbreak." As reported in the *Sydney Morning Herald,* November 18.

Rogers, P. 1993. "Integrated urban water resources management." *Natural Resources Forum,* pp. 33–42 (February).

Rogers, P. 1997. "Water for big cities: Big problems easy solutions?" Draft paper, Harvard University, Cambridge, Massachusetts.

Rosenberg, C.E. 1962. *The Cholera Years.* University of Chicago Press, Chicago.

Rouse, H., and S. Ince. 1957. *History of Hydraulics.* Iowa Institute of Hydraulic Research, Iowa State University, Ames, Iowa.

Ruiz-Tiben, E., D.R. Hopkins, T.K. Ruebush, and R.L. Kaiser. 1995. "Progress toward the eradication of Dracunculiasis (Guinea Worm Disease): 1994." *Emerging Infectious Diseases,* Vol. 1, No. 2, April–June.

Shimada T., A. Balakrish, G. Nair, and B.C. Deb. 1993. "Outbreak of *Vibrio cholerae* non-O1 in India and Bangladesh." Letters. *Lancet,* Vol. 341, p. 1347.

Snow, J. 1936. *Snow on Cholera.* The Commonwealth Fund. Oxford University Press, London.

Tauxe, R.V., E.D. Mintz, and R.E. Quick. 1995. "Epidemic cholera in the New World: Translating field epidemiology into new prevention strategies." *Emerging Infectious Diseases,* Vol. 1, No. 4, pp. 141–146.

United Nations. 1977. "Report of the United Nations Water Conference, Mar del Plata, March 14–25, 1977." United Nations Publications E.77.II.A.12 (New York).

United Nations. 1992. Chapter 18 "Protection of the Quality and Supply of Freshwater Resources: Application of Integrated Approaches to the Development, Management and Use of Water Resources" of Agenda 21. United Nations Publications, New York.

United Nations. 1997. *Comprehensive Assessment of the Freshwater Resources of the World.* World Meteorological Organization and the Stockholm Environment Institute, Geneva, Switzerland.

UNICEF. 1997. *The Progress of Nations 1997.* United Nations, Division of Communication, New York.

USA Today. 1997. "Tanzania epidemic." *USA Today,* November 22 (International Asia edition).

White G.F., D.J. Bradley, and A.U. White. 1972. *Drawers of Water: Domestic Water Use in East Africa.* University of Chicago Press, Chicago, Illinois.

World Health Organization. 1986. "Major parasitic infections: A global review." *World Health Statistical Quarterly,* Vol. 39, pp. 145–160. World Health Organization Parasitic Diseases Programme, Geneva, Switzerland.

World Health Organization. 1992a. "Cholera in 1991." *Weekly Epidemiological Record,* Vol. 67, pp. 253–260.

World Health Organization. 1992b. *Guidelines for Cholera Control.* Programme for Control of Diarrhoeal Disease, Publication No. WHO/CDD/SER/80.4, rev. 4. World Health Organization, Geneva, Switzerland.

World Health Organization. 1992c. *8th Programme Report: 1990–1991.* Programme for Control of Diarrhoeal Disease. World Health Organization, Geneva, Switzerland.

World Health Organization. 1993. "Dracunculiasis: Global surveillance summary, 1992." *Weekly Epidemiological Record,* Vol. 68, pp. 125–131.

World Health Organization. 1994a. "Dracunculiasis: Global surveillance summary, 1993." *Weekly Epidemiological Record,* Vol. 70, pp. 121–128.

World Health Organization. 1994b. "Cholera strain on Rwanda-Zaire border identified." WHO Press Release WHO/61—25 July 1994. Geneva, Switzerland.

World Health Organization. 1995a. "Guinea worm wrap-up." No. 47, WHO Collaborating Center for Research, Training, and Eradication of Dracunculiasis at the U.S. Centers for Disease Control and Prevention, March/April 1995.

World Health Organization. 1995b. *Community Water Supply and Sanitation: Needs, Challenges and Health Objectives.* 48th World Health Assembly, A48/INF.DOC./2, 28 April, Geneva, Switzerland

World Health Organization. 1996. "Water supply and sanitation sector monitoring report: Sector status as of 1994." WHO/EOS/96.15. Geneva, Switzerland.

World Health Organization. 1997a. *Dracunculiasis: Disease Sheet.* Division of Control of Tropical Diseases. Geneva, Switzerland. Via *http://www.who.org.*

World Health Organization. 1997b. "Dracunculiasis: Global surveillance summaries. Annual." *Weekly Epidemiological Record.* Vols. 57, 60, 61, 66, 68, 69, 70, 72. Geneva, Switzerland.

World Resources Institute (WRI). 1994. *World Resources 1994–95: A Guide to the Global Environment.* Oxford University Press, New York.

The Status of Large Dams:
The End of an Era?

Dams and reservoirs have been an integral part of human development and water management from the earliest days of civilization. Dams serve many functions: they store water in wet periods for use in dry periods for agriculture and cities; they produce electricity by tapping the energy of falling water; they reduce the risks of disastrous flooding; they create reservoirs used for recreation and play; and they assist navigation by leveling and increasing flows in low-flow periods. However, dams have recently become the focus of intense international debate because of their negative, and often ignored, impacts on both people and natural ecosystems.

Humans have been building dams since the earliest days of human civilization. There are examples over 5,000 years old from Mesopotamia and Egypt. Legend has it that the earliest known dam across a river, the Sadd el-Kafara, was built in the Middle East 5,000 years ago (Gleick 1994). In the late 1600s B.C., the grandson of Hammurabi, Abi-Eshuh, dammed the Tigris River during a regional war (Hatami and Gleick 1993). Examples of ancient dams can be found in China, South Asia, Europe from the Middle Ages, and North America before the arrival of Europeans. Earthen dams were built in Sri Lanka 2,500 years ago. In A.D. 833, the Chinese built a 30-meter high dam that is still used for irrigation diversions on the Abang Xi River, though its reservoir filled with silt long ago (Petts 1984). The Anasazi Indians built irrigation dams in Colorado 800 years ago (Ortiz 1979).

Despite this long history, the widespread construction of really big dams did not begin until the middle of the twentieth century, when improvements in engineering and construction skills, hydrologic analysis, and technology made it possible to build them safely. Much of the initial big dam construction occurred in the United States, where the government built about 50 large dams between 1902 and 1930. Between 1930 and 1980 it built a thousand more, together with tens of thousands of smaller ones. (See Box 3.1 for definitions of "large dams.") The first of the truly massive dams was Hoover Dam, built by the U.S. Bureau of Reclamation on the Colorado River in the 1930s. While the Bureau of Reclamation had built other dams, the size of Hoover Dam exceeded all its previous dams combined. The administration of President Franklin Roosevelt completed the construction of the five largest dams on earth by 1945—Hoover, Bonneville, Fort Peck, Shasta, and Grand Coulee. Outside of the arid western United States, the Tennessee Valley Authority

---

BOX 3.1

What Is a "Large Dam"?

There is no single definition of a "large dam." According to criteria set by the *International Journal on Hydropower and Dams,* there are more than 300 "major dam projects" that meet at least one of the following criteria:

- dam height exceeding 150 meters;
- dam volume exceeding 15 million cubic meters;
- reservoir volume exceeding 25 billion cubic meters; or
- installed electrical capacity exceeding 1,000 megawatts.

The International Commission on Large Dams (ICOLD) offers a less restrictive set of criteria for defin-

ing a "large dam." According to ICOLD criteria, there are about 40,000 large dam and over 800,000 small ones (McCully 1996). A large dam is one whose

- height is 15 meters or higher; or whose
- height is between 10 and 15 meters if it meets at least one of the following conditions:
  - a crest length of not less than 500 meters;
  - a spillway discharge potential of at least 2,000 cubic meters per second; or
  - a reservoir volume of not less than 1 million cubic meters.

---

built 20 large dams in the east in 20 years. In 1944, Congress authorized nearly 300 dams in the Missouri Basin, including the massive Garrison, Oahe, and Fort Randall dams. At about that same time, the U.S. Army Corps of Engineers began a 40-year building spree that resulted in the construction of over 400 dams. Of the ten largest reservoirs in the United States, nine were completed before 1970 (USCOLD 1996). Today, the reservoirs behind U.S. dams store 60 percent of the entire average annual river flow of the United States (Hirsch et al. 1990). In the Colorado River basin, reservoir storage is equal to nearly five years of average runoff.

Other countries also saw large dams as vital for national security, economic prosperity, and agricultural survival. In the USSR, Stalin dreamed of transforming that country's massive rivers into controlled projects to provide electricity for Soviet industries and "transform nature" into a machine for the communist state. By the 1970s the total area flooded by dams in the USSR greatly exceeded that of the United States, and the world's tallest dam—the 300-meter-tall Nurek—was completed in 1980 in Tadjikistan. In India, dam construction initiated by the British colonial government was quickly adopted by the independent government after 1947, and McCully (1996) estimates that 15 percent of India's total national expenditures between 1947 and 1980 went to the construction of thousands of dams and associated facilities. Following the Chinese revolution in 1949, more than 600 dams were built every year for three decades, with both wondrous and disastrous results.

Today, nearly 500,000 square kilometers of land worldwide are inundated by reservoirs capable of storing 6,000 cubic kilometers of water (Shiklomanov 1993, 1996; Collier et al. 1996). This redistribution of fresh water is so large that scientists recently reported that it is responsible for a small but measurable change in the orbital characteristics of the Earth (Chao 1995). Today's dams have nearly 640,000 megawatts of installed hydroelectric capacity and produce nearly 20 percent of the world's total supply of electricity. Figure 3.1 shows the breakdown of hydroelectric production by region for 1997. Only a small fraction is generated in Africa and Oceania, and much of the hydropower generation is concentrated in a few countries. Figures 3.2 and 3.3 show that Canada, the United States, Brazil, and China dominate in terms of both installed capacity and hydropower generation. In 63 countries, hydropower supplies more than 50 percent of total electricity supply; it supplies between 90 and 100 percent in 23 countries (see Table 3.1, Map 3.1, and Table 14 of the data section at the back of the book).

The construction of large dams continues today. In Asia, China is pursuing an ambitious hydropower development program. China's current installed capacity is 53,000 megawatts, which is expected to increase to 70,000 megawatts over just the next three years. India has 10,000 megawatts approved or under construction and another 28,000 megawatts planned. Laos, Nepal, Malaysia, Russia, and the Philippines all have major construction projects under way. In Latin America, Brazil has the most ambitious plans, with more than 10,000 megawatts under construction and plans for another 20,000 megawatts. Honduras, Mexico, and Ecuador all have major projects under consideration. In Europe, projects for new or rehabilitated hydroplants are being pursued in Albania, Bosnia, Croatia, Greece, Iceland, Macedonia, Portugal, Slovenia, and Spain. In North

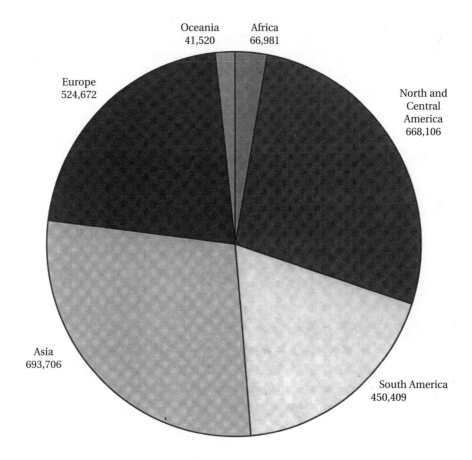

FIGURE 3.1   HYDROELECTRIC PRODUCTION BY REGION.
Total hydroelectric production by region in 1997 is shown here in gigawatt-hours. *Source:* International Journal on Hydropower and Dams, 1997, *World Atlas and Industry Guide,* Aqua-Media International, Surrey, England.

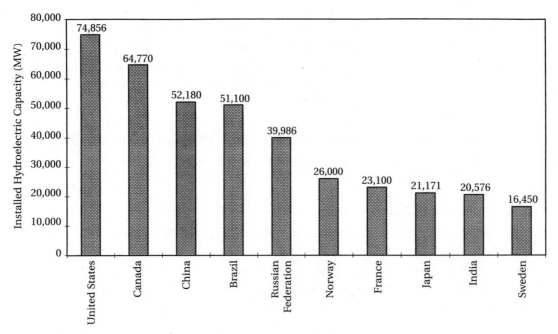

FIGURE 3.2   COUNTRIES WITH LARGEST INSTALLED HYDROELECTRIC CAPACITY, 1997.
Ten countries account for 62 percent of global installed hydroelectric capacity, for which the installed capacities in megawatts (MW) are shown here. China, Brazil, the Russian Federation, and India have developed lower fractions of their technically and economically feasible hydroelectric potential than have the United States, Canada, Norway, France, Japan, and Sweden. *Source:* International Journal on Hydropower and Dams, 1997, *World Atlas and Industry Guide,* Aqua-Media International, Surrey, England.

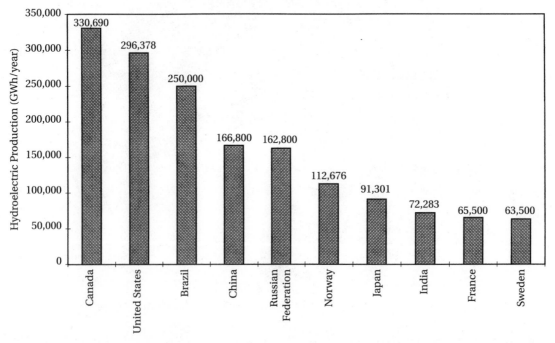

FIGURE 3.3   COUNTRIES WITH HIGHEST HYDROELECTRIC PRODUCTION, 1997.
Ten countries generate approximately 65 percent of all hydroelectric production. The countries are shown here, along with their 1997 hydroelectric production in gigawatt-hours per year (GWh/year). *Source:* International Journal on Hydropower and Dams, 1997, *World Atlas and Industry Guide,* Aqua-Media International, Surrey, England.

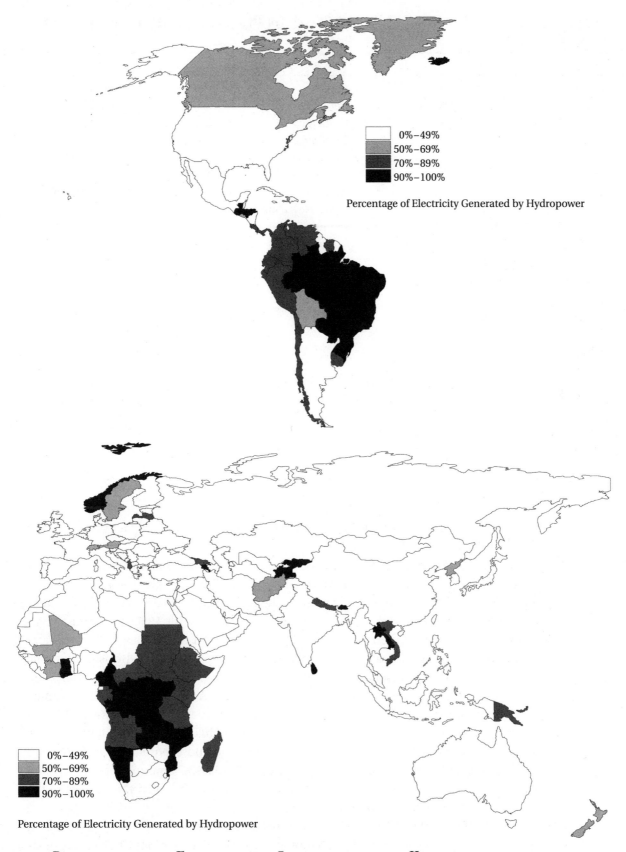

Percentage of Electricity Generated by Hydropower

0%–49%
50%–69%
70%–89%
90%–100%

Percentage of Electricity Generated by Hydropower

MAP 3.1 PERCENTAGE OF ELECTRICITY GENERATED WITH HYDROPOWER, 1997
Source: International Journal on Hydropower and Dams, 1997, *World Atlas and Industry Guide,* Aqua-Media International, Surrey, England.

TABLE 3.1   CATEGORIZATION OF COUNTRIES BASED ON PERCENTAGE OF ELECTRICITY GENERATED FROM HYDROPOWER

90–99%	80–89%	70–79%	60–69%	50–59%
Armenia	Albania	Angola	Afghanistan	Bolivia
Bhutan	Central African Republic	Colombia	Austria	El Salvador
Brazil	Chile	Costa Rica	Canada	Greenland
Burundi	Ecuador	Dominica	Côte d'Ivoire	Haiti
Cameroon	Ethiopia	Latvia	Korea DPR	São Tomé and Príncipe
Congo	Fiji	Madagascar	Mali	Sweden
Congo, DR	Gabon	Panama	New Zealand	Switzerland
Ghana	Georgia	Papua New Guinea	Reunion	
Guatemala	Kenya	St. Vincent		
Honduras	Nepal	Sudan		
Iceland	Peru	Venezuela		
Kyrgyzstan	Suriname	Vietnam		
Laos	Tanzania			
Malawi	Uruguay			
Mozambique				
Namibia				
Norway				
Paraguay				
Rwanda				
Sri Lanka				
Tajikistan				
Uganda				
Zambia				

America, Canada plans to build over 3,000 megawatts of additional capacity over the next 15 years (Hydropower and Dams 1997).

In addition to current projects, dam designers have for years dreamed of building even larger projects to harness the immense resources of the world's greatest water systems. Perhaps the most grandiose water-transfer scheme ever conceived was the North American Water and Power Alliance, or NAWAPA, proposed in the early 1960s by the Ralph Parsons Company, a construction-engineering firm. NAWAPA would have collected water from the Fraser, Yukon, Peace, Athabasca, and other rivers of Alaska, British Columbia, and the Yukon territory and transferred this water throughout Canada, the western and midwestern United States, and northern Mexico. NAWAPA is an example of the ultimate 1960s fantasy of water engineers, effectively replumbing the entire face of western North America with 369 massive projects costing hundreds of billions of dollars. It would have provided over 5,000 cubic kilometers of water storage and eventually transferred 136 cubic kilometers per year southward (Sewell 1985). Massive amounts of water would have had to be lifted over the Rocky Mountains, which lie between the sources of the water and its ultimate destinations. The centerpiece of the project would have been the damming of the Rocky Mountain Trench, an 800-kilometer long gorge in the Canadian Rockies adjacent to Banff and Jasper National Parks (U.S. Senate 1964). The massive environmental and economic costs of this project guarantee that it will never be built, but the design stands as a monument to what we are willing to consider when water supplies are perceived to be limited.

Similar grandiose projects have been discussed for every major river of the world. In Canada, 160 kilometers of proposed dikes across the end of James Bay would turn it into a freshwater reservoir the size of Lake Superior at a cost exceeding $100 billion. In Russia, a 20,000-megawatt dam at Turukhansk on an eastern tributary of the Yenisei River in Siberia has been proposed. Other enormous Siberian river schemes have been proposed, and canceled, many times, in many forms. One form would have diverted 120 cubic kilometers per year from the Ob, Irtysh, Yenisei, Onega, Pechora, and Dvina rivers away from the Arctic Ocean toward central Asia and other more populated regions of the country. Five to ten thousand megawatts of electrical capacity would have been needed to pump the water over various mountain ranges (Voropaev and Velikanov 1985).

In South America, there has been talk of damming the world's mightiest river, the Amazon, with a 64-kilometer-long dam, resulting in a 190,000 square kilometer reservoir larger than the nation of Uruguay and a generating capacity up to 80,000 megawatts. The Congo, the largest river in Africa, has 40,000 megawatts of hydroelectric potential at Inga Falls, which has been considered as a site for a dam. And perhaps the ultimate project would be a dam across the Straits of Gibraltar, turning the Mediterranean Sea into a separate closed sea.

Environmental and Social Impacts of Large Dams

Much has been written about the environmental impacts of large dams (see, for example, Goldsmith and Hildyard 1986; White 1988; Covich 1993; McCully 1996). Several types of impacts, though, merit special attention because of the difficulty of properly measuring, mitigating, or reversing them. These impacts include loss of riparian habitat, effects on aquatic species, and reservoir-induced seismicity, as well as the social impacts on people who must be uprooted and resettled. (See Box 3.2 for the different impacts of large versus small dams.)

Only a small fraction of land area worldwide has been lost to reservoirs, but this land is often of high quality and special value as fertile farmland, riparian woodland, or wildlife habitat. River floodplains are among the world's most diverse ecological systems, balancing aquatic and terrestrial habitat, species, and dynamics. In addition to the area lost to the reservoir behind a dam, there are often secondary impacts on surrounding lands, as humans are required to move to build new homes and farms. Wildlife populations displaced by the dam must seek new habitat. Access roads built in remote areas where many dams are built also bring in loggers and others searching for short-term financial gains (McCully 1996).

Dams can also destroy dramatic scenery when they are located in mountainous terrain. The Itaipú Reservoir inundated the spectacular waterfall of Sete Quedas on the Paraná River. The dam that provides water for the city of San Francisco flooded the Hetch Hetchy Valley in California, which was considered comparable in beauty to the famous Yosemite Valley. The beauty of Glen Canyon along the Colorado River was known to only

---

BOX 3.2
Environmental Impacts of Large versus Small Dams

While much has been written about the environmental impacts of large dams, there is still considerable uncertainty about either the impacts of single small dams or the cumulative impacts of many small dams. Moreover, there are many different types of dams built and operated for many different purposes in a variety of environments, making any generalizations unreliable and misleading.

Single large dams have attracted the most attention from the environmental science and environmental advocacy communities because they often create enormous reservoirs, flood large land areas, displace large populations, and form significant barriers to aquatic species and navigation. Smaller dams may be built on smaller river systems, be designed to either create a reservoir or operate in "run-of-river" mode without storing water, generate small amounts of electricity, or divert small amounts of irrigation water. They thus tend to avoid the more obvious environmental and social disruption of mega-projects, but they produce smaller benefits as well. Proper comparison of impacts is therefore extremely difficult.

Appropriate measures of the environmental consequences of dams must include not only their size but how they are designed and operated (Gleick 1992). In a comparison of large and small dam systems in the western United States, multiple small dams were shown to have some impacts that exceeded those of a single large dam. In particular, while many large hydroelectric dams with large reservoirs flood a substantial amount of land and lose water to evaporation, certain kinds of small dams may actually flood more land or lose more water to evaporation per unit of energy produced. Another factor relevant to determining net impacts is how dams are operated. Dams operated in "run-of-river" mode—independent of size—have less impact on natural riverine ecosystems than dams that store water in large reservoirs and alter the timing, quality, and character of streamflow.

This area requires further analysis. No comprehensive study has been done, for example, on the numbers of people displaced by dams as a function of energy produced, land flooded, or water supplied. Few analyses of the net impact to natural ecosystems have been done as a function of either the size of a dam or the number of dams built on a single river. And information is especially limited on impacts in developing countries, where baseline environmental data, reliable social indicators, and environmental assessments are less common or less readily available.

---

a few before it was lost under the rising waters of Lake Powell created by the Glen Canyon Dam. Indeed, massive public opposition to additional dams along the Colorado resulted in large part from the desire to prevent the destruction of the Grand Canyon, one of the most important symbols of natural beauty in the United States.

Because dams so radically alter natural free-flowing river systems, they put large pressures on aquatic ecosystems and species. Even at the turn of the century, there was evidence that dam construction led to the disappearance of species (Stanford and Ward 1979). There is often an immediate decline in diversity of native fishes after dam construction because the new environment is so different from the original river ecosystem or because nonnative species are introduced that outcompete native species. Dams interrupt the seasonal upstream migration of many species and alter the flow, sediment regime, dissolved oxygen content, and temperature of the habitat, all of which affect reproduction and survival (Covich 1993). They alter the hydrology of a basin and the dynamics of riverine ecosystems through changes in nutrient retention, water levels, and water chemistry.

Reservoirs also differ significantly from the free-flowing ecosystems they replace. Large fluctuations in reservoir levels during droughts or seasonal drawdowns alter the habitat along the margins where fish feed and spawn. Nutrient cycling is also greatly affected by these fluctuations. Temperature differences in reservoirs often result in completely different aquatic flora and fauna.

The ecological impacts of dams on rivers extend all the way to the sea. Many of the world's major fisheries depend on the volume and timing of freshwater and nutrient flows from large rivers into estuaries or the ocean. Almost all of the fish caught in the Gulf of Mexico, the Grand Banks of Newfoundland, the Caspian Sea, the eastern Mediterranean, and off the west coast of Africa depend on river discharges. The vast salmon fisheries of the eastern Pacific along Canada and the United States rely on the ability of the fish to spawn in freshwater streams. Dams in all of these areas, combined with overfishing and fisheries mismanagement, have led to great declines in many anadromous fish populations.

Partly as a consequence of large dams, more than 20 percent of all freshwater fish species are now considered threatened or endangered, and many of the most severe impacts have been felt by amphibians, insects, waterfowl, and plants. The west coast of the United States used to boast 400 separate salmon and steelhead stocks, but a century of habitat loss, dam construction, changes in flow regimes, and overfishing have eliminated nearly 50 percent of them. Of the remainder, 169 are at high or moderate risk of extinction (McCully 1996). Similar impacts have been felt in anadromous fisheries on both sides of the Atlantic. Other examples of aquatic species threatened by dams include several species of river dolphins in Latin America and Asia; sturgeon in North America, China, and Russia; and commercial and noncommercial fisheries at the mouth of almost every major river, including the Nile, Colorado, Volga, and Indus, and in the Black, Azov, and Caspian seas. Table 18 in the Data section lists endangered and threatened aquatic species worldwide.

Another unusual consequence of some dam construction is reservoir-induced seismicity, in which earthquakes are caused by the filling of a large reservoir. For most cases of reservoir-induced seismicity, the intensity of seismic activity increases with the size of a reservoir and the speed the reservoir is filled. The strongest shocks normally occur as the reservoir approaches maximum volume and earthquakes may continue for a few years after initial filling. Each dam that experiences induced seismicity, however, has unique characteristics that make any sort of prediction extremely difficult.

Reservoir-induced seismicity has occurred near at least 200 reservoirs around the world. Of the 11 reported cases of earthquakes greater than 5.0 on the Richter scale associated with filling of reservoirs, 10 of them occurred in dams higher than 100 meters. Fifteen percent of all reported cases had quakes larger than 4.0 on the Richter scale (McCully 1996). The most severe example of reservoir-induced seismicity is thought to be a 6.3 quake at the Koyna Dam in India in 1967. Reservoirs can both increase the frequency of earthquakes in seismically active areas and cause earthquakes to happen in areas previously thought to be seismically inactive.

More recently, one of the most serious concerns about large dams has been the involuntary displacement and resettlement of people living in areas to be flooded by reservoirs. According to Robert Goodland (1994), the World Bank's senior environment advisor, "Involuntary resettlement is arguably the most serious issue of hydro projects nowadays." Addressing the resettlement issue is complicated; it involves restoring and improving living standards as well as trying to resolve a host of problems usually ignored by traditional economic solutions. Often entire villages are destroyed, with generations of history and culture lost, churches and burial grounds flooded, and homes and farmlands inundated.

Limited and often contradictory data are available on populations displaced by dam construction and the creation of reservoirs. Table 15 in the Data section presents the best assessment of displaced populations, culled from case studies, World Bank literature, and resettlement overviews, but it only represents a small fraction of all people thought to have been displaced (van Wicklin, World Bank, personal communication, 1997). According to a World Bank study (1994), approximately 40 million people were displaced by dams between 1986 and 1993, an average of about 5 million people a year. This estimate is a "back-of-the-envelope" calculation and is likely to be a serious overestimate, though it may be approximately correct as an estimate for the total population displaced during the entire century of dam construction (McCully, personal communication, 1997). Figure 3.4 shows the countries that have displaced the most people by dam construction, using reported data in Table 15 in the Data section.

The vast majority of people displaced are in India and China, two countries with extremely high population densities and aggressive dam-development programs. In China, over 10 million people were officially displaced by water projects between 1960 and 1990, and the massive Three Gorges Dam project now under construction will displace well over 1 million more. Even more people have been relocated in India (McCully 1996; Goodland 1997). Figure 3.5 displays those projects completed, planned, or under construction that displace the largest number of people.

A wide range of complexities are involved in resettlement: conflicts between those being resettled and the communities into which they are moving; inadequate implementation of resettlement plans; insufficient financial or institutional support; lack of empowerment of local communities and populations; loss of social resilience when communities dissolve; and so on. The World Bank has issued formal guidelines in this area that are very influential. These guidelines require that the goal of all resettlement plans should be the improvement of pre-removal living standards; at a minimum those standards should be restored. Some observers consider these goals insufficient, difficult to accomplish, and rarely achieved in practice. Scudder (1997), among others, argues that restoration of income and living standards is not enough and that "mere restoration can be expected to increase . . . impoverishment." He supports this conclusion with several observations:

- Living standards tend to drop in the years immediately following resettlement;
- Development levels in project areas tend to be lower than surrounding areas to begin with;
- Expenses following resettlement are often greater than before, especially for food purchases and agricultural production;
- There is a general tendency to underestimate and understate people's incomes in project areas; and
- Merely restoring living standards does not compensate those resettled for negative health impacts and "sociocultural trauma."

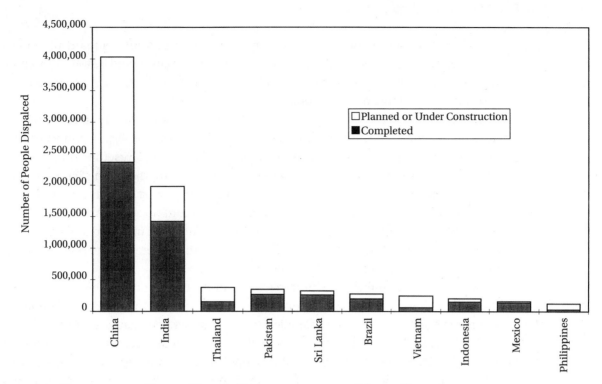

FIGURE 3.4   COUNTRIES THAT HAVE DISPLACED THE MOST PEOPLE BY DAM CONSTRUCTION.

The ten countries displacing the largest number of people with both completed projects and projects that are planned, under construction, or postponed are shown here. *Sources:* See those listed for Table 15 in the Data section at the back of the book.

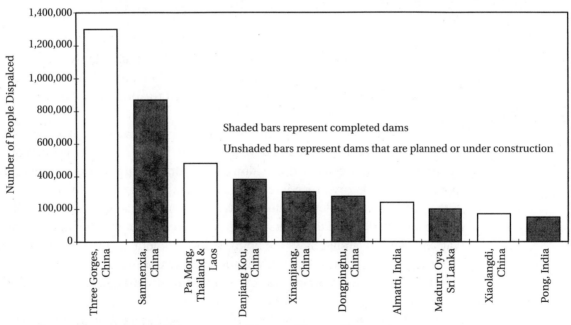

FIGURE 3.5   COMPLETED AND PROPOSED DAMS THAT DISPLACE THE LARGEST NUMBER OF PEOPLE.

The ten completed and proposed dams that displace the largest number of people are shown here. Projects in India and China displace large numbers of people in part because of the high population densities of those countries. Upper estimates from Table 16 are used, if a range was presented. *Sources:* See those listed for Table 15 in the Data section at the back of the book.

As a result of these concerns, simply maintaining a current standard of living is increasingly considered an inadequate goal. More appropriate would be the requirement that displaced populations be made better off than before the project began, an even more difficult but more equitable objective.

Resettlement also causes problems in the host population, with whom resettled populations compete for economic opportunities, land, social services, and political power. Few resettlement projects have been documented where host–resettler conflicts have not appeared. Ten years after resettlement in connection with the Kpong Dam in Ghana, deadly fighting broke out over rights to limited lands for grazing and farming, and political competition continued between the original hosts and populations resettled in the area. Scudder (1997) recommends that the host population be included in the improved social services and economic development opportunities offered to the resettled population, but he notes that this rarely happens.

Another serious concern is the impact on rural populations downstream of large dams. While these populations are often assumed to benefit from the project outputs, evidence suggests there are often significant, underestimated negative impacts (Scudder 1997). While few detailed studies have been done, work on the Manantali Dam project in the Senegal River identified over 500,000 people adversely affected by changes in river and nutrient flows downstream of the dam. The costs of the Bakolori Dam to downstream villagers are estimated to exceed the benefits realized from the project (Adams 1993).

New Developments in the Dam Debate

One consequence of the negative environmental and social impacts resulting from major dam projects is growing opposition to new ones. Few major dams are now moving forward free of public scrutiny and analysis, and several major projects have been canceled or postponed. Many more have been redesigned to address public concerns and interests.

While many new construction plans are well along and may fill a vital need, serious obstacles will limit the actual level of development and lead to the cancellation of some current projects. Growing public opposition on environmental and social grounds, more comprehensive regulatory procedures, and new complexities in financing major dam projects are all working to slow development. In the United States, federal and state laws have protected over 16,000 kilometers of undammed rivers from further development. In Norway and Sweden, similar laws are limiting development on pristine rivers, as local sentiment against big dams spreads. Even more striking was the news in late 1997 that for the first time the U.S. government was ordering the destruction of a hydroelectric dam because of the damage it was doing to migratory fish.

Organized opposition to dams in the United States goes back nearly 100 years. John Muir, a famous American environmentalist, led the first national battle around the turn of the century to prevent the construction of the Hetch Hetchy Dam in Yosemite National Park in the mountains of California. In a famous exchange between Muir and Gifford Pinchot, the two positions—protection versus use—were clearly defined:

Muir: "These Temple destroyers, devotees of ravaging commercialism, seem to have a perfect contempt for Nature, and instead of lifting their eyes to the God of the mountains, lift them to the almighty dollar."

Pinchot: "The injury . . . by substituting a lake for the present swampy floor of the valley . . . is altogether unimportant compared with the benefits to be derived from its use as a reservoir."

Muir and those in favor of preserving the valley were unsuccessful, but this well-publicized fight led to the creation of the Sierra Club, one of the largest grassroots environmental organizations in the United States. It also set the groundwork for later, more successful, public opposition to major dam projects (McPhee 1971; Reisner 1986; Ehrlich and Ehrlich 1987).

Public debate and opposition to major dam projects have appeared in other countries, such as Nepal, India, several countries in Latin America, and even China, where active dissent against government policies is a dangerous and courageous act. Many dam planners tend to blame outside factors for opposition to major projects. In the 1950s, British colonial administration blamed external political organizers in favor of independence for organizing resistance to Kariba Dam on the Zambezi. In 1990, Botswana blamed international environmentalists and outside safari firm interests for opposition to the Southern Okavango Integrated Water Development Project. In the early 1990s, India blamed outside parties for organizing opposition to the Narmada Dam Project. In 1996, South Africa water planners blamed international activists for opposition to the Lesotho Highlands Project. Although each of these cases did attract international attention and opposition, they also suffered from early and serious local resistance.

International nongovernmental organizations (NGOs) do play a role in advocating for people affected by major projects and in opposing large dams. These organizations, such as Probe International and the International Rivers Network, collect and disseminate information about projects, lobby governments and international financial organizations, and work to raise public opinion and opposition. They unabashedly feel their mission is to stop large dams and will go to great lengths to do so. These groups have forced major dam developers and funders to reevaluate their participation in many questionable projects and to redefine standards to protect natural ecosystems and local populations. The growing sophistication of such NGOs suggests that major dam projects will continue to undergo outside independent scrutiny and attention.

Opposition is also due, however, to true grassroots local concerns. As more and more large projects are proposed, and as more evidence of the social and environmental impacts of those projects accumulates, local communities are no longer willing to trade guaranteed immediate social and environmental disruption for a promise of hypothetical future benefits. Moreover, the frequent disregard for the civil rights of people living in rural areas affected by *some* dam projects has led to a widespread distrust of *all* dam projects.

One of the major factors driving opposition to large dams has been the inadequate assessment and mitigation of adverse social impacts. Until the 1970s and 1980s, few major dam projects included any assessment of the environmental and social impacts of construction and operation. Scudder (1997) argues that the short-term and cumulative adverse impacts of dam construction have been routinely and serious underestimated, lowering the living standards of millions of local people. Even today, the assessment of these problems is irregular and inadequate, in part because of the massive nature of the problem and in part because of the difficulty in evaluating the severity of some impacts that are simply unquantifiable.

Major projects should go forward only if local populations are fully consulted, involved, and represented in decisions about a project and its alternatives. Yet planning for almost all major dam projects is still usually conducted behind closed doors by a small number of engineers and water experts. In most countries, dam developers solicit little or no public feedback, and decisions are made by a small group of water managers or governmental organizations, together with international funding agencies (Oud and Muir 1997). Such lack of consideration for the public has also contributed to growing opposition to dams around the world. While the United Nations and others (Gleick et al. 1995) argue that public participation is a fundamental principle for sustainable water planning and management, many project proponents are still nervous about opening plans to public scrutiny. The Mekong River Commission, sponsored by the United Nations Development Programme, was widely reported to be resisting public participation and transparency during 1994 to 1996 (Goodland 1997).

With few exceptions, all major dam projects have been designed, developed, and owned by governments and national utilities. In the developed world, funding for major projects has traditionally come from internal financial sources. In the developing world, funding has come from multilateral and bilateral aid agencies and from export–import banks in major industrialized countries. Increasingly, however, the extremely high costs of new infrastructure, combined with the difficulty of raising capital for large projects, are shifting development toward private-sector funding.

There are some serious potential problems with a shift to private-sector funding. Oud and Muir (1997) note that increased private funding leads to:

- An emphasis on financial project efficiency that leads to cost cutting in the areas of operations and maintenance;

- Externalization of as much of the indirect costs associated with the project as possible;

- Levying of water rates that provide an attractive financial return; and

- Off-loading of as much risk as possible onto other parties, particularly governments.

These problems may also lead to inequitable distribution of costs among the project beneficiaries or sacrifices in the quality of materials used in construction. While these may be reasonable outcomes from the point of view of private funders, they demonstrate the need to maintain clear regulation and control over safety and workmanship and over the considerable nonmarket costs of large dams, including the environmental and social impacts.

Partly in response to organized opposition to large dams funded by international aid organizations, the World Bank published a review of its experience with large dam development in August 1996. This review was heavily criticized as inadequate, and in early 1997 the Bank and the World Conservation Union (IUCN) sponsored a meeting with dam designers and builders, NGOs, people affected by dams, academics, and others to debate new principles to guide future assessments and dam development. Out of these efforts came a proposal for a World Commission on Dams (WCD), constituted in late 1997 and early 1998 to work for two years to evaluate dam projects and develop principles to guide dam efforts. This meeting and the goals of the WCD are described more completely in Chapter 6.

There are other signs that the philosophy of dam construction is slowly changing. In late 1997, the U.S. government for the first time refused to relicense a hydroelectric dam and ordered the structure destroyed on the grounds that its costs as a barrier to migratory fish significantly exceed its hydroelectric benefits. The 16-meter high Edwards Dam, on the Kennebec River in Maine, was built 160 years ago and produces one-tenth of 1 percent of Maine's electricity. The dam blocks spawning grounds for striped bass, shad, sturgeon, Atlantic salmon, and herring. In 1986, a change in Federal law required the Federal Energy Regulatory Commission (FERC), which is responsible for licensing hydroelectric dams, to balance conservation, recreation, and other environmental values with electricity generation in decisions to renew licenses. The Edwards Dam is the first instance in which FERC has exercised its power to deny a license (Goldberg 1997; FTGWR 1997b). About 550 dams are up for relicensing in the next 15 years, and many of them block spawning habitat of threatened and endangered fish.

Two other dams are facing destruction in the northwestern United States, the Elwha and Glines Canyon dams on the Elwha River in Washington State. Elwha Dam is over 30 meters high; Glines Canyon Dam is over 50 meters high. Both dams were built by a logging company to supply electrical power to a wood products plant. These dams block spawning runs of endangered salmon that are of great cultural and social importance to local tribes. These values are increasingly considered more important than the limited economic benefits of the dams and their reservoirs. The National Park Service is seriously considering dismantling these two dams and restoring the Elwha River to a more natural condition. The estimated cost of dismantling the dams is between $60 and $100 million; the cost would be less if the Elwha River is allowed to redistribute trapped sediment naturally (Collier et al. 1996).

The Edwards, Elwha, and Glines Canyon dams are not huge facilities and they have particularly severe environmental and social impacts. However, they symbolize many of the problems facing all dam projects, and the precedence of their destruction may herald a whole new generation of hydrologic engineers and ecological scientists whose field of expertise is removal of large dams. As a step in this direction, in mid-1997, a leading U.S. environmental group proposed that one of the largest dams in the United States, the Glen Canyon Dam on the Colorado River, be decommissioned and its reservoir drained. In this proposal, the dam itself would be left as a monument, but the river in this portion of its run would be returned to a free-flowing condition.

It may be many years before any sizeable dam is intentionally decommissioned or destroyed because of its environmental or social consequences, but many major projects are already being delayed or protested. On the pages that follow are two case studies of major projects now under way that have attracted substantial local and international attention and opposition. The issues that have become important in these projects are issues common to many others, and water planners and dam builders may learn relevant lessons by studying them.

The Three Gorges Project, Yangtze River, China

The Three Gorges Dam will be the largest and most powerful dam ever built. The dam will stretch 2 kilometers across the Yangtze (or Chang Jiang) River; it will be 185 meters tall and will create a 600-kilometer-long reservoir with a total storage capacity approaching 40 billion cubic meters. The dam is being built in a 200-kilometer stretch of canyons formed by immense limestone cliffs, known as Three Gorges (see Map 3.2). The gorges—the Xiling, Wu, and Qutang—offer some of the most scenic landscape anywhere in the world and are classic symbols of China itself. The beauty of the region has inspired the work of Chinese poets and artists for centuries, including much of the work of Li Bai (A.D. 701–762), considered by many Chinese to be the world's greatest poet. Ironically, an opera has been written about the scenic Three Gorges to mark construction of a dam that will flood the site celebrated for centuries by poets and painters. According to the Xinhua News Agency (1997), an opera company in Chongqing is producing "The Goddess of the Gorges" as part of the China International Opera and Dance Year. The Three Gorges Dam is designed to have an installed hydroelectric capacity approaching 18,000 megawatts, well above the 12,600 megawatts produced by Brazil's Itaipu Dam, currently the world's largest hydroelectric dam. The turbines will generate up to 85 billion kilowatt-hours (kwh) of electricity per year (one-ninth of China's total electricity supply), provide flood protection on the tempestuous Yangtze River, and improve river navigation for thou-

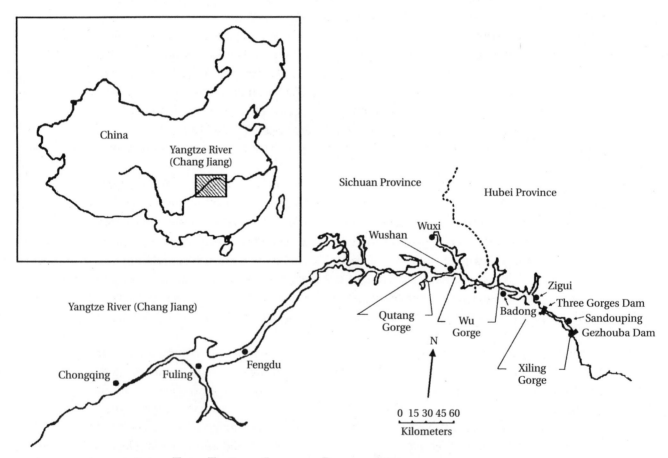

MAP 3.2   THE THREE GORGES RESERVOIR

sands of kilometers. Historically, people living in the middle and lower reaches of the Yangtze River have suffered tremendous losses from flooding. In 1931, some 145,000 people were drowned and over 300,000 hectares of agricultural land flooded. Currently, 15 million people and 1.6 million hectares of land along the Yangtze River are seriously threatened by flooding.

The project has generated enormous international controversy because of its magnitude, the serious impacts it will have on the ecology and environment in the region, and the massive human resettlement required. The reservoir would permanently submerge tens of thousands of hectares of farmland and forest, as well as 160 towns, 1,500 factories, and hundreds of archaeological sites. Official estimates are that 1.3 million people will need to be resettled—the largest population displacement in the history of dam construction—and recently some sources have suggested that the number may be as high as 1.9 million (Zich 1997).

The idea of building a gigantic dam on the Yangtze River in the Three Gorges area was proposed more than 70 years ago by Sun Yat-sen. Preparatory work and hydrologic investigations began decades ago, but the development of the project remained a dream because of lack of resources, the Cultural Revolution, and other political events. Box 3.3 offers a chronology of important events. The U.S. Bureau of Reclamation was one of the first international organizations to help Chinese engineers identify a possible site in the 1940s. Soviet engineers performed similar services in the 1950s. Massive flooding along the Yangtze in 1954 left 30,000 people dead and one million people homeless, and it brought a renewed sense of urgency to damming the Three Gorges. Chairman Mao Tse-tung vowed to speed up preparations for the dam, and since that time, hundreds of Chinese government agencies, bureaucracies, and academic bodies have produced detailed studies on all aspects of the project. China's elaborate decision-making process, ideological struggles, the Cultural Revolution, economic troubles, and prolonged government debate over the feasibility of the project all led to substantial delays.

In 1986, the Chinese Ministry of Water Resources and Electric Power asked the Canadian government to finance a $14 million feasibility study to be conducted by a consortium of Canadian firms. The consortium, known as CIPM Yangtze Joint Venture, included three private companies (Acres International, SNC, and Lavelin International) and two state-owned utilities (Hydro-Quebec International and British Columbia Hydro International). The World Bank was asked to supervise the feasibility study to ensure that it would "form the basis for securing assistance from international financial institutions" (Probe International 1997). In the spring of 1992, the National People's Congress officially approved the construction of the project and in December 1994, the Chinese government began construction.

The current phase in construction is one of the most difficult. Flood-discharge dikes, power-plant facilities, and actual construction of the main dam involve excavating approximately 8 million cubic meters of earth, pouring 12 million cubic meters of concrete, and installing 150,000 tons of equipment and metal structures (Reuters 1997b). This work is expected to last until the installation of the first power equipment in 2003.

In August 1997, China awarded a contract for 14 power-generating units to GEC Alsthom, ABB, and an industrial consortium formed by Germany's Voith and Siemens and General Electric Canada (VGS). The members of the consortium have provided

BOX 3.3

Three Gorges Dam Project Chronology of Events*

1919   First mention of the Three Gorges project in Sun Yat-sen's "Plan to Develop Industry."

1931   Massive flooding along the Yangtze River kills 145,000 people.

1932   Nationalist government proposes building a low dam at Three Gorges.

1935   Massive flooding kills 142,000 people.

1940s   The U.S. Bureau of Reclamation helps Chinese engineers identify a site.

1947   Nationalist government terminates all design work.

1949   Communist revolution in China.

1953   Mao Tse-tung proposes building a dam at Three Gorges to control flooding.

1954   Flooding along the Yangtze leaves 30,000 people dead and one million people homeless.

1955   Soviet engineers play a role in project planning and design.

January 1958   Mao appoints Chou Enlai to begin planning for the project.

May 1959   Yangtze Valley Planning Office (YVPO) identifies Sandouping site for dam.

1966   All work halted by the Cultural Revolution (1965–1975).

1976   Planning recommences.

February 1984   Ministry of Water Resources and Electric Power recommends immediate commencement of construction.

Spring 1985   The National People's Congress delays a decision on the project until 1987 because of economic difficulties.

1986   The Chinese Ministry of Water Resources and Electric Power asks the Canadian government to finance a feasibility study.

August 1988   Canadian-World Bank "Three Gorges Water Control Project Feasibility Study" is completed and recommends construction at "an early date."

February 28, 1989   *Yangtze! Yangtze!*—a book opposing dam construction—is released. The book is edited by Dai Qing and includes contribu-

* For a more detailed chronology of the internal political debates between 1955 and 1992, see Dai Qing (1989, 1994)

export credits and 15- to 20-year commercial loans. The 14 generators and turbines will have a capacity of 700 megawatts each and are scheduled to go on line between 2003 and 2006. The project will need another 12 generators and turbines, which would be built in China (Reuters 1997a; BBC 1997). China's State Development Bank signed a loan package in September 1997 with three German commercial banks. This package includes both export credits and a $200 million commercial loan (FTGWR 1997a). The first electricity is scheduled to be produced in 2003 and the project is to be completed in 2009 (China 1996).

Financing for the Project

The Three Gorges Dam project will be enormously expensive, but—as is the case with almost all massive infrastructure projects—no one knows with any certainty just how expensive it will actually be or how to measure some of the costs. In addition, tremendous cost overruns are endemic to major construction projects in China—a problem

tions from many Chinese scholars, researchers, and activists.

April–June 1989 Democracy movement sweeps through China. Dai Qing is jailed.

February 1992 Politburo Standing Committee agrees to the construction of the project.

April 3, 1992 China's National People's Congress (NPC) formally approves the "Resolution on the Construction of the Yangtze River Three Gorges Project"; 177 delegates oppose the project, 644 abstain, 1,767 approve.

April 27, 1992 The Canadian government cancels development assistance for the project.

May 1992 Some 179 members of the Democratic Youth Party reportedly detained in connection with their protests against the Three Gorges project in Kai County, Sichuan (HRW 1995).

January 1993 An armed fight involving over 300 persons occurs in the vicinity of the dam (HRW 1995).

December 14, 1993 The U.S. Bureau of Reclamation terminates agreements for technical services because of economic and environmental impacts.

Early 1994 The full resettlement program begins in earnest.

Mid-1994 Excavation and preparation of the dam's foundations begin at Sandouping.

December 14, 1994 Premier Li Peng formally declares the project under construction.

May 1996 The U.S. Ex-Im Bank's board votes unanimously to withhold support for the project and voices serious reservations about the dam's environmental and social impacts and its economic viability.

August 1997 China awards a contract for 14 power-generating units to GEC Alsthom, ABB, and an industrial consortium formed by Germany's Voith and Siemens and General Electric Canada (VGS).

September 1997 The State Development Bank of China signs a loan package with Germany's Kreditanstalt Fur Wiederaufbau, Dresdner Bank, and DG Bank that includes both export credits and a $200 million commercial loan.

November 8, 1997 Yangtze River is diverted out of its bed; construction on main dam begins.

2003 The first electricity is scheduled to be produced.

2009 The project is scheduled to be completed.

found elsewhere as well. In 1990, the official estimate for the project was $10.7 billion (Probe International 1993). More recent estimates for the project range from a low of $25 billion to a high of $60 billion (Dai Qing 1994; China 1996; JPN 1996; McCully 1996; Reuters 1997b). There are also additional costs of the project that cannot be estimated accurately. For instance, many environmental and social impacts are unquantified or unquantifiable. Serious siltation in the reservoir and the subsequent lack of siltation downstream make the lifetime of the dam uncertain and will lead to additional direct and indirect costs. Significant amounts of equipment are to be imported from abroad, with uncertain prices and currency exchange rates. The collapse of Asian currency and financial markets in late 1997 is also likely to have an impact on the costs. And the total investment will increase with the duration of the construction period, which itself depends on a wide variety of uncertain variables.

The project is being funded by both internal and external sources. China has identified four internal sources of funds: the State Three Gorges Construction Funds, power revenues from the existing Gezhouba Hydropower Plant, power revenues from the Three

Gorges Project when the first electricity is produced (scheduled for the year 2003), and loans and credits from the State Development Bank. In spite of the above sources, the Chinese government estimates that there will be a substantial gap in funding, which it hopes to make up through export credits, overseas commercial loans, and issuance of bonds (HRW 1995). Shortage of money has not apparently affected the progress of the project to date.

Several international export credit agencies have agreed to participate in the project. Export credit agencies are governmental bodies that provide financing and insurance to companies bidding on foreign projects. Canada's Export Development Corporation and Germany's export–import bank, Hermes-Buergschaften, have agreed to grant loan guarantees for the project totaling hundreds of millions of dollars. Most export credit agencies have no environmental or human rights regulations on their books.

Some important major dam financiers and builders are withholding support or are refusing to participate, including the World Bank, the U.S. Bureau of Reclamation, Bechtel Enterprises, and Canadian utilities such as BC Hydro and Ontario Hydro. The U.S. Bureau of Reclamation has stated that Three Gorges is not "environmentally or economically feasible." The World Bank has warned that the current design of the project is "not an economically viable proposition." Former British Columbia premier Mike Harcourt instructed BC Hydro not to bid on Three Gorges contracts for environmental reasons. Maurice Strong, former Ontario Hydro Chairman and organizer of the 1992 Earth Summit, stated that Ontario Hydro would become involved in the project "over my dead body" (Probe International 1997).

The U.S. Ex-Im Bank has also decided to avoid the project. In 1992, the U.S. Congress required the Ex-Im Bank to develop environmental review procedures and authorized the Bank's directors to grant or withhold financing support after taking into account beneficial and adverse environmental effects of proposed transactions. Guidelines and procedures to be used in evaluating Ex-Im Bank transactions were drafted and issued in February 1995. After the Chinese began development of Three Gorges, the Ex-Im Bank was approached by several U.S. corporations interested in participating as suppliers of equipment for the project. These exporters requested that the Ex-Im Bank issue Letters of Interest indicating the Bank's willingness to provide financing to the Three Gorges Development Authority for its purchases of U.S.-made equipment.

As required by its guidelines, the Bank reviewed financial, technical, and environmental issues raised by the Three Gorges Project and consulted with the National Security Council, nongovernmental organizations, Chinese officials, U.S. companies, and members of Congress and Congressional staff. In May 1996, the Ex-Im Bank's board voted unanimously to withhold support and at the same time voiced serious reservations about the dam's environmental and social impacts. The bank requested further information from the Chinese government on a wide-ranging set of issues, including water pollution, sedimentation problems, threats to endangered aquatic and terrestrial species, destruction of cultural resources, and forcible resettlement (Ex-Im Bank 1996).

This refusal was unusual and indicates the depth of concern that some have for the financial—as well as the social—repercussions of the project. China is the Ex-Im Bank's largest customer in Asia, and the Ex-Im Bank has recently maintained an aggressive outreach effort to support U.S. exporters doing business there. The Ex-Im Bank has supported $3.8 billion in U.S. exports to China during the last four years and is considering billions in future financing for transactions in various sectors, including power, aircraft, airports, telecommunications, chemical plants, and project finance. In the power sector alone, between 1994 and 1996, the Ex-Im Bank financed nearly $640 million worth of U.S. exports for both hydroelectric and nuclear power projects (Ex-Im Bank 1996).

The private sector is also split over the project. Some commercial banks have offered financing for the sale of equipment to China. German banks are lending some $270 million for the purchase of turbines and generators for the project. The State Development Bank of China signed a loan package with Germany's Kreditanstalt Für Wiederaufbau, Dresdner Bank, and DG Bank in late September 1997. This package includes both export credits and a $200 million commercial loan. In addition to the German and Canadian money, other project financing is coming from Switzerland, Brazil, France, Spain, and Norway (FTGWR 1997a).

Other private financiers are wary of the project. *The Wall Street Journal* noted that Three Gorges is so politically risky that few bankers are willing to be associated with it (Smith 1995; Cooper 1996). Planned international bond issues were canceled after Beijing consulted with several U.S. investment banks. Whether or not financing problems act as a constraint on the project in the future remains to be seen.

Impacts of the Project

One of the strongest arguments in favor of the project is that the electricity produced by the dam would otherwise be produced by Chinese coal-burning power plants, which have serious environmental impacts. The Chinese government estimates that if the electricity to be generated by Three Gorges were instead produced with Chinese coal, 50 million more tons of coal would be burned annually, producing 100 million tons of carbon dioxide, 1.2–2 million tons of sulfur dioxide, 10,000 tons of carbon monoxide, and large quantities of particulates (China 1996). The Chinese people suffer most from the serious environmental impacts associated with coal combustion, but there are also international effects. With the growing debate over the emission of greenhouse gases, and efforts to develop and implement a formal international treaty limiting emissions, there is great concern about the actions and role of China, which depends on coal for a substantial part of its electricity supply. China argues that Three Gorges will help it continue to grow economically without emitting as many greenhouse gases.

Dam proponents also claim that a dam at Three Gorges will protect millions of people living along the middle and lower reaches of the river from disastrous floods and transform the Yangtze into a smooth navigable waterway for ocean-going vessels (see Map 3.2). Currently, barges larger than 1,000 tons can travel only within a limited path of 2,500 kilometers, mostly in the middle and the lower reaches of the Yangtze. In the upper river, navigation is almost the only means of long-distance freight transportation. For Chongqing, the major port city in Sichuan province 600 kilometers upstream of the dam site, 90 percent of material is transported by water, but the navigation condition of the upper Yangtze is rather poor and most tributaries allow transportation only of barges smaller than 100 tons. Because the Three Gorges Project will create a huge lake, it will dramatically increase the depth of water and improve navigation up to Chongqing.

These large "benefits" will be accompanied by large "costs." The destruction caused by the dam will be massive: more than a million people will lose their homes and livelihoods; fertile agricultural lands will be destroyed; rare and endangered fish species will be threatened with extinction; and important archaeological and historical sites will be forever lost. Land with a population of 14 million people will be affected by the reservoir. More than 100 towns will be submerged. Fourteen thousand hectares of agricultural land will be drowned, as will more than 100 archaeological sites, some dating back over 12,000 years. Other impacts include irreversible changes to the hydrology of the river for thousands of kilometers, the destruction of commercial fish stocks, modification of the complex downstream agricultural floodplain, and a possible increase, not decrease, in certain flood risks (Probe International 1997). Ironically, the project has stimulated an enormous temporary increase in tourism in the region as many international tour operators advertise the last chance to see the beauty of the area.

In its analysis of the environmental impacts of the Three Gorges Project, the Chinese Academy of Science (1988, 1995) considered the large-scale resettlement of people and inundation of land to be the most devastating aspects of the project. Resettlement has been a serious problem for every large dam built in China, especially for projects such as Sanmenxia, Xinanjiang, and Danjiangkou reservoirs, each of which involved relocation of more than 300,000 people (see Table 15 in the Data section). But the scale of relocation associated with Three Gorges is unprecedented. According to the feasibility report of the Chinese Academy of Sciences (1988, 1995), the dam will submerge parts of 19 counties, including the cities of Wanxian and Fuling, which have cultural histories extending back more than 1,000 years. Parts of Chongqing may also be flooded. Total resettlement will exceed one million and may approach 1.9 million (Zich 1997). Unlike many other dam resettlement projects, urban populations will account for a substantial fraction of the total population displaced, making the resettlement from Three Gorges even more complicated and expensive than usual (Fearnside 1988). Resettlement will be further complicated by the scarcity of good land. Due to China's huge population, good land is already occupied and cultivated. Most resettlements will be squeezed to the more marginal higher elevations of the reservoir area, which have already been overpopulated in recent years.

As might be expected from plans of this magnitude, there will be enormous ecological impacts. The fish resources of the Yangtze River are abundant and quite vulnerable. Major changes in fish populations are likely because the dynamics of the river, the chemical and temperature composition of the water, and the character of the natural habitat and food resources available for these fish species will be altered. The dam itself will block migration of fish and spawning grounds for up to 172 different fish species. A number of species will not be able to adapt to the new environment and may suffer a dramatic reduction in numbers. In particular, the project will seriously affect the fish species in the middle reach of the Yangtze River, which is a major breeding area for four rare native fishes. Of special concern are the Chinese sturgeon and Chinese freshwater dolphin, which inhabit only the middle and lower reaches of the Yangtze River. The breeding of sturgeon has already been affected by the Gezhouba Dam, and the Chinese dolphin has been reduced to a few hundred in number (Perrin and Brownell 1987; Fearnside 1988). Concern has also been expressed for the Siberian crane, which is endangered and depends on overwintering habitat in the middle and lower Yangtze that will be affected by the dam (Fearnside 1988).

A complex set of impacts will be associated with sedimentation in the river. The Yangtze River is one of the most silt-laden rivers in the world. Every year, it carries between 500 million and a billion tons of silt (Milliman and Meade 1983; Leopold 1997). Although it is difficult to predict the rate of sedimentation into the reservoir, it is generally agreed that under normal operating conditions more than 70 percent of the silt load of the river will be trapped behind the dam every year. This implies that, within 100 years, the sediments of the Yangtze River will silt up all the dead storage behind the dam.

The deposition of sediments in the reservoir will have effects downstream. These sediments have been a source of nutrients for downstream agriculture and fisheries in the Yangtze Delta for thousands of years. Opponents of the project warn that siltation may be especially rapid near the tail of the reservoir, which would seriously affect navigation of the upper reach of the Yangtze River during the dry season. The accumulated siltation near the tail will raise the riverbed in this area. Intensified cultivation of unsuitable hilly land would produce serious soil erosion, which could cause additional sedimentation and further reduce the life of the reservoir.

One problem is the inability of designers to forecast the rate of sediment accumulation in a reservoir. These uncertainties may lead to significantly lower financial benefits than currently expected (Leopold 1997). Even when the records of sediment inflow are reliable, the deposition rate and location are often unanticipated. For the multipurpose reservoirs in India, Murty (1989) states that the "annual loss rates of siltation in most reservoirs are 145 percent to 875 percent of the figures assumed at the time of construction." Leopold (1997) points out that the Three Gorges Dam is designed to operate under conditions practically untested in the world. China has about 330 major reservoirs, and sediment deposition in 230 of them has become a significant problem, resulting in a combined loss of 14 percent of the total storage capacity. In some, more than 50 percent of the storage capacity has been lost in less than 10 years because of poor analysis and bad design (Abu Zeid and Biswas 1990; Hu Chunhong 1995).

Many other diverse impacts may occur. The reservoir is located in a region where conditions are conducive to outbreaks of waterborne diseases such as malaria and schistosomiasis. Increased evaporation in the reservoir will reduce the flow of water downstream of the dam. Opponents of the project are worried about the reliability of the multiple ship locks required to open navigation around the dam. The navigation of the Yangtze River may be interrupted if any one of the five complex integrated locks has a problem. Similar problems have already occurred with the ship lift of the Gezhouba Dam, which only has one lock and is technically much simpler than that of the Three Gorges design. The free-flowing river will be transformed into a slow-moving reservoir and concern has been expressed about the release of toxic substances and pollutants from the land and residential areas that will be submerged. The regulated flow of water will increase the chance of saltwater intrusion into the Shanghai estuary and coastal groundwater aquifers, affecting water supply for the city. Ultimately, the full scope of the impacts of the dam cannot be predicted in advance, but if the dam is completed, these impacts will be felt for decades, or centuries, to come.

Chinese and International Opposition to the Dam

Throughout the protracted debate over the Three Gorges Dam, numerous objections and challenges to the project have been mounted by environmentalists, social scientists, geologists, sedimentation experts, hydraulic power engineers, military planners, and other Chinese specialists concerned about the dam's likely consequences.

In early 1989, shortly after Chinese and Canadian feasibility studies recommended dam construction, an extraordinary political event took place in China. Dozens of prominent citizens, scientists, intellectuals, and journalists released *Yangtze! Yangtze!*, an independently published collection of essays critical of the project (Dai Qing 1989, 1994). *Yangtze! Yangtze!*'s independent publication alone was extraordinary, but it had the further effect of influencing delegates attending the National People's Congress to take a stand against the project. At the meeting where Three Gorges finally was to be approved, nearly one-third of the delegates voted against it or abstained, an unprecedented display of political dissent. *Yangtze! Yangtze!* was subsequently banned in China.

Public opposition caused China's leaders to postpone the start of construction for five years. Not long thereafter, the pro-democracy demonstrations in Tiananmen Square began. As part of the subsequent political crackdown, the Communist Party ordered the arrest of some of those who opposed construction of the dam, including Dai Qing, *Yangtze! Yangtze!*'s chief editor. She was detained without trial in a maximum security prison for almost a year and threatened with execution for her part in opposing Three Gorges (Probe International 1993, 1997). Little information is available about current efforts within China, whether official or unofficial, to halt the project, though some experts continue to speak out. In late 1997, one of China's most respected hydrologists Huang Wanli, emeritus professor at Qinghua University, called for construction to be stopped because of concern over increased flood risks upstream of the dam (Caufield 1997).

In an unusual public criticism, the National Environmental Protection Agency of China (NEPA) in Beijing issued a report in March 1997 critical of construction practices and environmental protection programs (WWEE 1997). The report said that construction at Three Gorges has destroyed "vast tracts of soil and vegetation without a proper study of the project's impact." NEPA added that environmental protection programs were not being implemented or were lagging behind as rapid construction at the dam site went forward. Many environmental problems were "still outstanding" the agency said, in the sharpest criticism of the project by a state agency since construction began. Though most criticism of the project has been silenced in China, international organizations have stepped up their activities. In 1989, Probe International obtained and reviewed the Canadian feasibility study of the dam (CIDA 1988) and in 1993 published a detailed critique of it (Probe International 1993). Major international opposition to the dam still exists, and efforts continue to try to convince the Chinese government to stop or slow construction.

The Three Gorges Dam project is continuing to move forward. In November 1997, the Yangtze River was blocked and diverted in preparation for actual construction on the dam. The Xinhua news agency quoted Premier Li Peng as saying the giant Three Gorges dam project had proceeded smoothly in the past five years, the design for blocking the river was feasible, and the resettlement of local residents had gone well (*Financial Post* 1997). In many ways the project is emblematic of China itself, a massive development project being built to try to address massive development problems. Whether Three

Gorges is ever able to meet its objectives or becomes a symbol of the worst kind of human development of the twentieth century remains to be seen.

The Lesotho Highlands Project, Senqu River Basin, Lesotho

The Kingdom of Lesotho, one of the world's poorest countries, is completely surrounded by the Republic of South Africa, the continent's most affluent nation. All of Lesotho is over 1,500 meters in elevation, including the highest point in southern Africa at over 3,800 meters. Lesotho also receives the highest rainfall in the region and the country is the origin of many of the region's rivers, including the Senqu (Orange), the Mohokare (Caledon), and the Tugela (see Map 3.3).

Lesotho is almost completely devoid of exportable natural resources, but it has been able to market one asset, its water. A transfer of water from Lesotho to South Africa was first seriously discussed in the early 1980s, but in 1983, guerilla violence erupted in Lesotho over the possibility (Khits'ane 1997). In 1986, a treaty was signed by South Africa's apartheid government and Lesotho's military regime, establishing the Lesotho Highlands Water Project, currently the continent's largest civil engineering works. The project consists of many phases (named, practically enough, "1A," "1B," "2," "3," and "4"). The overall goal is to divert most of the major tributaries of the Senqu River, which naturally flows south out of Lesotho into South Africa (where it becomes the Orange), north into the Gauteng Region, which is the industrial heartland of South Africa. If the project is brought to completion, nearly 50 percent of Lesotho's water will be diverted for use in a different watershed in South Africa. More than 90 percent of the construction works are located in Lesotho.

The project is managed by the Lesotho Highlands Development Authority (LHDA), a joint Lesotho–South African organization, which has responsibility for construction, environmental protection, and all resettlement and compensation issues. In South Africa, the project is overseen by the Department of Water Affairs and Forestry and the Trans-Caledon Tunnel Authority (TCTA), established by South Africa under the authority of the 1956 South African Water Act (Act 54 of 1956) to implement its treaty responsibilities with Lesotho. A Joint Permanent Technical Commission (JPTC) was established to represent both countries. The JPTC has monitoring and advisory powers over the administrative, technical, and financial activities of the project (TCTA 1995).

The Lesotho Highlands Project consists of several major and minor dams, a series of water-transfer tunnels dug through the mountains, and various associated infrastructures, including hydroelectric generators and pumping plants. These components are summarized in Box 3.4. The heart of the system is Katse Dam and its reservoir. Katse Dam—a 185-meter concrete arch dam—is the highest ever built in Africa and blocks the south-flowing Malibamats'o River, a tributary of the Senqu. Water backs up to the northern end of the reservoir, from which it will flow into a series of diversion tunnels bored out of rock that will move the water out of the Senqu watershed and north into South

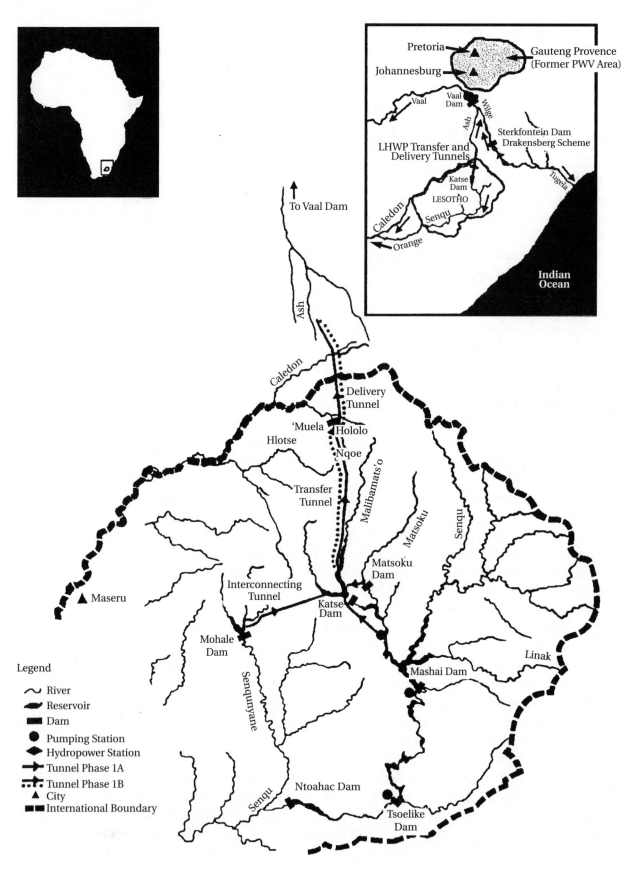

To Vaal Dam

Ash

Caledon

Delivery
Tunnel

'Muela   Hololo

Hlotse

Nqoe

Transfer
Tunnel

Malibamats'o

Matsoku

Senqu

Matsoku
Dam

Interconnecting
Tunnel

Katse
Dam

Maseru

Mohale
Dam

Mashai Dam

Linak

Senqunyane

Senqu

Ntoahac Dam

Tsoelike
Dam

Inset map:

Pretoria

Johannesburg

Gauteng Provence
(Former PWV Area)

Vaal

Vaal
Dam

Wilge

Ash

Sterkfontein Dam
Drakensberg Scheme

LHWP Transfer and
Delivery Tunnels

Katse
Dam

Tugela

Caledon

LESOTHO

Senqu

Orange

Indian
Ocean

Legend

〜 River
▬ Reservoir
▭ Dam
● Pumping Station
◆ Hydropower Station
➤ Tunnel Phase 1A
•••➤ Tunnel Phase 1B
▲ City
▰▰▰ International Boundary

MAP 3.3 THE KINGDOM OF LESOTHO

Africa via the Vaal River Basin to the Gauteng region. This water will be sold for use by Rand Water, the region's wholesaler. Below Katse Dam, a series of other dams would create other reservoirs, whose contents would then be pumped into Katse, for eventual delivery north.

South Africa will receive all of the water from the project, a total of 18 cubic meters per second from Phase 1A and an additional 12 cubic meters per second from Phase 1B. In return, Lesotho gets an annual payment, consisting of a base sum plus a sum contingent on the amount of water delivered. The project will also generate a small amount of hydroelectric power for Lesotho. Lesotho also gets secondary benefits associated with the project. Massive infrastructure, in the form of major roads and transmission lines, has been constructed. Housing for workers and contractors has been built and will be turned over to the government on completion of the project.

Financing for the Project

During the early years of the project, international economic sanctions were imposed on the apartheid regime of South Africa, making it difficult for South Africa to receive international support for construction. As a result, South Africa entered into an agreement with Lesotho to raise the funds on behalf of South Africa for Phase 1A, bypassing the economic and trading sanctions. South Africa undertook indirectly to repay all foreign loans. Following the change in government in South Africa in 1994 and the country's acceptance by the international community of nations, funding arrangements were made through international monetary markets (Lesotho Highlands Project 1997).

Funding for Phase 1A was secured from several sources, the largest amount being raised by loans from the industrial center of South Africa. Financial institutions in South Africa, Lesotho, and Swaziland also have participated. Export credit agreements were negotiated by foreign contractors involved in the project, and other funding arrangements were made with the Development Bank of South Africa, the African Development

---

BOX 3.4

Components of the Lesotho Highlands Project

Phase 1A: Katse Dam and reservoir (1,950 million cubic meters)
Transfer and delivery tunnels (82 kilometers)
'Muela power station (72 megawatts)
'Muela tailpond

Phase 1B: Mohale Dam and reservoir (958 million cubic meters)
Tunnel from Mohale to Katse (32 kilometers)
Matsoku weir and transfer tunnel (6.4 kilometers)

Phase 2: Mashai Dam and reservoir (3,306 million cubic meters)
Parallel transfer and delivery tunnels to Ash River
Pumping plants from Mashai to Katse

Phase 3: Tsoelike dam and reservoir (2,224 million cubic meters)
Pumping plants from Tsoelike to Mashai

Phase 4: Ntoahae Dam and reservoir
Pumping plants from Ntoahae to Tsoelike

Source: TCTA 1995.

Bank, the European Union Development Fund, and bilaterally from some of its member countries including the United Kingdom, Sweden, France, and Germany. The Development Bank of South Africa contributed 6 percent of the costs of building roads and other necessary infrastructure through the project area. Concessionary loans were raised from both multilateral and bilateral aid agencies such as the Commonwealth Development Corporation of the United Kingdom. The World Bank made a US$110 million commitment to Phase 1A to help finance project design, set up the financial package, and pay for part of construction (Lesotho Highlands Project 1997). While this contribution is relatively small, it has had the effect of influencing other international banks and funding agencies to support the project (Lesotho Highlands Project 1997). Deutsche Morgan Grenfell is serving as financial adviser to the LHDA to help raise capital to fund construction of Phase 1B. The entire project is expected to cost US$8 billion, with Phase 1A estimated to cost $2.5 billion and Phase 1B an estimated $1.5 billion.

Under treaty obligations, South Africa will pay for all capital costs related to the purpose of water transfer—approximately US$2.67 billion. These include costs incurred in Lesotho and funded by loans raised by the Lesotho Highlands Development Authority. Lesotho is responsible for approximately $23 million in capital costs for the 'Muela hydroelectric element of the scheme, a modest contribution toward its overall costs (Lesotho Highlands Project 1997).

South Africa is responsible for the payment of interest, the redemption of capital costs of the full water-delivery component, and water royalties to Lesotho. South Africa finances these payments through the sale of water to the consumers in the Vaal supply area and by a levy imposed on Rand Water consumers within the Vaal River System supply area starting in April 1988. The revenue generated over the life of the project is supposed to be sufficient to pay for the construction, maintenance, operating, royalty, and finance costs of the water-delivery component. Since being introduced, the levy has raised more than R1.2 billion (over US$250 million) toward project costs.

The South African Department of Water Affairs and Forestry (DWAF), which made the decision to construct the dam and is responsible for South Africa's participation in the project, sets the price at which water is to be sold and the project is to be repaid. Rand Water's rates went up 25 percent in 1996 and it predicts that rates will continue to rise 25 percent per year for the next four years as the costs of Phase 1A come on line. Rand Water does not know what the impacts of these cost increases will be; there is no certainty about the size of the increases; and it is unprepared at present to evaluate clearly the comparative costs of conservation alternatives. Some senior executives of Rand Water expressed serious concern in October 1996 about several aspects of the project, particularly the economic and the "consultative" aspects. They also explicitly believe that the costs of Phase 2 and beyond are likely to be prohibitive (Rand Water, personal communication, 1996). In late February 1998, Rand Water increased rates a further 9 percent, largely to cover LHP costs (SAPA 1998). Between March 1997 and April 1998, the price of water for Rand Water consumers almost doubled.

New studies suggest that considerable water conservation is possible in the Gauteng region, where water losses due to old infrastructure often reach 50 percent or more and inefficient mining and industrial processes waste water. Substantial amounts of water are also being used to irrigate the gardens of urban middle and upper middle class homes (Horta and Coverdale 1996). Demand management and efficient technologies would delay, probably reduce, and possibly eliminate the need for Lesotho Highlands Project water from the uncompleted phases.

Impacts of the Project

As with all major dam projects, there are a number of costs that will be borne by the local people and environment that have not been clearly evaluated, anticipated, or mitigated. Some of these have led to international attention and opposition. Rural development in the Lesotho Highlands area, which might have been brought about to a greater extent by careful planning, is being poorly and incompletely handled. The lives of more than 20,000 people will be disrupted by the project. The resettlement scheme was rated one of the worst in Africa in a 1995 World Bank report because of inadequate compensation for displaced people and increases in local health problems, including AIDS (World Bank 1995; Horta 1995; Horta and Coverdale 1996). Major environmental disruption of the Senqu River system is likely, but poorly studied. Several rare and endangered plants, animals, birds, and fish will be affected both in Lesotho and further downstream in South Africa.

Part of the project is an agreement to provide basic services to populations affected by the dam. All villages within the watershed of the project were to receive basic drinking-water supplies, consisting of standpipes within 150 meters of homes providing a standard of 30 liters per person per day. Pit toilets were also supposed to be provided for sanitation. Despite the fact that Phase 1A is almost completed and that billions of dollars have been spent, even this modest rural development is far behind schedule. Moreover, development plans do not include providing electricity to all residents in the affected area, despite the fact that hydroelectricity is being produced by the project and will be exported to other parts of Lesotho.

The treaty signed by Lesotho and South Africa specifies that people affected by the project should be able to maintain a standard of living equal to that which existed prior to the project and that they should be compensated for their losses. Phase 1A necessitated the relocation of 71 households to other sites. Approximately 240 other houses were replaced because of earthworks, powerlines, road construction, and other project-related activities. In addition, a total of 1,600 hectares of arable land and 3,200 hectares of rangeland has been lost due to Phase 1A.

Localized reservoir-induced seismicity caused by the rapid filling of the Katse Reservoir started occurring soon after impoundment commenced. These events caused structural damage to houses and other mud and stone buildings and have led to fear and apprehension among the local population. At the village of Mapeleng, seismic activity was accompanied by a 1.5-kilometer-long crack in the earth's surface, terrorizing many villagers who has now requested to be relocated.

In the area to be flooded by Phase 1B, more people and more arable land will be affected because of a higher population density and more intense agricultural activities.

Official estimates suggest that 46 of the 83 villages within the Phase 1B scheme area will suffer tangible and measurable losses in terms of physical structures and land. This includes nearly 1,000 households. Thirteen villages will have to be completely resettled. Of a total of 700 households in the remaining affected villages, nearly 400 will lose a substantial proportion of their land (Lesotho Highlands Project 1997).

Other social disruptions have occurred, including labor disruptions caused by the large inflow of foreign workers, the "boomtown" problems associated with big construction projects, and the introduction of diseases formerly unknown in the area, including AIDS. Labor disputes between Basotho and South African workers and between workers and the Lesotho Highlands Development Authority deteriorated into violence in late 1996, leading to at least five deaths and many injuries. In part, this violence resulted from controversy over the different wage scales paid to local and international workers, alleged discrimination by South African management against Basotho employees, and layoffs occurring at the end of Phase 1A, which led to unemployment and social disruptions. The World Bank called for the creation of an international independent commission to look into the violence, though no such commission was ever created (L. Pottinger, personal communication, 1997).

Some of the opposition stems from concerns about environmental impacts. For the Lesotho Highlands Project, these concerns include impacts to sensitive species of fish and birds. Prior to the completion of Phase 1A, minimum flows in the river below the dam site never fell below 3.5 cubic meters per second (cms) and maximum flows often exceeded 50 cms. After Phase 1A goes into operation, releases below the dam will be limited to a mere 0.5 cms as the water is diverted north. No studies appear to have been done on the minimum flow requirements needed to support instream ecosystems, and no provision has yet been made to provide such requirements, in large part because Phase 2 of the project will result in a reservoir that will flood all the way to the foot of Katse Dam. Because of the possibility that Phase 2 will never be built, such studies are now needed.

An endangered minnow, the Maluti minnow (*Pseudobarbus quathlambae*), occurs only in Lesotho and only in seven localities in the Phase 1 area. Of the seven known populations, four will be destroyed by Phases 1A and 1B. Environment officials from the Ministry of Water Affairs and Forestry have estimated that approximately 15 percent of the average flow of the Malibamats'o would need to be released from the Katse Dam in order to maintain the ecosystems of the Lesotho Highlands riverine system. This would reduce the economic viability of the project by decreasing both hydroelectric production and the volume of the water projected to flow north into South Africa. Endangered and threatened birds also live in the region, including the bearded vulture, bald-faced ibis, and black stork. Nesting sites of these birds will be eliminated by the reservoirs of Phases 1A and 1B (Horta 1995).

Impacts of the project will be felt outside of Lesotho, but little information on downstream impacts of the project is available (Horta and Coverdale 1996). In addition to the Lesotho portions of the basin, the Senqu/Orange River Basin includes parts of central and western South Africa and Namibia. Recently, these parties have begun to question openly the further development of the project and call for reevaluation of environmental and social impacts downstream. Specifically, there are concerns that the Lesotho Highlands Project will deprive other Orange River water users of sufficient water for their own needs. Namibia has only agreed to the initial stages of the project and renegotiations between Lesotho, Namibia, and South Africa will be necessary in order to proceed with the next phases.

Project Update

Phase 1, which consists of two parts (1A and 1B), is currently under construction and Phase 1A has virtually been completed (see Box 3.5 for project highlights). In Phase 1A, the Malibamats'o River was dammed by the Katse Dam. The dam was closed in October 1995 and the reservoir began filling. In late October 1997, a test of the water-delivery tunnels was successfully completed (G. van der Merwe, personal communication, 1997). Construction delays forced spilling of some water through the diversion tunnels, but the remaining details of the first phase were completed, the reservoir filled, and the first water deliveries made to South Africa in early 1998. Ironically, heavy rains filled South Africa's reservoirs in late 1997 and early 1998, and the first Lesotho Highlands Project water deliveries were diverted to low-valued agricultural uses.

Infrastructure development for Phase 1B has begun. Phase 1B will consist of a second large dam (Mohale Dam) and reservoir and another tunnel to transfer water from Mohale to Katse Reservoir (see Box 3.4). Phase 1B will boost total water deliveries to the Vaal system in South Africa from 18 cubic meters per second to over 30 cubic meters per second. Phase IB was given tentative approval in late 1997 and final approval in early 1998. Bids for the construction of the Mohale Dam and the tunnel linking Mohale to Katse were received in late 1997 and the contract was awarded in March 1998. The advanced infrastructure at Mohale has been completed and consists of a social platform, commercial platform, and housing platform, which required the leveling of mountain tops in the vicinity to provide flat areas for structures (G. van der Merwe, personal communication, 1997).

In March 1998, new opposition to the project emerged from within the heart of South Africa itself. Residents of Alexandra, a major black township outside of Johannesburg, challenged the project on the grounds that they were not consulted and that agencies associated with the project failed to consider alternatives that may be cheaper and more effective. These alternatives include reduction of massive leaks in the water distribution system that may lose as much as half of all water supposed to be delivered to the townships. The opposition from civic groups in South Africa is unusual because it comes from the project's intended beneficiaries rather than the rural communities affected by the development itself, where most of the opposition has occurred. These townships fear the burden of rising water costs to pay for a project on which they were never consulted. John Roome, the World Bank official in charge of the Lesotho Highlands Project, was quoted as saying: "They have a point. They have not really been consulted very well. While it is a good thing, it is not in our current guidelines that we have to consult with the end-users. That is government's job" (Aslam 1998). In late April 1998, these civic groups filed a formal claim with the World Bank Inspection Panel requesting investigation of possible harm to them arising from the project, despite heavy political pressure within South Africa not to file such a claim (Aslam 1998; Pottinger, personal communication, 1998).

BOX 3.5
Lesotho Highlands Project Highlights

October 24, 1986:   Treaty is signed between South Africa and Lesotho agreeing to the project.

1988:   First contract is awarded for construction of the Northern Access Road.

1989:   Second contract is awarded for construction of the Northern Access Road.

1991:   Main contracts are awarded for construction of Katse Dam and the tunnels.

1993:   First concrete is placed on Katse Dam.

1994:   First contract is awarded for construction of the Phase 1B access roads.

1995:   Impounding of water behind Katse Dam starts on October 20.

1996:   Construction of Delivery Tunnel North is completed. Water level in Katse Dam reaches elevation necessary for royalty payments to Lesotho. Construction of Delivery Tunnel South is completed.

August 1997:   'Muela Dam (55 meters high) is completed; 'Muela hydropower plant (72 megawatts) is 90 percent complete.

October 29, 1997:   First water test of tunnels and delivery system.

January 1998:   First water is delivered to South Africa.

April 1998:   Construction begins on Mohale tunnels. Contracts awarded for Phase 1B.

Despite internal and external opposition, the new government of South Africa is intent on continuing with the rest of Phase 1. The South African Water Minister has publicly supported the project:

> We in South Africa have one of the widest gaps in the world between the haves and the have-nots, and this is obviously apparent in the relative availability of water to the people. . . . The central truth is that the [Lesotho Highlands Water] Project is about people. It is a successful project which is mutually beneficial to the people of South Africa and Lesotho . . .
>
> Professor Kader Asmal, MP, Minister of Water Affairs and Forestry, 18 November 1996, Johannesburg, at the issuance of two new water bonds by the TCTA, in association with the LHDA

Despite the rhetoric, the Lesotho Highland Project cannot yet be declared "successful." For it to be considered truly successful, given international standards for water developments set out by international financial and environmental organizations and by the Dublin and Earth Summit conferences, major changes will be required. Far better attention will have to be paid to the social and environmental impacts than has been given so far. More participation in decision making by affected Basotho people and the Johannesburg area civic organizations should be encouraged rather than discouraged. More rapid and effective restoration and improvement of quality of life for displaced populations should be achieved. Independent and comprehensive evaluation of environmental impacts and alternatives for protecting resources of concern should be commissioned. And the true benefits of the project—specifically the water and electricity—should be priced to reflect actual project costs.

Finally, given the problems that occurred during Phase 1A, Phases IB, 2, 3, and 4 should be carefully reconsidered. There is already evidence that Phase 1B could prove too expensive for the project beneficiaries and the impacts too great for local populations. If this phase goes ahead, reevaluation and mitigation of the impacts should be done to international standards. Phases 2 to 4 appear even less attractive at this point given their anticipated environmental, social, and economic costs and the limited additional benefits they offer. Until a comprehensive independent reevaluation can be done, these phases should be delayed or canceled.

REFERENCES

Abu Zeid, M., and A.K. Biswas. 1990. "Impacts of agriculture on water quality." *Water International,* Vol. 15, No. 3, pp. 160–167.

Adams, W.M. 1993. "Development's deaf ear: Downstream users and water releases from the Bakalori Dam, Nigeria." *World Development,* Vol. 21, No. 9, pp. 1405–1416.

Aslam, A. 1998. "South Africans clash with World Bank over water." Inter Press Service (IPS), March 10, Washington, D.C.

British Broadcasting Service (BBC). 1997. "China: Foreign consortia win contracts for China's Three Gorges Project." BBC Monitoring Service (August 25).

Canadian International Development Agency (CIDA). 1988. "Three Gorges water control project feasibility study, People's Republic of China." Canadian International Development Agency, Toronto (March).

Caufield, C. 1997. "Rough sailing at Three Gorges Dam." *World Rivers Review,* Vol. 12, No. 6 (December), pp. 1, 10.

Chao, B.F. 1995. "Anthropological impact on global geodynamics due to water impoundment in major reservoirs." *Geophysical Research Letters,* Vol. 22, pp. 3533–3536.

China. 1996. "The Three Gorges Project: A brief introduction." Official government fact sheet. Chinese Embassy, Washington, D.C. Via *http://www.china-embassy.org/Press/gorges.htm.*

Chinese Academy of Sciences. 1988. *Environmental Impact Statement for the Three Gorges Project.* Chinese Academy of Sciences, Nanjing Geography Institute.

Chinese Academy of Sciences. 1995. *Environmental Impact Statement for the Yangtze Three Gorges Project (A Brief Edition).* Environmental Impact Assessment Department, Chinese Academy of Sciences and the Research Institute for Protection of Yangtze Water Resources. Science Press, Beijing, China.

Collier, M., R.H. Webb, and J.C. Schmidt. 1996. *Dams and Rivers: Primer on the Downstream Effects of Dams.* United States Geological Survey (USGS), Circular 1126, Branch of Information Services, Tucson, Arizona, and Denver, Colorado.

Cooper, H. 1996. "Ex-Im Bank snubs Chinese dam project." *The Wall Street Journal* (May 31), p. A3.

Covich, A.P. 1993. "Water and ecosystems," in P.H. Gleick (ed.), *Water in Crisis: A Guide to the World's Fresh Water Resources.* Oxford University Press, New York, pp. 40–55.

Dai Qing. 1989. *Yangtze! Yangtze!* Chinese edition. Guizhou People's Publishing House, China.

Dai Qing. 1994. *Yangtze! Yangtze!* English edition. P. Adams and J. Thibodeau (eds.), Probe International, Earthscan Publications Limited, United Kingdom.

Ehrlich, A.H., and P.R. Erhlich. 1987. *Earth.* Methuen London, Ltd., London.

Ex-Im Bank. 1996. "Three Gorges Dam in China: Transcript of press briefing on Board Meeting." May 30, Washington, D.C. Via *http://www.exim.gov/t3gorges.html*.

Fearnside, P.M. 1988. "China's Three Gorges Dam: Fatal project or step toward modernization?" *World Development*, Vol. 16, No. 5, pp. 615–630.

Financial Post. 1997. "China to block Yangtze River in November." October 15, 1997, Via *http://www.nextcity.com/main/article/pi/97–10-15-FinancialPost.htm*

Financial Times Global Water Report. 1997a. "Germans cement water links." *Financial Times Global Water Report*, Issue 31, September 26, p. 7.

Financial Times Global Water Report. 1997b. "Washington orders dam destruction." *Financial Times Global Water Report*, Issue 36, December 4, p. 9.

Gleick, P.H. 1992. "Environmental consequences of hydroelectric development: The role of facility size and type." *Energy*, Vol. 17, No. 8, pp. 735–747.

Gleick, P.H. 1994. "Water, war, and peace in the Middle East." *Environment*, Vol. 36, No. 3, pp. 6–42. Heldref Publishers, Washington.

Gleick, P., P. Loh, S. Gomez, and J. Morrison. 1995. *California Water 2020: A Sustainable Vision.* Pacific Institute Report. Pacific Institute for Studies in Development, Environment, and Security, Oakland, California.

Goldberg, C. 1997. "Fish are victorious over dam as U.S. agency orders shutdown." *The New York Times*, November 26, p. A12.

Goldsmith, E., and N. Hildyard (eds.). 1986. *The Social and Environmental Impacts of Large Dams.* Wadebridge Ecological Centre, Cornwall, United Kingdom.

Goodland, R. 1994. "Ethical Priorities in Environmentally Sustainable Energy Systems: The Case of Tropical Hydropower." World Bank Environment Working Paper 67.

Goodland, R. 1997. Environmental sustainability in the hydro industry: Disaggregating the debates," in T. Dorcey, A. Steiner, M. Acreman, and B. Orlando (eds.), *Large Dams: Learning from the Past, Looking at the Future.* IUCN and the World Bank Group Workshop Proceedings. Gland, Switzerland 11–12 April 1997, Washington, D.C., pp. 69–102.

Hatami, H., and P. Gleick. 1993. *Chronology of Conflict Over Water in the Legends, Myths, and History of the Ancient Middle East.* Pacific Institute for Studies in Development, Environment, and Security, Oakland, California.

Hirsch, R.M., J.F. Walker, J.C. Day, and R. Kallio. 1990. "The influence of man on hydrologic systems," in *The Geology of North America*, Vol. O-1, *Surface Water Hydrology*, Geological Society of America, Boulder, Colorado, pp. 329–359.

Horta, K. 1995. "The Mountain Kingdom's white oil: The Lesotho Highlands Water Project." *The Ecologist*, Vol. 25, No. 6, pp. 227–231.

Horta, K., and S. Coverdale. 1996. "Dams and distress for kingdom in the sky." *People and the Planet*, Vol. 5, No. 3, pp. 24–25.

Hu Chunhong. 1995. "Controlling reservoir sedimentation in China." *International Journal of Hydropower and Dams* (March), pp. 50–52.

Human Rights Watch/Asia (HRW). 1995. "The Three Gorges Dam in China: Forced resettlement, suppression of dissent and labor rights concerns." *Human Rights Watch/Asia*, Vol. 7, No. 2, p. 1.

Hydropower and Dams. 1997. "1997 World Atlas and Industry Guide." *The International Journal on Hydropower and Dams.* Surrey, United Kingdom.

Japan Press Network (JPN). 1996. "MITI and Ex-Im Bank consider aid for Three Gorges Dam." (J. Tofflemire, May 20).

Khits'ane, M. 1997. "Work in the areas affected by Lesotho Highlands Water Project (LHWP)." Highlands Church Action Group (HCAG) (March). Via *http://www.irn.org/*.

Leopold, L.B. 1997. "Sediment problems at Three Gorges Dam." University of California, Berkeley. Via *http://irn.org/*.

Lesotho Highlands Project. 1997. South African governmental website for the Lesotho Highlands Project. Via *http://www.gov.za/dwaf/web-pages/Watres/LHWP/lhwpframe. htm.*

McCully, P. 1996. *Silenced Rivers: The Ecology and Politics of Large Dams.* Zed Books, London.

McPhee, J. 1971. *Encounters with the Archdruid.* Farrar, Straus, and Giroux, New York.

Milliman, J.D., and R.H. Meade. 1983. "World-wide delivery of river sediment to the oceans." *Journal of Geology,* Vol. 91, No. 1, pp. 1–21.

Murty, K.S. 1989. *Soil Erosion in India, River Sedimentation,* Vol. 1. International Research and Training Center on Erosion, Beijing, China.

Ortiz, A. 1979. *Handbook of North American Indians. Vol. 9: Southwest.* Smithsonian Institute, Washington, D.C.

Oud, E., and T.C. Muir. 1997. "Engineering and economic aspects of planning, design, construction, and operation of large dam projects," in T. Dorcey, A. Steiner, M. Acreman, and B. Orlando (eds.), *Large Dams: Learning from the Past, Looking at the Future.* IUCN and the World Bank Group Workshop Proceedings. Gland, Switzerland 11–12 April 1997, Washington, D.C., pp. 17–39.

Perrin, W.F., and R.L. Brownell, Jr. (eds.). 1987. *Report of the Workshop on Biology and Conservation of the Plantanistoid Dolphins.* October 28–30. Wuhan, People's Republic of China. International Union for the Conservation of Nature and Natural Resources (IUCN), Gland, Switzerland.

Petts, G.E. 1984. *Impounded Rivers: Perspectives for Ecological Management.* John Wiley and Sons, New York.

Probe International 1993. *Damming the Three Gorges: What Dam Builders Don't Want You to Know,* M. Barber and G. Ryder (eds.), Earthscan Publications Limited, United Kingdom.

Probe International. 1997. "Financing disaster: China's Three Gorges Dam." Web site http://www.nextcity.com/ProbeInternational/ThreeGorges/index.html.

Reisner, M. 1986 (1993 revision). *Cadillac Desert: The American West and its Disappearing Water.* Penguin Books, New York.

Reuters. 1997a. "China: Three Gorges Dam contract official—No extra for GEC Alsthom." Reuters News Service (August 25).

Reuters. 1997b. "China firms win $807 million contracts on Three Gorges." Reuters News Service (August 19).

SAPA. 1998. "Rand Water announces increase from April." South African Press Agency, February 26, Johannesburg, South Africa.

Scudder, T. 1994. "Recent experiences with river basin development in the tropics and subtropics." *Natural Resources Forum,* Vol. 18, No. 2, pp. 101–113.

Scudder, T. 1997. "Social impacts of large dam projects," in T. Dorcey, A. Steiner, M. Acreman, and B. Orlando (eds.), *Large Dams: Learning from the Past, Looking at the Future.* IUCN and the World Bank Group Workshop Proceedings. Gland, Switzerland 11–12 April 1997, Washington, D.C., pp. 41–68.

Sewell, W.R.D. 1985. "Inter-basin water diversions: Canadian experiences and perspectives," in G.N. Golubev and A.K. Biswas (eds.), *Large Scale Water Transfers: Emerging Environmental and Social Experiences.* United Nations Environment Program, Water Resources Series Volume 7, Tycooly Limited, Oxford, United Kingdom, pp. 7–35.

Shiklomanov, I. 1993. "World fresh water resources," in P.H. Gleick (ed.), *Water in Crisis: A Guide to the World's Fresh Water Resources.* Oxford University Press, New York, pp. 13–24.

Shiklomanov, I.A. 1996. "Assessment of the water resources and water availability in the world." Draft report to the Comprehensive Assessment of the Freshwater Resources of the World. State Hydrological Institute, St. Petersburg, Russia.

Smith, C.S. 1995. "China dam project is hard sell abroad." *The Wall Street Journal* (May 3), p. A10.

Stanford J.A., and J.V. Ward. 1979. "Stream regulation in North America," in J.V. Ward and J.A. Stanford (eds.), *The Ecology of Regulated Streams*. Plenum Press, New York, pp. 215–236.

Trans-Caledon Tunnel Authority (TCTA). 1995. *Lesotho Highlands Water Project*, Volumes 1–3. Laserline Pensord Press, United Kingdom.

United States Committee on Large Dams (USCOLD). 1996. Register of Dams. Via *http://www.uscold.org/~uscold/uscold_s.html*

United States Senate. 1964. "Western water development: A summary of water resources projects, plans, and studies relating to the western and midwestern United States." Committee on Public Works, Special Subcommittee on Western Water Development, U.S. Government Printing Office, Washington, D.C.

Voropaev, G.V., and A.L. Velikanov. 1985. "Partial southward diversion of northern and Siberian rivers," in G.N. Golubev and A.K. Biswas (eds.), *Large Scale Water Transfers: Emerging Environmental and Social Experiences*. United Nations Environment Program, Water Resources Series Volume 7, Tycooly Limited, Oxford, United Kingdom, pp. 67–83.

White, G. 1988. "The environmental effects of the High Dam at Aswan." *Environment*, Vol. 30, No. 7, p. 4.

World Bank. 1994. *Resettlement and Development: The Bankwide Review of Projects Involving Involuntary Resettlement 1986–1993*. World Bank, Washington, D.C.

World Bank. 1995. *Resettlement Remedial Action Plan for Africa*. World Bank, Washington, D.C.

World Water and Environmental Engineering (WWEE). 1997. "Vast tracts of soil destroyed by Three Gorges Project." July 1997. Via *http://www.nextcity.com/main/article/pi/97-07-01-worldwater.htm*.

Xinhua News Agency. 1997. "Dam opera." China (August 21).

Zich, A. 1997. "Before the flood: China's Three Gorges." *National Geographic*, Vol. 192, No. 3, pp. 2–33.

Conflict and Cooperation Over Fresh Water

In the past decade, there has been a major shift and rethinking in the field of international security analysis. With the end of the Cold War and the breakup of the Soviet Union, the world has moved away from a focus on political and military conflict between two major superpowers to a growing concern over regional conflicts; civil, religious, and ethnic wars within regions; and the connections between environmental degradation, scarcity of resources, and regional and international politics and disputes. Traditional political and ideological questions that have long dominated international discourse are now becoming more tightly woven with other variables that loomed less large in the past, including population growth, transnational pollution, resource scarcity, and inequitable access to resources and their use. These issues have been loosely termed "environmental security," and interest in them has engendered a wide range of innovative and controversial analyses and studies.

This change results from a growing recognition that new and large-scale environmental threats with political ramifications are beginning to dominate international discourse about environmental concerns. During the 1970s and early 1980s, environmental concerns focused primarily on local or regional air or water pollution problems. In the mid-1980s, however, new environmental problems with global ramifications became more evident, including global climate change, the depletion of atmospheric ozone, population pressures on limited resources, and inequities in resource use between developed and developing countries. Many of these new problems affect more than one region or nation and involve problems that cross political borders.

While many of the past, present, and future causes of conflict and war may seem to have little or no direct connection with the environment or with resources, a strong argument can be made for linking certain resource and environmental problems with the prospects for political frictions and tensions, or even war and peace. History shows that access to resources has been a proximate cause of war, resources have been both tools and targets of war, and environmental degradation and the disparity in the distribution of resources can cause major political controversy, tensions, and violence.

At the center of the ongoing debate is the assertion that resource scarcity and certain forms of environmental degradation are important factors contributing to political instability or violent conflict at local, regional, and interstate levels. A related argument is that as population growth continues and certain large-scale environmental problems worsen, cases of environmentally related instability are likely to proliferate in the future (Myers 1989; Gleick 1990, 1991, 1993; Homer-Dixon 1991, 1994; Homer-Dixon and Percival 1996; Libiszewski 1995; Dabelko and Dabelko 1996; Carnegie Commission 1997).

In short, there is a growing perception that local, regional, and global environmental deficiencies or resource scarcities will increasingly produce conditions that may lead to conflict. The academic disputes over definitions and terms will, no doubt, continue, but a wide variety of recent comments by senior diplomats and policymakers leave no doubt that issues related to environmental security have moved to the highest levels.

In 1987, the General Secretary of the Soviet Union, Mikhail Gorbachev, stated:

> [The world] is not secure in the direct meaning of the word when currents of poison flow along river channels, when poisonous rains pour down from the sky, when an atmosphere polluted with industrial and transport waste chokes cities and whole regions, when the development of atomic engineering is justified by unacceptable risks. . . . The relationship between man and the environment has become menacing. Problems of ecological security affect all—the rich and the poor. What is required is a global strategy for environmental protection and the rational use of resources. (Gorbachev 1987)

In 1989, then United States Secretary of State James Baker said:

> The strategic, economic, political, and environmental aspects of national security and global well-being are, today, indivisible. (Baker 1989)

In November 1989, Prime Minister Margaret Thatcher gave a speech to the United Nations General Assembly saying:

> While the conventional, political dangers—the threat of global annihilation, the fact of regional war—appear to be receding, we have all recently become aware of another insidious danger. It is as menacing in its way as those more accustomed perils with which international diplomacy has concerned itself for centuries. It is the prospect of irretrievable damage to the atmosphere, to the oceans, to earth itself.

In 1994, United Nations Ambassador Madeleine K. Albright (now U.S. Secretary of State) stated:

> We believe that environmental degradation is not simply an irritation, but a real threat to our national security. . . . Left unaddressed, it could become a kind of creeping Armageddon . . . it could, in time, threaten our very survival.

In April 1997, as United States Secretary of State, Ms. Albright went on to say:

> Not so long ago, many believed that the pursuit of clean air, clean water, and healthy forests was a worthy goal, but not part of our national security. Today environmental issues are part of the mainstream of American foreign policy. (U.S. Department of State 1997)

The argument and focus of debate has now shifted from "whether" there is a connection to "when," "where," and "how" environmental and resource problems will affect regional and international security. There is also a long way to go before the United States and the rest of the world produce a common policy agenda or set of initiatives that truly incorporate environmental and resources issues into approaches to reduce the risk of regional and national conflicts. Nevertheless, there is a new framework under construction that is permitting scholars and policymakers to apply new tools, to set new priorities, and to organize responses to a range of environmental threats to peace and security.

This framework is particularly well developed in the area of water resources. In the past several years there has been considerable progress in both understanding the nature of the connections between water resources and conflict and in evaluating regional cases where such connections may be particularly strong (Gleick 1993, 1994, 1996; Klötzli 1993, 1994; Böge 1993; Libiszewski 1995; Kelly and Homer-Dixon 1995; Wolf 1997). There has also been progress in trying to identify policies and principles for reducing the risks that freshwater disputes will lead to conflict and to better understand mechanisms for promoting cooperation and collaboration over shared freshwater resources.

Progress has been more than academic. In October 1994, for example, Israel and Jordan signed a peace treaty that explicitly addressed water allocations, sharing water information, and joint management policies for the Jordan River Basin. In 1996, India and Bangladesh signed a formal treaty that moves toward resolving their longstanding dispute over the Farraka Barrage and flows in the Ganges/Brahmaputra system. In 1997, the International Law Commission, after nearly three decades of negotiations, drafting, and discussion, finalized the Convention on the Non-Navigational Uses of International Watercourses, which was approved by the General Assembly in April (UN 1997a). And countries like Brazil, South Africa, and Zimbabwe are incorporating mechanisms and principles for resolving conflicts over shared waters in their new water laws.

Conflicts Over Shared Water Resources

Fresh water is integral to all ecological and societal activities, including the production of food and energy, transportation, waste disposal, industrial development, and human health. Yet freshwater resources are unevenly and irregularly distributed, and some regions of the world are extremely short of water. As we move into the twenty-first century, water and water-supply systems are increasingly likely to be both instruments of political conflict and the objectives of military action as human populations grow, standards of living improve, and global climatic changes make water supply and demand more problematic and uncertain.

There are four major links between water and conflict. Water has been used as a military and political goal. Water has been used as a weapon of war. Water-resources systems have been targets of war. And inequities in the distribution, use, and consequences of water-resources management and use can be a source of tension and dispute (Gleick 1993). These links can occur at local, subnational, and international levels. Appendixes A and B to this chapter present some historical and ongoing disputes and conflicts over freshwater resources or their allocation and use. These appendixes also summarize the nature of the conflict and whether or not it led to military action or violence.

Water as a Military and Political Goal

Where water is scarce, competition for limited supplies can lead groups, communities, and even nations to see access to water as a matter of highest concern. This situation falls into the most traditional Cold War/realpolitik framework where water can be a defining factor in the wealth and power—and in the economic and political strength—of a nation. Access to water resources may serve as a focus of dispute or provide a justification for actual conflict. While it may never be the sole reason for conflict, history suggests that it has already proven to be an important factor.

Four important conditions influence the likelihood that water will be the object of military or political action: (1) the degree of water scarcity; (2) the extent to which the supply is shared by two or more groups; (3) the relative power of those groups; and (4) the ease of access to alternative sources.

Water is unevenly distributed throughout the world. Some regions receive enormous amounts of rainfall or river flow, while others are extremely dry and suffer from "absolute" scarcity. Human factors, such as high population densities or intensive industrial development, may cause conditions of "relative" scarcity. In the past, absolute scarcity was a more important factor in regional water issues. Increasingly, however, relative scarcity is the problem leading to water disputes, as population pressures come up against resource limits (Falkenmark 1986; Gleick 1993; Engelman and LeRoy 1993; UN 1997b). If water is abundant, conflict over access to adequate supplies is unlikely, though other factors, such as hydropower allocations, transportation access, or water quality, may also cause disagreements.

The problem of shared resources complicates the problem of scarcity. When water is extensively shared, misunderstandings or lack of agreement about allocations are more likely. Fresh water is very widely shared because political borders rarely coincide with watershed boundaries. At the international level, over 220 river basins are shared by two or more nations (UN 1978). But even countries with few or no internationally shared rivers or aquifers often have internal water disputes among states, ethnic groups, or economic classes trying to gain access to additional water supplies.

If there are great disparities in the economic or military strength of the parties involved, unilateral and inequitable decisions are more likely. A weaker party will rarely provoke or initiate military action—and even more rarely prevail—against a stronger adversary, but if a weaker nation either controls a water source or is dependent on water from an outside source, disputes and conflicts may occur. When adversaries are equally matched economically or militarily, negotiation and cooperation are more common outcomes (Wolf 1997).

Finally, if there are few technologically or economically attractive alternative sources of supply, the potential for conflict is higher. If water is scarce and shared, but there exist alternative sources such as other rivers, groundwater aquifers, or even expensive desalination, conflicts are less likely to occur. There is a high economic, social, and political cost to conflicts; they are likely to be avoided if acceptable substitutes can be found.

Over the course of human history these different factors have come together many times to produce a wide range of disputes over access to shared freshwater resources (see Appendixes A and B). Forty-five hundred years ago, the control of irrigation canals vital to survival was the source of conflict between the states of Umma and Lagash in the ancient Middle East. Twenty-seven hundred years ago, Assurbanipal, King of Assyria from 669 to 626 B.C., seized control of wells as part of his strategic warfare against Arabia. In the modern era, the Jordan River Basin has been the scene of a wide variety of water disputes. In the 1960s, Syria tried to divert the headwaters of the Jordan away from Israel, leading to air strikes against the diversion facilities (Falkenmark 1986; Lowi 1992). The 1967 war in the Middle East resulted in Israel winning control of all of the headwaters of the Jordan as well as the groundwater of the West Bank. In these cases, water was certainly not the sole issue precipitating conflict, but it was an important factor in both pre- and post-1967 border disputes.

Water remains an important factor in the politics of the region. The multilateral and bilateral peace talks conducted in the 1990s, which led to the interim agreement between the Israelis and the Palestinians and to the peace treaty between Israel and Jordan, explicitly included negotiations and agreements on the shared water resources of the Jordan River. Israeli and Syrian concerns over the Banias, which originates in the Golan Heights, remain an important unresolved issue. Jordanian concerns about Syrian dams on the Yarmouk, the major tributary to the Jordan, are still unanswered.

Disputes over the allocation of water occur at the subnational level as well and have the potential to turn violent. In California in the mid-1920s, farmers repeatedly destroyed an aqueduct taking water from their region to the urban centers of southern California. The governor of Arizona called out the local militia in the 1930s to protest the construction of water diversion facilities on the Colorado River between Arizona and California (Reisner 1986). This dispute was eventually resolved in court.

Court decisions do not always successfully end disputes. An interim court decision in India to allocate additional waters from the Cauvery River—which originates in the state of Karnataka—to Tamil Nadu actually precipitated violent conflicts resulting in the deaths of over 50 people (Gleick 1993). In 1997, the World Court issued a decision in the dispute between Hungary and Slovakia over the Gabcikovo-Nagymaros project on the Danube, effectively refusing to decide and returning the dispute to the two parties for more negotiation (Klötzli 1993; Lipschutz 1997). As populations continue to grow, regional water scarcity may lead to more frequent examples of this kind of water dispute and conflict.

Water as an Instrument or Tool of Conflict

The usual tools and instruments of war are military weapons. But the use of water as both offensive and defensive weapons has a long history. One of the earliest accounts is an ancient Sumerian myth from 5,000 years ago, which parallels the biblical account of the great flood. In this myth, the Sumerian deity Ea punishes humanity's sins by causing a great flood. In 695 B.C., Sennacherib completed the destruction of Babylon by diverting irrigation canals to wash over the ruins of the city. Herodotus wrote in 400 B.C. about Cyrus the Great's successful invasion of Babylon in 539 B.C. by diverting the Euphrates River into the desert and entering the city along the dry riverbed. In 1503, Leonardo da Vinci, in one of history's oddest collaborations, worked with Machiavelli on an unsuccessful project to divert the Arno River away from Pisa during the war between Florence and Pisa (Honan 1996).

In the last several years, two examples suggest that the use of water as a weapon continues to be considered. North Korea announced plans in 1986 to build a major hydroelectric dam on the Han River upstream of South Korea's capital, Seoul. The project would provide much-needed electricity to the North, but South Korea sees only its potential as a military weapon. South Korean hydrologists calculated that the destruction of the dam by the North and the sudden release of the reservoir's contents would destroy most of Seoul. While the project currently remains on hold due to serious political and economic difficulties in North Korea, South Korea has built a series of levees and check dams above Seoul to defend against any such threat (Chira 1986; Koch 1987).

Another use of a large dam and reservoir as a weapon of war was proposed during the Persian Gulf war. The allied coalition arrayed against Iraq discussed the possibility of using the Ataturk Dam in Turkey on the Euphrates River to shut off the flow of water to Iraq, which is highly dependent on flows in the Euphrates for water supply. No formal request to Turkey was ever made, and Turkey subsequently stated that it would never use water as a means of political pressure, but the possibility remains a concern in the region (Gleick 1993). Both Syria and Iraq continue to have an ongoing dispute with Turkey over the operation of Ataturk and the level and quality of flows in the Euphrates reaching both downstream countries.

In 1997, a fresh dispute arose between Singapore and Malaysia. Singapore has never been self-sufficient in water because of its high population density and small size; it depends on piped water from Malaysia for nearly half of all its needs (Zachary 1997). In addition, Singapore imports water from Malaysia that it then treats and sells back under an agreement signed in 1965 (Dupont 1998). Relations between the two countries have long been clouded by economic competition, and religious, political, and ethnic differences that have flared periodically since their separation in 1965. In early 1997, these relations soured again after comments about mutual concerns were exchanged by senior politicians, leading Ahmed Zamid Hamidi, the head of the youth wing of Malaysia's ruling party, to urge the government to review the basis of water agreements with Singapore (Lee 1997). The chief minister of Johor State in Malaysia went further, suggesting that they appropriate two of the three water-purification plants operated by Singapore in Johor (Jayasankaran and Hiebert 1997). Singapore is clearly worried that Malaysia might use water as a political and strategic weapon against Singapore—a point made to Malaysian Prime Minister Mahathir Mohamad by Singapore's Prime Minister Goh Chok Tong: "An agreement by Malaysia to meet Singapore's long-term water needs beyond the life of the present water agreements would remove the perception in Singapore that water may be used as a leverage against Singapore" (*Straits Times* 1997; Dupont 1998). Singapore has also launched a campaign to increase water supplies and to reduce consumption through an aggressive conservation program. Among their supply plans are new desalination plants that would produce water at about eight times the cost of current supplies (Zachary 1997). Whether this high price moves Singapore toward more diplomatic negotiations with Malaysia remains to be seen.

Water and Water Systems as Targets of Conflict

Where water resources and water-supply systems are a strategic resource with economic, political, or military importance, they become targets during wars or conflicts. The city of Babylon in ancient times was often a subject for conquest, as described above, and around 720 B.C. Sargon of Assyria destroyed the irrigation systems of the Haldians of Armenia. In modern times, dams and hydroelectric facilities were regularly bombed as strategic targets during World War II and the Korean War. The United States targeted irrigation levees in North Vietnam. Syria tried to destroy Israel's National Water Carrier while it was under construction in the 1950s. The Persian Gulf war saw several examples of this problem: the Iraqis intentionally destroyed the water desalination plants of Kuwait and in turn suffered from the destruction of their water-supply system by the allied forces assembled to liberate Kuwait.

Inequities in Water Distribution, Use, and Development

Tensions and conflicts may also result from such indirect factors as the inequitable distribution, use, and development of water resources. Many rivers, lakes, and groundwater aquifers are shared by two or more nations, and many more cross other kinds of political borders. More than half the land area of the world, and perhaps 70 percent of inhabitable land area, is in an international watershed, where river flows or lakes are shared. These include the Nile, Jordan, Tigris, Euphrates, and Orontes rivers in the Middle East; the Indus, Mekong, Ganges, and Brahmaputra rivers in Asia; the Great Lakes, and the Colorado, Rio Grande, Amazon, and Paraná rivers in the Americas; and Lake Chad and the Congo, Zambezi, Niger, Senegal, Okavango, and Orange rivers in Africa, to name only a few. This geographical fact has led to the geopolitical reality of disputes over the uneven distribution of shared waters. Table 4.1 lists internally available renewable water resources and total water withdrawals for countries in four major shared river basins, revealing the enormous disparities in resource supply within specific regions. In particular, note the countries whose water use exceeds internally available water resources. These countries, including Egypt, Iraq, and Syria, have a particularly strong interest in continuing to receive water from river flows that originate outside of their borders.

Equally uneven is the level of water use. Many industrialized nations and nations with extensive irrigated agriculture withdraw more than 1,500 cubic meters of water per person annually for all uses. At the other extreme, nations with limited supplies or low levels of economic development may use fewer than 100 cubic meters per person per year. Table 4.2 lists the 15 countries that use the most water per capita and the 15 countries that use the least. Such a low level of water use has direct and undesirable human consequences, including adverse impacts on health (see Chapter 2), the inability to grow sufficient food for local populations, and constraints on industrial and commercial activities. Table 2 in the Data section lists data on water use by country. Table 2.1 in Chapter 2 lists countries where domestic water use falls below the basic water requirement of 50 liters per person per day.

One of the most important water constraints facing many regions is insufficient water to grow food. Sandra Postel (1997) suggests that as annual water availability drops below 1,700 cubic meters per person, domestic food self-sufficiency becomes almost impossible and countries must begin to import water in the form of grain. Table 4.3 lists countries where annual renewable water supplies are below 1,700 cubic meters per person per year. The number of countries in this category will continue to rise with population growth and overall dependence on grain imports will deepen and spread. Food insecurity is a political concern and can lead to economic weakness and other regional problems.

Finally, there are often adverse consequences of water development and use, and people who do not receive the benefits from water projects may feel these consequences. Examples include contamination of downstream water supplies or groundwater

112

TABLE 4.1 WATER SUPPLY AND DEMAND BY COUNTRY IN FOUR MAJOR RIVER BASINS

River Basin	Countries	Renewable Internal Supply (km³/year)[a]	Water Withdrawals (km³/year)[b]
Nile	Sudan	35	17.8
	Ethiopia	110	2.2
	Egypt	1.8	55.1
	Uganda	39	0.2
	Tanzania	80	1.2
	Kenya	20.2	2.1
	Congo (Democratic Republic)	935	0.36
	Rwanda	6.3	0.8
	Burundi	3.6	0.1
	Eritrea	2.8	—[c]
Zambezi	Zambia	80.2	1.7
	Angola	184	0.5
	Zimbabwe	14.1	1.2
	Mozambique	100	0.6
	Malawi	17.5	0.9
	Botswana	2.9	0.1
	Tanzania	80	1.2
	Namibia	6.2	0.25
Tigris and Euphrates	Iraq	35.2	42.8
	Iran	128.5	70
	Turkey	196	31.6
	Syria	7	14.4

[a] These data include only internally available or generated water resources, not water flows into a country from outside its borders. For those data, see Table 1 in the Data section.

[b] Water withdrawals include water reported as taken from rivers, lakes, or groundwater for human use. See Table 2 in the Data section.

[c] Included in Ethiopia's withdrawals.

TABLE 4.2 COUNTRIES WITH THE SMALLEST AND LARGEST REPORTED PER-CAPITA WATER WITHDRAWALS[a]

Country	Withdrawals (m³/p/yr)	Country	Withdrawals (m³/p/yr)
Congo, DR (formerly Zaire)	9	Suriname	1,181
Bhutan	15	Pakistan	1,277
Guinea-Bissau	17	Australia	1,306
Solomon Islands	18	Bulgaria	1,600
Comoros	18	Chile	1,625
Burundi	20	Madagascar	1,638
Congo	20	Korea DPR	1,649
Uganda	20	Afghanistan	1,702
Papua New Guinea	25	Canada	1,752
Central African Republic	26	Tajikistan	2,065
Benin	28	United States of America	2,162
Togo	28	Iraq	2,367
Gambia	29	Kyrgyzstan	2,527
Cameroon	31	Turkmenistan	6,346[b]
Equatorial Guinea	31	Guyana	7,616[b]
Lesotho	31		

[a] These data include only reported water use. Data on rainfall used in agriculture and unreported water use are not available.

[b] These numbers are reported by FAO but may be in error.

TABLE 4.3 COUNTRIES WITH PER-CAPITA ANNUAL RENEWABLE WATER SUPPLIES BELOW 1,700 CUBIC METERS PER PERSON PER YEAR (AS OF THE MID-1990S)

Country	Population (millions)	Per-Capita Availability (m³/p/year)
Kuwait	2.04	10
Malta	0.35	46
United Arab Emirates	1.59	94
Libya	4.55	132
Qatar	0.37	143
Saudi Arabia	14.13	170
Jordan	4.01	219
Singapore	2.72	221
Bahrain	0.52	223
Yemen Democratic Republic	11.7	350
Israel	4.6	467
Tunisia	8.18	504
Algeria	24.96	573
Oman	1.5	657
Burundi	5.47	658
Djibouti	0.41	732
Cape Verde	0.37	811
Rwanda	7.24	870
Morocco	25.06	1,197
Kenya	24.03	1,257
Belgium	9.85	1,269
Cyprus	0.7	1,286
South Africa	35.28	1,417
Poland	38.42	1,463
Korea, Republic of	42.79	1,542
Egypt	52.43	1,656
Haiti	6.51	1,690

aquifers, dislocation of people because of dam construction, and the destruction of fishery resources that support local populations.

What is the connection between these issues and conflict? For the most part, inequities will lead to poverty, shortened lives, and misery, but perhaps not to direct conflict. But in some cases, they will increase local, regional, or international disputes; create refugees that cross, or try to cross, borders; and decrease the ability of a nation or society to resist economic and military aggression (Gleick 1993; Homer-Dixon 1994).

Reducing the Risk of Water-Related Conflict

The ultimate goal of any assessment of the risks of conflict over water resources must include efforts to reduce those risks. Various regional and international approaches exist for reducing water-related tensions. Among the approaches are legal agreements, the application of proper technology, institutions for dispute resolution, and innovative water management. Unfortunately, these mechanisms have never received the international support or attention necessary to resolve many conflicts over water. Efforts by the United Nations, international aid agencies, and local communities to ensure access to clean drinking water and adequate sanitation can reduce the competition for limited water supplies and the economic and social impacts of widespread waterborne diseases. Improving the efficiency of water use in agriculture can extend limited resources, increase water supplies for other users, strengthen food self-sufficiency, reduce hunger, and lower expenditures for imported food. In regions with shared water supplies, third-party participation in resolving water disputes can also help end conflicts.

The Role of Law

International law has an important role to play. There are typically two forms of international codes of conduct for resolving disputes over water: general principles of international behavior and law, and specific bilateral or multilateral treaties. Several broad legal instruments prohibit environmental warfare or the use of the environment as a weapon of war, including the Environmental Modification Convention of 1977, the 1977 Bern Geneva Convention, and the 1982 World Charter for Nature. In this area, the problem is not lack of law but lack of enforcement and acceptance by the international community (Gleick 1993).

International law in the area of shared water resources is both well advanced and, in what may appear to be a contradiction, largely ineffective. More than 30 years of negotiations and discussions have occurred since the original statement of the 1966 Helsinki Rules governing international waters, and the new 1997 Convention on the Non-Navigational Uses of International Watercourses took 27 years to formalize (McCaffrey 1993; United Nations 1997a). The negotiations and the final product reflect the difficulty of integrating legal principles with the complexities of the hydrologic cycle.

The full text of this Convention is published in the Water Briefs section of this book (pp. 211–230). Among the most relevant portions are Article 7 of the Convention, which obliges States to take all appropriate measures to prevent harm to other States from their use of water; Article 8, which obliges watercourse States to cooperate on the basis of equality, integrity, mutual benefit, and good faith in order optimally to use and protect shared watercourses; and Article 33, which offers provisions for the peaceful settlement of disputes by negotiation, mediation, arbitration, or appeal to the International Court of Justice. Among the weaknesses of the Convention are the inherent conflict between equitable uses and the obligation not to cause appreciable harm. These principles are not binding and offer little concrete guidance to countries trying to allocate scarce water resources, but they may prove effective at encouraging states to negotiate water disputes and to implement joint management activities.

In addition to basic principles of international law, as represented by the Convention, there are hundreds of different bilateral and multilateral treaties signed by parties to international rivers that allocate water, regulate navigation and power, monitor and control water quality, and affect all other aspects of joint management. Each of these treaties is individually negotiated. Some have been highly effective at reducing water-related conflicts. Historical evidence suggests that these forms of agreements tend to be more effective, though they also have their limitations. Since 1814, approximately 300 water-related treaties have been negotiated dealing with management, water allocation, or flood control and hydropower. Wolf (1997) has identified 145 treaties dealing only with water in the twentieth century alone.

Despite the effectiveness of some legal tools, there are concerns that existing international water law may be unable to handle the strains of ongoing and future problems. It is also possible that international water law may simply be an inappropriate mechanism for addressing some of these problems, such as subnational and local disputes. Many political entities claiming water rights or allocations will not be served by the International Court of Justice, or the 1997 Convention, or other international bodies.

In spite of these concerns, good progress has been made recently to resolve long-standing water disputes, such as a peace treaty between Jordan and Israel that includes water provisions, as well as a new treaty between India and Bangladesh over the Farakka Barrage and the Ganges River. These positive developments, described on the following pages, offer the hope that cooperation will continue to be the preferred mode of resolving water-related disputes. At the same time, new unresolved problems are arising in new regions, such as southern Africa. While diplomatic and institutional solutions are being explored here as well, the appearance of new international and intra-regional water disputes due to inequities driven by population growth or economic competition raises concerns for the future.

The Israel—Jordan Peace Treaty of 1994

Much has been written about water in the Middle East, especially in the past few years as peace negotiations there have waxed and waned (see, for example, Gleick 1994; Biswas 1994; Rogers et al. 1994; Lonergan and Brooks 1994; Isaac and Shuval 1994; Spiegel and Pervin 1995; Wolf 1995; Biswas et al. 1997). Many seemingly irreconcilable issues complicate the political landscape of the region, including the equitable distribution, management, and use of fresh water. Yet in 1994, Israel and Jordan signed a peace treaty that, among other things, explicitly resolves a variety of contentious water issues over the Jordan River basin.

The water resources of the Middle East are limited and poorly distributed, particularly in the Jordan River basin—a surprisingly small river given its historical, hydrological, and political importance. The Jordan River Basin includes portions of Israel, Jordan, Syria, Lebanon, and Palestine. While many different estimates have been made of the water resources and the levels of water use for each country, there is agreement that water use in Israel, Jordan, and the territories of Palestine is approaching, if not already past, the limit of the available renewable supply (see, for example, Kolars 1992; Gleick 1994, 1997; Biswas 1994; Hillel 1994; Brooks 1997). Lebanon is one of the few countries of the region with a current surplus of surface water and groundwater, though there are disputes over the Orontes River, which is also shared by Turkey and Syria. Syria is sometimes considered relatively water rich; but as Table 4.1 shows, this is true only if it continues to have access to large flows of both the Euphrates and the Tigris, a doubtful prospect given Turkey's massive upstream development activities. There are also shortages in the more populated southern regions where Syria shares the Yarmouk River, a tributary of the Jordan, with Jordan. Syria has recently built many small dams on tributaries of the Yarmouk, creating unresolved problems with Jordan.

After many years of hostile relations between Israel and Jordan, and many years of negotiations over issues of concern, the two countries signed a peace treaty on October 26, 1994. As part of this treaty, specific articles and agreements address water policies, allocations, data sharing, and joint management. In particular, *Article 6: Water* makes five important points:

1. The Parties agree mutually to recognize the rightful allocations of both of them in Jordan River and Yarmouk River water and Araba/Arava groundwater in accordance with the agreed acceptable principles, quantities and quality as set out in Annex II, which shall be fully respected and complied with.

2. The Parties, recognizing the necessity to find a practical, just and agreed solution to their water problems and with the view that the subject of water can form the basis for the advancement of cooperation between them, jointly undertake to ensure that the management and development of their water resources do not, in any way, harm the water resources of the other party.

3. The Parties recognize that their water resources are not sufficient to meet their needs. More water should be supplied for their use through various methods, including projects of regional and international cooperation.

4. In light of paragraph 3 of this article, with the understanding that cooperation in water-related subjects would be to the benefit of both parties, and will help alleviate their water shortages, and that water issues along their entire border must be dealt with in their totality, including the possibility of transboundary water transfers, the Parties agree to search for ways to alleviate water shortage and to cooperate in the following fields:

   a. development of existing and new water resources, increasing the water availability including cooperation on a regional basis, as appropriate, and minimizing wastage of water resources through the chain of their uses;

   b. prevention of contamination of water resources;

   c. mutual assistance in the alleviation of water shortages;

   d. transfer of information and joint research and development in water-related subjects, and review of the potentials for enhancement of water resources development and use.

5. The implementation of both Parties' undertakings under this article is detailed in Annex II.

Annex II lays out details for allocating water to both countries from the Yarmouk and Jordan rivers, makes provisions for guaranteeing flows during different seasons, lays the groundwork for the construction of a diversion/storage dam on the Yarmouk, and sets up a Joint Water Committee for documenting water uses, monitoring implementation, and resolving water disputes.

Other Water Issues in the Middle East

Despite this agreement, arguments over water in the Middle East have a long history and are wrapped up in rancorous disagreements extending into the religious, political, economic, and social realms. Long-term faithful cooperation is going to be hard to achieve. In the long run, specific international agreements will have to be supplemented by real reform activities in the domestic water policies of each of the countries in the region to increase the efficiency and productivity of water use, to restructure the water-allocation systems, and to change the way water is priced and managed.

The use of water for agricultural production in the region is a particularly important part of the problem, and few observers believe that large-scale agricultural water use can long continue. Much of the water demand in all the countries in the Middle East today is consumed by efforts to feed local populations from local production with subsidized water and artificially supported farm sectors. In the long run, agriculture in the Jordan Basin and elsewhere in the Middle East will have to be reconfigured and restructured. Highly efficient agriculture, high-value crops, and production that uses reclaimed water or water not fit for direct human consumption will survive. Additional food needs will have to be provided by international markets and food purchases. Israel and Jordan are already moving along this path.

A critical component of a peaceful solution will necessarily bring economics more directly into decisions about agricultural water pricing. As the 1992 Dublin Principles proclaimed (see Box 6.3 in Chapter 6), water has an economic value and should be treated as an economic good, which requires that it be assigned a price more related to its true value. While Israel and Jordan have taken steps to reduce agricultural subsidies and to price water more rationally, others have not. Egypt continues to supply almost all water to its farmers without charge. Syria not only subsidizes water but also pays premiums over the international price for the production of certain crops (Khaldi 1992; Biswas et al. 1997). One study showed that increasing the price of agricultural water in Egypt to $0.04 per cubic meter could reduce agricultural water demand by 20 percent with little change in agricultural income (Hazell et al. 1995).

There are many factors that militate against reducing agricultural water subsidies, including powerful political interests, concerns about rural employment, worries about vulnerability to international food boycotts, and economic difficulties in purchasing food on the international market. While none of these barriers is insurmountable, they will require serious attention by water policymakers and others.

In 1997, Jordan began a radical transformation of its water sector, raising prices closer to real costs, initiating privatization of the water supply and distribution system in Amman, and pressuring water users to pay for water and water services. The old pricing system of declining or flat block rates has been partly replaced with a low initial rate followed by an increasing block rate. Efforts are also under way to reduce the losses in the water-supply system, which have been estimated to be as high as 50 percent (FTGWR 1997). Some of these changes were stimulated by the agreement with Israel, which removed some of the uncertainties about guaranteed water supplies. Among the issues resolved has been the siting of a diversion dam on the Yarmouk River near Adassiyah to help divert river flows to Jordanian agricultural projects.

As domestic reforms accelerate and as international agreements are implemented, the chances improve that real conflict over water in the Middle East will either be resolved or become less of a factor in regional politics. But many problems remain, with no sign of improvement. In particular, most countries have not even begun the difficult domestic reforms that are needed. Syria, Iraq, and Turkey have unresolved disputes over water quantity and quality in the Tigris and Euphrates systems and are holding almost no discussions or negotiations. Egypt is struggling with rapid population growth and an increasingly constrained external supply. And the progress made in the mid-1990s between the Palestinians and Israelis over groundwater and other issues has been halted. Water-related disputes have been seen in this region for 5,000 years, and it may be too much to expect them to disappear soon.

The Ganges—Brahmaputra Rivers: Conflict and Agreement

South Asia has several great rivers, including the Indus, the Ganges, and the Brahmaputra. These rivers have been at the heart of water discussions and disputes throughout the centuries, and they contribute to both the well-being and the impoverishment of hundreds of millions of people. The Ganges and Brahmaputra basins together contain an estimated 400 million people living at a standard of living as low as anywhere on Earth (Rogers et al. 1994). There has been little regional cooperation over these waters, particularly over how to allocate flows, address the disastrous flooding that occurs, or tap the hydropower potential of the upstream riparians. In the 1980s, the South Asian Association for Regional Cooperation (SAARC) was formed, but it expressly excluded water as a topic for discussion, leaving that issue for bilateral or multilateral political talks. Over the years, various interim agreements have been signed; these were followed by renewed disputes and chronic lack of agreement, and international discussions repeatedly failed to resolve the conflicts. Moreover, while the riparians include Nepal, China, India, Bhutan, and Bangladesh, most discussions have excluded China, Bhutan, and Nepal. Yet at the very end of 1996, India and Bangladesh signed a water-sharing agreement that may have finally put to rest the most contentious water issues between the two countries—the allocation of flows of the Ganges between India and Bangladesh, and the operation of the Farakka Barrage.

The Farakka Barrage was built by India during the late 1960s and early 1970s across the Ganges River just upstream of the Bangladeshi border to divert water into the Hooghly River for irrigation use and to improve navigation. This dam was built without international agreement and it seriously affects dry-season flows to Bangladesh. In the early 1990s, dry-season flow in the Ganges reaching Bangladesh fell to 10 percent of mean dry-season flow, causing serious problems for agriculture and food production.

When Bangladesh became independent in 1971, the Farakka Barrage was still under construction. Because of the good relations between India and Bangladesh at the time, an interim agreement for operating Farakka was signed in 1974. These good relations, however, soon soured after changes in government, and Bangladesh brought the issue to the UN General Assembly in 1976, asking for help in resolving their concerns. A change in government in India in 1977 led to another agreement (The 1977 Ganges Waters Agreement), which guaranteed a minimum flow to Bangladesh and committed both governments to find a way to increase flows in the dry season. The agreement over water sharing held for five years without conflict, but ultimately expired (Crow 1985; Crow et al. 1995).

Other attempts at agreement were made during the 1980s, with a "memorandum of understanding" signed in 1982. This memorandum itself expired in 1984 and a subsequent 1988 agreement on flows from India to Bangladesh was negotiated and quickly lapsed without a renewal (Wolf 1997). Finally, in December 1996, India and Bangladesh signed a water-sharing accord. What had seemed intractable over nearly three decades was suddenly possible due to changes in government and political will on both sides, pushed along by unofficial dialogues between the two countries (RCSS 1997). In addition to specifying water allocations in normal and dry periods, both governments agreed to conclude water-sharing treaties and agreements over more than 50 other shared rivers and to find solutions to augmenting the flow of the Ganges in the dry season. The full text of the Treaty can be found in the Water Briefs section (pp. 206–209).

The treaty certainly doesn't mark the end of problems in the region over water—much depends on the future implementation of the accord. Yet it offers a promising beginning to a final resolution and a framework in which future disputes can be addressed. In the spirit of cooperation that followed this agreement, new avenues for resolving other outstanding issues between India and Bangladesh have opened up, showing that cooperative agreements over resource disputes can play a broader role in peace and security arrangements in South Asia.

Water Disputes in Southern Africa

Southern Africa is a diverse region encompassing Angola, Botswana, Lesotho, Malawi, Mozambique, Namibia, South Africa, Swaziland, Tanzania, Zambia, and Zimbabwe. The climate of southern Africa is extremely variable and the water resources are often unreliable. The entire region is largely dependent on rainfall and river runoff for water supply, and every major perennial river (a river that flows all year) in the region is shared by two or more nations (Heyns 1995; Conley 1996).

Growing populations and economic development are putting increasing pressure on these resources and access to water has long been a concern and focus of controversy. At the same time, recent political changes have improved the chances that negotiations and cooperation over water can reduce the risks of water-related conflicts and promote regional cooperation. Effective joint management will require joint basin control over international rivers, the integration of environmental and social factors into estimates of the benefits and costs of physical infrastructure for water supply, and the more efficient use and allocation of existing supplies. Strong efforts are now being made in these areas, and cooperative management of shared waters has the potential to benefit all parties.

At the same time, several regional disputes remain unresolved. In particular, Mozambique is still suffering the consequences of a long civil war and has disputes over rivers shared with South Africa. Another dispute is brewing between Botswana, Namibia, and Angola over development and allocation of the Okavango River. These problems could benefit from sustained and organized negotiation among the parties.

Mozambique's Shared Water

Mozambique's rivers all originate outside of its borders in South Africa, Zimbabwe, and Swaziland, and by the time they reach Mozambique they are substantially depleted or contaminated. Mozambique has become concerned that its neighbors, particularly South Africa, are contributing to their difficulty in growing food and to the problems in maintaining even minimal flows in the shared rivers. There is growing discussion in southern Mozambique about whether or not Mozambique can increase the reliability of flows in its rivers through more effective international agreements.

One unusual aspect of the dispute involves the connection between the dispute over the shared rivers between the two countries and those rivers that flow through the Kruger National Park in South Africa along the border with Mozambique. Of particular concern to the Mozambican government is the operation of a number of South African dams and increased agricultural withdrawals on tributaries of the Limpopo, Injaka, and Incomati rivers. During the dry season, many of these rivers effectively cease flowing by the time they reach the border.

Rivers of Kruger National Park: Tributaries to the Limpopo and Incomati Rivers

Six major rivers flow east from the Republic of South Africa into Kruger National Park and then into Mozambique, where they ultimately join the Limpopo or Incomati rivers (see Box 4.1 and Table 4.4). All six international tributaries are highly utilized outside of the park and are threatened by growing populations and utilization. Kruger is the most important center for tourism in the southern African region, with over 7 million visitors annually (Venter and Deacon 1995), and it is one of the largest and most important centers of biodiversity in the area.

South Africa knows about these concerns, and Mozambique has even threatened to take its concerns to the World Court for resolution if South Africa doesn't more adequately address its needs. According to Peter van Niekerk, a senior South African water planner,

> South Africa is very aware that the Mozambicans are unhappy about the quantity of water reaching them. In fact, the Mozambicans were so unhappy that they declined to sign a proposed Memorandum of Understanding [MoU]—they called it a 'feel-good document.' (Arenstein 1996)

The threatened legal action is seen as a political lever to force South Africa to address Mozambique's concerns, and even if no formal legal action is taken, the threat itself could harm South Africa's image both within the subcontinent and in the Organisation for African Unity (OAU). The recent World Court decision on the dispute between Hungary and Slovakia over the Gabcikovo Project on the Danube, however, indicates that the Court's strong preference is for disputing river basin parties to come up with negotiated settlements directly (Lipschutz 1997).

Instead of a formal MoU, Mozambique and South Africa recently signed a draft document calling for the creation of a joint water commission between the two countries, a considerable improvement over the attitudes that prevailed during the recent war in Mozambique and the attitudes of the apartheid government in South Africa. A more for-

TABLE 4.4 TRIBUTARIES TO THE LIMPOPO AND
INCOMATI RIVERS SHARED BY SOUTH AFRICA AND
MOZAMBIQUE

River Basin	Basin Area (km²)	Natural Flow (mcm/yr)
Luvuvhu River	3,568	395
Letaba River	13,400	553
Shingwedzi River	5,600	78
Olifants River	54,575	1,950
Sabie River	7,096	762
Crocodile River	10,526	1,238

Source: Breen et al. 1994.

mal treaty between South Africa and Mozambique was scheduled to be signed in late 1997 (Arenstein 1996), and South Africa has called for a review of all of its international water agreements under its new water law (see Chapter 5).

In 1983, South Africa, Mozambique, and Swaziland created a technical committee — the Tripartite Permanent Technical Committee (TPTC)—to address issues of concern on the Limpopo, Incomati, and Maputo rivers shared by the three nations. This committee has not functioned well in the past and may not be capable in its current form of addressing the environmental issues related to flows through Kruger National Park. The end of war in Mozambique and recent improvements in relations between Mozambique and South Africa, however, offer the possibility that water issues can now be addressed explicitly and directly in the TPTC or a new institution to be created. Allocations of water to Mozambique should be negotiated with the impacts to Kruger National Park considered a fundamental constraint. Recently, there has been discussion of creating a "mirror" park to Kruger on the Mozambique side of the border. Water for that park should also be guaranteed through formal agreement between the two countries.

The Okavango River

The Okavango River is shared by Angola, Namibia, and Botswana and is the largest river in southern Africa that doesn't drain to the ocean. Most of the flow originates in Angola, flows southeast to Namibia, along the Namibian–Angolan border, and then turns south into Botswana. The river empties into the Okavango Delta—a world-renowned ecosystem and a major tourist destination, with a diversity of plant and animal life unrivaled in Africa. The Okavango Delta has been classified as a World Heritage Site. Inflow to the delta averages about 10,000 million cubic meters (mcm) per year.

Several years ago, Botswana proposed a major project called the Southern Okavango Integrated Water Development Project. Its main objective was to provide water for irrigators, urban users, livestock, and, in particular, a large mine. International concern about the environmental impacts of this diversion project and the quality of the environmental assessment led to an outside analysis by the International Union for the Conservation of Nature (IUCN), which was skeptical about the need for the project, concerned about its economics, and critical about its environmental implications. Following this analysis

BOX 4.1

International Tributaries to the Incomati and Limpopo Rivers Shared by South Africa and Mozambique

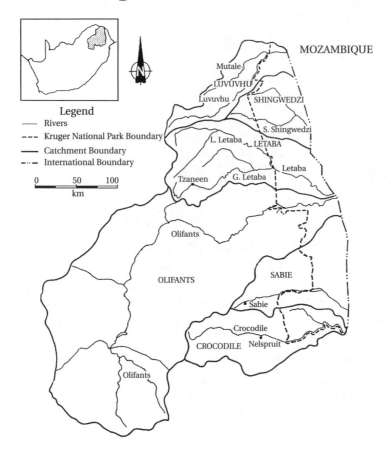

and the subsequent publicity, the project was put on hold, though it remains of interest to Botswana (Conley 1996).

More recently, the relationship between Namibia and Botswana has been strained by Namibian plans to construct a 250-kilometer pipeline to divert water from the Okavango River to eastern Namibia and its capital of Windhoek. Namibia intends to build an "emergency" pipeline to connect their Eastern National Water Carrier with the Okavango to help deal with severe drought and anticipated water shortages. This development would extract about 20 mcm of water from the Okavango for urban water needs, though some plans have called for as much as 100 mcm per year (James 1996; Leitch 1997). While an Okavango River Basin Commission (OKACOM) comprised of Angola, Botswana, and Namibia was formed in September 1994, there is no long-term agreement over management or allocation of the river and no agreement that this project

The Letaba River: The Letaba River is highly utilized and has become an ephemeral, rather than a perennial, river. Substantial amounts of water are taken for irrigation and the support of exotic species of trees in plantations. Two fish species normally found in the river no longer occur within the park. The Letaba joins the Limpopo River in Mozambique.

The Olifants River: The Olifants is the largest river flowing through Kruger National Park and is heavily developed. There are few rivers in the world that have been as heavily manipulated, dammed, and polluted as the Olifants. Nearly 2.5 million people live within its catchment, and coal and asbestos mining activities are widespread. Thirty dams have been built on portions of the watershed, with the water used for primary water supply or irrigation. So much water is taken from the river that many rural settlements cannot develop sufficient supplies, and richer homesteads nearby have water trucked in every day. High salinity, heavy metals contamination, and high silt loads resulting from industrial activities in the basin are also major problems. The Olifants ultimately flows through Kruger and into Mozambique, where it joins the Limpopo River. By the time it reaches Mozambique, it has been described as little more than a "trickle of effluent."

The Sabie River: The Sabie River rises on the eastern slopes of the Transvaal Drakensberg and runs through the southern portion of the park by the village of Skukuza. This is the only river in the park that has never stopped flowing, and it remains in good condition with a high biotic diversity, though threats from exotic plant species and upstream development exist. The Sabie River is part of the international system of the Incomati River.

The Crocodile River: The Crocodile rises in the mountainous areas southwest of Kruger National Park and forms the southern boundary of the park to the Mozambique border. Across the border it joins the Incomati River. Substantial amounts of water are withdrawn for intensive and inefficient irrigation in South Africa, especially of sugarcane and citrus. Flows are highly unnatural, variable, and contaminated by agricultural runoff.

The Luvuvhu River: The Luvuvhu River rises in the Soutpansberg region and forms part of the northwestern border of the park. This river is one of the most beautiful in the park, with enormous environmental assets. In recent years, great pressures have been exerted on the water by farmers in the region and the Luvuvhu has become a seasonal rather than a perennial river. The river flows into the Limpopo River along the northern boundary of the park.

The Shingwedzi River: This small river in the northern section of the park is experiencing increasing abstractions from development west of the park. It is joined by the Letaba River and ultimately flows into the Limpopo River in Mozambique.

should go forward (Heyns 1995). Negotiations and discussions between Namibia and Botswana are continuing, however, and the two countries have signed an agreement to maintain the flow of information, to share detailed feasibility studies, and to apply common principles to those studies (Communiqués, 6/27/96 and 10/15/96).

Some reports have appeared suggesting that Botswana is increasing the size of its defense forces in response to this dispute, and there is concern in South Africa about the potential for conflict between the two countries over this issue (Leitch 1997; Samson and Charrier 1997). While neither country would comment on recent Botswana arms purchases, Peter Mokaba, Deputy Minister for Environment Affairs in South Africa, has said that a change in the balance of power in the region could lead to an arms race (Leitch 1997).

Good rains fell in the region in the first few months of 1997. This had the effect of making the need for the project less immediate and providing time to explore alternative water supply and demand management options. Nevertheless, Namibia is strongly considering proceeding with the extension of the Eastern National Water Carrier despite objections from Botswana. In mid-1997, Dr. Vaino Shivute of the Ministry of Agriculture, Water and Rural Development in Namibia categorically stated that the construction of the pipeline will go ahead (Leitch 1997).

Such an action could be a violation of the new Convention on Non-Navigational Uses of Watercourses, which lays out principles governing shared water, and which all three basin states voted for (UN 1997a). These principles call for joint basin management, cooperative agreements over allocations, and joint assessments to determine the environmental consequences of watershed development. Furthermore, a unilateral decision to build would set a bad precedent for future water withdrawals and allocations from the Okavango by Angola and Botswana. Ironically, these international principles are being used as a justification for the project. According to Leitch (1997), a spokesman for the Namibian Ministry of Agriculture, Water and Rural Development says that the new Convention gives Namibia an inalienable, international riparian right to abstract water from the Okavango, in consultation with other states and in an environmentally acceptable way.

OKACOM successfully petitioned the Global Environment Facility to fund a study that would lay out terms of reference for an environmental assessment with the goal of producing an integrated management plan for the entire basin and all three riparians. Part of this plan would be a treaty on the equitable allocation of Okavango water to each party. If the next rainy season continues to relieve the immediate pressures on Namibia for new supply, diplomatic efforts may permit Namibia, Botswana, and Angola to work through the OKACOM or another mutually acceptable forum to come up with a satisfactory joint basin management plan. In any case, adequate environmental impact assessments must be done to evaluate not only the impacts of construction but also the consequences of water withdrawals downstream during high and low flows and during both routine and emergency operations.

Summary

Conflicts are caused by many factors, including religious animosities, ideological disputes, arguments over borders, and economic competition. After more than a decade of academic arguments, political rethinking, and increased participation in security discussions by environmental scientists and analysts, it is widely accepted that resource and environmental factors can play a tangled but definite role in local, regional, and even international disputes. Not all water-resources disputes will lead to violent conflict; indeed most lead to negotiations, discussions, and nonviolent resolutions. But in certain regions of the world, water is a scarce resource that is becoming increasingly important for economic and agricultural production and increasingly scarce due to growing population and resource degradation. In these regions, shared water is evolving into an issue of high politics and the probability of water-related disputes is increasing. Policymakers must be alert to the likelihood of disagreements over water resources and to possible changes in international water law, regional political arrangements, and patterns of use that could minimize the risk of conflict.

Date (B.C.)	Parties Involved	Basis of Conflict	Violent Conflict/ Military Involvement?	Description	Sources
3000	Gods and Sumerians	Tool of war	No	An ancient Sumerian legend recounts the deeds of the deity Ea, who punishes mankind for its sins by inflicting the earth with a six-day storm. The Sumerian myth generally parallels the Biblical account of Noah and the Deluge, although some details differ. According to the Sumerians, the patriarch, Utu, speaks with Ea, who warns him of the impending flood and orders him to build a large vessel filled with "all the seeds of life." Like Noah's Ark, Utu's ship is grounded on a mountain peak, surrounded by water.	Burstein 1978
2500	Lagash–Umma Border Dispute	Military/political goal; tool of war; target of war	Yes	**2500** The dispute over the Gu'edena ("Edge of Paradise") region begins during the reign of Akurgal, King of Lagash. The northwestern city-state of Umma invades. Water is not a source of conflict at this point. **2450** Enatum of Lagash fights two wars over region. Umma is defeated, but resumes fighting. Lagash crushes rebellion. Umma must comply with the treaty, which requires payment for use of region. Entemena, Enatum's successor, constructs a feeder canal from Tigris, leading to over-irrigation. Urlama, King of Lagash 2450–2400, diverts water from region to boundary canals, drying up boundary ditches to deprive Umma of water. **2400** Urlama's son, Il, "the field thief," engages in improper use of irrigation and intends to shift levees but fails. Il cuts off Girsu's (city in Umma) water supply. Agriculture in Lagash is dependent on two sources of water, what are today the Shatt al-Hai and the Tigris.	Cooper 1983; Lambert 1965; Jacobsen 1969; Kramer 1956
1720–1684	Abi-Eshuh vs. Iluma-Ilum	Tool of conflict	Yes	Grandson of Hammurabi, Abish (or Abi-Eshuh), dams the Tigris in order to prevent the retreat of rivals from the southern marshes of Mesopotamia (Sea Country). Led by Iluma-Ilum, who had declared the Sea Country's independence of Babylon, the rebels head south from a northern expedition along the Tigris. Abi-Eshuh attempts to trap and flood the retreating forces by damming the Tigris. While he succeeds in damming the river, he fails to stop Iluma-Ilum. This failed attempt marks the decline of the Sumerians, who had reached their apex under Hammurabi.	King 1915; Saggs 1962
circa 1200	Moses floods (Biblical account)	Tool of conflict	No	From Exodus. Moses dams a tributary of the Nile to Egyptian Plains to prevent the Egyptians from reaching the Jews as they retreat through the Sinai.	Honor 1926
circa 1200	Moses floods Egyptian Plains (Biblical account)	Tool of conflict	No	When Moses and the retreating Jews find themselves trapped between the Pharaoh's army and the Red Sea, Moses miraculously parts the waters of the Red Sea, allowing his followers to escape. The waters close behind them and cut off Egyptians.	
720–705	Sargon II (Assyrian) Destroys Armenian Waterworks	Target of conflict	Yes	After a successful campaign against the Haldians of Armenia, Sargon destroys their intricate irrigation network and floods their land. The destruction, like that of Sennacherib's against Babylon, was both defiant and defensive, ensuring the complete submission of the area. Olmstead argues that the Assyrians later replicated the irrigation system under Sargon II's son. While Armenian wars are recorded, the chronology of events varies according to historian. The Haldians had constructed canals from the eastern deserts to provide potable water that the brackish Lake Van could not.	Drower 1954; Olmstead 1975

continued

125

Date (B.C.)	Parties Involved	Basis of Conflict	Violent Conflict/ Military Involvement?	Description	Sources
705–682	Sennacherib (Assyrian) and the fall of Babylon	Tool of conflict	Yes	Sennacherib builds a series of aqueducts to provide adequate water from the Gomel River. In quelling rebellious Assyrians (695), he razes Babylon and "even diverted one of the principal irrigation canals so that its waters washed over the ruins."	Lloyd 1961; Olmstead 1975; Winkler 1907
	Sennacherib and Hezekiah (Biblical account)	Tool of conflict	Yes	As recounted in Chronicles 32:3, Hezekiah digs a well outside the walls of Jerusalem and uses a conduit to bring in water. Preparing for a possible siege by Sennacherib, he cuts off water supplies outside of the city walls, and Jerusalem survives the attack.	
681–669	Esarhaddon (Assyrian) and the Siege of Tyre	Tool of conflict	Yes	Esarhaddon refers to an earlier period when gods, angered by insolent mortals, create a destructive flood. According to inscriptions recorded during his reign, Esarhaddon besieges Tyre, cutting off food and water.	Luckenbill 1927
668–626	Assurbanipal, Siege of Tyre, Drying of Wells	Tool of conflict	Yes	Assurbanipal's inscriptions also refer to a siege against Tyre, although scholars attribute it to Esarhaddon. In campaigns against both Arabia and Elam (645), Assurbanipal, son of Esarhaddon, uses wells to defeat his enemy. During the final attack against Elam, Assurbanipal dries up wells to deprive Elamite troops, while he guards wells from Arabian fugitives in an earlier Arabian war. On his return from victorious battle against Elam, Assurbanipal floods the city of Sapibel, an ally of Elam. According to inscriptions, he dams the Ulai River with the bodies of dead Elamite soldiers and deprives dead Elamite kings their food and water offerings.	Luckenbill 1927; Smith 1871
612	Fall of Nineveh in Assyria and the Khosr River	Tool of conflict	Yes	A coalition of Egyptian, Median (Persian), and Babylonian forces attack the capital of Assyria and destroy the stronghold. Nebuchadnezzar's father, Nebopolassar, leads the Babylonians. The converging armies divert the Khosr to create a flood, allowing them to elevate their siege engines on rafts. The Assyrian brothers Ashur-etil-ili and Sin-shar-Ishkun ruled during the time of the invasion.	Lloyd 1961; Saggs 1962
605–562	Nebuchadnezzar Uses Water to Defend Babylon	Tool of conflict	Yes	Nebuchadnezzar builds an immense wall around Babylon, using the Euphrates and canals as defensive "moats" surrounding the inner castle. Herodotus attributes the canals and wall to Queen Nitocris rather than Nebuchadnezzar. Berossus describes Nebuchadnezzar's plan to create an impregnable city, stating, "He arranged it so that besiegers would no longer be able to divert the river against the city by surrounding the inner city with three circuits of walls..."	Herodotus 1859 ed.; Saggs 1962; Burstein 1978
558–528	Cyrus the Great: The 360 Canals	Target of conflict	Yes	On his way from Sardis to defeat Nabonidus at Babylon, Cyrus faces a powerful tributary of the Tigris (probably the Diyalah). According to Herodotus' account, the river drowns his royal white horse and presents a formidable obstacle to his march. Cyrus, angered by the "insolence" of the river, halts his army and orders them to cut 360 canals to divert the river's flow. Other historians argue that Cyrus needed the water to maintain his troops on their southward journey, while another asserts that the construction was an attempt to win the confidence of the locals. According to the latter perspective, the ability to provide water legitimized his rule.	Herodotus 1859 ed.; Lane 1923; DeLattre 1888
539	Cyrus the Great: Invasion of Babylon	Tool of conflict	Yes	According to Herodotus, Cyrus invades Babylon in 539 by diverting the Euphrates above the city and marching troops along the dry river bed. This legend describes a midnight attack coinciding with a Babylonian feast. The drunken Babylonians, caught off guard, immediately fell to the Persian advance. In a less dramatic telling, the Persians divert the Tigris to enter a protected region north of Babylon. After a battle against Babylonian troops at Opis, Cyrus turns to Babylon, and the besieged city falls after "some months."	Herodotus 1859 ed.; Sykes 1921; Tatum 1987

continued

Date (B.C.)	Parties Involved	Basis of Conflict	Violent Conflict/ Military Involvement?	Description	Sources
424–390?	Artaxerxes and the Retreat of the Ten Thousand	Tool of conflict	Yes	Cyrus the Younger (424–401), son of Darius I, determined to crush his brother, whom he believes usurped the throne, marches from Asia Minor to Babylon with both Greek and Persian soldiers. Cyrus engages in battle with Artaxerxes' forces at Cunaxa (401), where Cyrus dies in the surprise attack. After an agreement with Artaxerxes, the remaining Greek soldiers retreat north along the Tigris led by Xenophon and Clearchus. Clearchus orders bridges built along old canals for fear Artaxerxes would breach a northerly dam.	Lane 1923
355–323	Alexander the Great: Destruction of Persian Dams	Target of conflict	Yes	Returning from a triumphant battle against Darius I and the razing of Persepolis, Alexander proceeds to India. After the Indian campaigns, he heads back to Babylon via the Persian Gulf and Tigris, tearing down defensive weirs that the Persians had constructed along the river. Arrian describes Alexander's disdain for the Persians' attempt to block navigation who saw them as "unbecoming to men who are victorious in battle."	Arrian 1893; Lloyd 1961; Lane 1923; Pearson 1960
	Alexander Maintains Old Babylonian Dams	Tool of conflict	Yes	While sailing north on the Euphrates, Alexander clears previously blocked canals, most notably the Pallacopas which ran east of the Euphrates. Alexander may have cleared them as a sign of friendship toward the defeated lands.	Lane 1923; Pearson 1960; Arrian 1893

Note: This chronology was partly compiled by Haleh Hatami and Peter H. Gleick and produced as "Chronology of Conflict Over Water in the Legends, Myths, and History of the Ancient Middle East." Report of the Pacific Institute for Studies in Development, Environment, and Security, Oakland, California (October 1992).

Sources

Arrian. 1893. *Arrian's Anabasis of Alexander and Indica,* trans. by Edward J. Chinnock. George Bell and Sons, London.

Burstein, S. 1978. "Babyloniaca Berossus." *Sources from the Ancient Near East.* Udena Publications, Malibu, California.

Cooper, J.S. 1983. "The Lagash-Umma Border Conflict." *Reconstructing History from Ancient Inscriptions.* Udena Publications, Malibu, California.

DeLattre, S.J. 1888. "Les Traveaux Hydrauliques en Babylonie." *Revue Des Questions Scientifiques.* Paris.

Drower, M.S. 1956. "Water-Supply, Irrigation, and Agriculture in Ancient Mesopotamia." *History of Technology,* Vol. 1, pp. 520–556.

Herodotus. 1859. *History of Herodotus.* G. Rawlinson (ed.). D. Appleton and Company, New York.

Honor, L. 1926. *Sennacherib's Invasion of Palestine.* Greenwood Press, New York.

Jacobsen, T. 1969. "A Survey of the Girsu (Telloh) Region." *Sumer,* Vol. 25, Nos. 1–2, pp. 103–108.

Jacobsen, T., and R.M. Adams. 1958. "Salt and Silt in Ancient Mesopotamian Agriculture." *Science,* Vol. 128, No. 3334, pp. 1251–1258.

King, L. 1915. *History of Babylon.* Frederick Stokes Publishers, New York.

Kramer, S.N. 1956. *From the Tablets of Sumer.* Falcon's Wing Press, Indian Hills, Colorado.

Lambert, M. 1965. "L'occupation du Girsu par urlumma roi d'umma." *Revue d'assyriologie,* Vol. 59, No. 2, pp. 81–84.

Lane, W.H. 1923. *Babylonian Problems.* John Murray, London.

Lloyd, S. 1961. *Twin Rivers.* Oxford University Press, London.

Luckenbill, D.D. 1927. *Ancient Records of Assyria and Babylonia.* Greenwood Press, New York.

Olmstead, A.T. 1975. *A History of Assyria.* University of Chicago Press, Chicago.

Pearson, L. 1960. *The Lost Histories of Alexander the Great.* William Clowes and Sons, Ltd., London.

Pritchard, J.B. (ed.) 1958. *The Ancient Near East.* Princeton University Press, Princeton, New Jersey.

Saggs, H.W.F. 1962. *The Greatness that was Babylon.* Hawthorn Books, Inc., New York.

Smith, G. 1871. *History of Assurbanipal.* Williams and Northgate, London.

Strabo. *Geography of Strabo* (from Lane 1923, appendix).

Sykes, P. 1921. *A History of Persia.* MacMillan and Co. Ltd., London.

Tabouis, G.R. 1931. *Nebuchadnezzar.* Whittlesey House, London.

Tatum, J. 1987. *Translation of Xenophon's 'Cycropedia'.* Garland Publishing, Inc., New York.

Winkler, H. 1907. *The History of Babylonia and Assyria,* J.A. Craig (ed.). Charles Scribner's Sons, New York.

Xenophon. *The Anabasis of Cyrus the Younger* (from Lane 1923, appendix).

Date	Parties Involved	Basis of Conflict	Violent Conflict/ Military Involvement?	Description	Sources
1503	Florence and Pisa	Military tool	No	Leonardo da Vinci and Machievelli plan to divert Arno River away from Pisa during conflict between Pisa and Florence.	Honan 1996
1642	China; Ming Dynasty	Military tool	Yes	The Huang He's dikes have been breached for military purposes. In 1642, "toward the end of the Ming dynasty (1368–1644), General Gao Mingheng used the tactic near Kaifeng in an attempt to suppress a peasant uprising."	Hillel 1991
1898	Egypt, France, and Britain	Military/ political goal	Military maneuvers	Military conflict nearly ensues between Britain and France in 1898 when a French expedition attempted to gain control of the headwaters of the White Nile. While the parties ultimately negotiated a settlement of the dispute, the incident has been characterized as having "dramatized Egypt's vulnerable dependence on the Nile, and fixed the attitude of Egyptian policy-makers ever since."	Moorhead 1960
1924	Owens Valley and Los Angeles, California	Political goal; water development	Yes	The Los Angeles Valley aqueduct/pipeline suffers repeated bombings in an effort to prevent diversions of water from the Owens Valley to Los Angeles.	Reisner 1986, 1993
1935	California and Arizona	Political goal; water development	Yes	Arizona calls out the National Guard and militia units to the border with California to protest the construction of Parker Dam and diversions from the Colorado River; dispute ultimately is settled in court.	Reisner 1986, 1993
1938	China and Japan	Military tool	Yes	Chiang Kai-shek orders the destruction of flood-control dikes of the Huayuankou section of the Huang He (Yellow) river to flood areas threatened by the Japanese army. West of Kaifeng, dikes are destroyed with dynamite, spilling water across the flat plain. The flood destroyed part of the invading army and its heavy equipment was mired in thick mud, though Wuhan, the headquarters of the Nationalist government was taken in October. The waters flooded an area variously estimated as between 3,000 and 50,000 square kilometers and killed Chinese estimated in numbers between "tens of thousands" and "one million."	Hillel 1991; Yang Lang 1989, 1994
1940–1945	Multiple parties	Military target	Yes	Hydroelectric dams routinely bombed as strategic targets during World War II.	Gleick 1993
1947 onward	Bangladesh and India	Inequitable distribution and use	No	Partition divides the Ganges River between Bangladesh and India; construction of the Farakka Barrage by India, beginning in 1962, increases tension; short-term agreements settle dispute in 1977–82, 1982–84, and 1985–88, and thirty-year treaty is signed in 1996.	Butts 1997; Samson and Charrier 1997
1947–1960s	India and Pakistan	Military/political goal, inequitable distribution and use	No	Partition leaves Indus basin divided between India and Pakistan; disputes over irrigation water ensue, during which India stems flow of water into irrigation canals in Pakistan; Indus Waters Agreement reached in 1960 after 12 years of World Bank–led negotiations.	Bingham et al. 1994; Wolf 1997
1948	Arabs and Israelis	Military tool	Yes	Arab forces cut off West Jerusalem's water supply in first Arab–Israeli war.	Wolf 1995, 1997
1950s	Korea and United States, others	Military tool	Yes	Centralized dams on the Yalu River serving North Korea and China are attacked during Korean War.	Gleick 1993
1951	Israel, Jordan, and Syria	Military/political goal; water development	Yes	Jordan makes public its plans to irrigate the Jordan Valley by tapping the Yarmouk River; Israel responds by commencing drainage of the Huleh swamps located in the demilitarized zone between Israel and Syria; border skirmishes ensue between Israel and Syria.	Wolf 1997; Samson and Charrier 1997
1953	Israel, Jordan, and Syria	Military/political goal; water development	Yes	Israel begins construction of its National Water Carrier to transfer water from the north of the Sea of Galilee out of the Jordan basin to the Negev Desert for irrigation. Syrian military actions along the border and international disapproval lead Israel to move its intake to the Sea of Galilee.	Samson and Charrier 1997

continued

Date	Parties Involved	Basis of Conflict	Violent Conflict/ Military Involvement?	Description	Sources
1958	Egypt and Sudan	Military/political goal; military tool	Yes	Egypt sends an unsuccessful military expedition into disputed territory amidst pending negotiations over the Nile waters, Sudanese general elections, and an Egyptian vote on Sudan–Egypt unification; Nile Water Treaty signed when pro-Egyptian government elected in Sudan.	Wolf 1997
1960s	North Vietnam and United States	Military target	Yes	Irrigation water supply systems in North Vietnam are bombed during Vietnam War.	Gleick 1993
1962–1967	Brazil and Paraguay	Military/political goal; water development	Military maneuvers	Negotiations between Brazil and Paraguay over the development of the Paraná River are interrupted by a unilateral show of military force in 1962 by Brazil, which invades the area and claims control over the Guaira Falls site. Military forces were withdrawn in 1967 following an agreement for a joint commission to examine development in the region.	Murphy and Sabadell 1986
1963–1964	Ethiopia and Somalia	Military/political goal	Yes	Creation of boundaries in 1948 leaves Somali nomads under Ethiopian rule; border skirmishes occur over disputed territory in Ogaden desert where critical water and oil resources are located; cease-fire is negotiated only after several hundred are killed.	Wolf 1997
1965–1966	Israel and Syria	Military/political goal	Yes	Fire is exchanged over "all-Arab" plan to divert the Jordan River headwaters and presumably preempt Israeli National Water Carrier; Syria halts construction of its diversion in July 1966.	Wolf 1995, 1997
1967	Israel and Syria	Military target and tool	Yes	Israel destroys the Arab diversion works on the Jordan River headwaters. During Arab–Israeli War, Israel occupies Golan Heights, with Banias tributary to the Jordan; Israel occupies West Bank.	Gleick 1993; Wolf 1995, 1997; Wallenstein and Swain 1997
1969	Israel and Jordan	Military target and tool	Yes	Israel, suspicious that Jordan is overdiverting the Yarmouk, leads two raids to destroy the newly built East Ghor Canal; secret negotiations, mediated by the United States, lead to an agreement in 1970.	Samson and Charrier 1997
1970s	Argentina, Brazil, and Paraguay	Political goal; water development	No	Brazil and Paraguay announce plans to construct a dam at Itaipu on the Paraná River, causing Argentina concern about downstream environmental repercussions and the efficacy of its own planned dam project downstream. Argentina demands to be consulted during the planning of Itaipu, but Brazil refuses. An agreement is reached in 1979 that provides for the construction of both Brazil and Paraguay's dam at Itaipu and Argentina's Yacyreta dam.	Wallenstein and Swain 1997
1974	Iraq and Syria	Military/political goal; water development	Military force threatened	Iraq threatens to bomb the al-Thawra Dam in Syria and masses troops along the border, alleging that the dam had reduced the flow of Euphrates River water to Iraq.	Gleick 1994
1975	Iraq and Syria	Military/political goal	Military maneuvers	As upstream dams are filled during a low-flow year on the Euphrates, Iraqis claim that flow reaching its territory is "intolerable" and asks the Arab League to intervene. Syrians claim they are receiving less than half the river's normal flow and pull out of an Arab League technical committee formed to mediate the conflict. In May, Syria closes its airspace to Iraqi flights and both Syria and Iraq reportedly transfer troops to their mutual border. Saudi Arabia successfully mediates the conflict.	Gleick 1993, 1994; Wolf 1997
1978 onward	Egypt and Ethiopia	Inequitable distribution and use; military/ political goal	No	Ethiopia's proposed construction of dams on the headwaters of the Blue Nile leads Egypt to repeatedly declare the vital importance of water. "The only matter that could take Egypt to war again is water" (Anwar Sadat—1979). "The next war in our region will be over the waters of the Nile, not politics" (Boutrous Ghali—1988).	Gleick 1991, 1994
1981	Iran and Iraq	Military target and tool	Yes	Iran claims to have bombed a hydroelectric facility in Kurdistan, thereby blacking out large portions of Iraq, during the Iran–Iraq War.	Gleick 1993
1982	Israel, Lebanon, and Syria	Military tool	Yes	Israel cuts off the water supply of Beirut during siege.	Wolf 1997

continued

129

Date	Parties Involved	Basis of Conflict	Violent Conflict/ Military Involvement?	Description	Sources
1986	North Korea and South Korea	Military tool	No	North Korea's announcement of its plans to build the Kumgansan hydroelectric dam on a tributary of the Han River upstream of Seoul raises concerns in South Korea that the dam could be used as a tool for ecological destruction or war.	Gleick 1993
1990	South Africa	Military tool	No	Pro-apartheid council cuts off water to the Wesselton township of 50,000 blacks following protests over miserable sanitation and living conditions.	Gleick 1993
1990	Iraq, Syria, and Turkey	Water development; military tool	No	The flow of the Euphrates is interrupted for a month as Turkey finishes construction of the Ataturk Dam, part of the Grand Anatolia Project. Syria and Iraq protest that Turkey now has a weapon of war. In mid-1990, Turkish president Turgut Ozal threatens to restrict water flow to Syria to force it to withdraw support for Kurdish rebels operating in southern Turkey.	Gleick 1993, 1995
1991	Karnataka, and Tamil Nadu (India)	Inequitable distribution and use	Yes	Violence erupts when Karnataka reacts to an Interim Order handed down by the Cauvery Waters Tribunal. The Tribunal was established in 1990 to settle two decades of dispute between Karnataka and Tamil Nadu over irrigation rights to the Cauvery River.	Gleick 1993; Butts 1997
1991	Iraq, Kuwait, and United States	Military target and tool	Yes	During the Gulf War, Iraq destroys much of Kuwait's desalination capacity during retreat. Baghdad's modern water supply and sanitation system is also intentionally targeted. Discussions are held about using the Ataturk Dam to cut off flows of the Euphrates to Iraq.	Gleick 1993
1992	Czechoslovakia and Hungary	Military/political goal; water development	Military maneuvers	Hungary abrogates a 1977 treaty with Czechoslovakia concerning construction of the Gabcikovo/Nagymaros project based on environmental concerns. Slovakia continues construction unilaterally, completes the dam, and diverts the Danube into a canal inside the Slovakian republic. Massive public protest and movement of military to the border ensue; issue taken to the International Court of Justice.	Gleick 1993
1992	Bosnia and Bosnian Serbs	Military tool	Yes	The Serbian siege of Sarajevo, Bosnia and Herzegovina, includes a cutoff of all electrical power and the water feeding the city from the surrounding mountains. The lack of power cuts the two main pumping stations inside the city despite pledges from Serbian nationalist leaders to United Nations officials that they would not use their control of Sarajevo's utilities as a weapon. Bosnian Serbs take control of water valves regulating flow from wells that provide more than 80 percent of water to Sarajevo; reduced water flow to the city is used to "smoke out" Bosnians.	Burns 1992; Husarska 1995
1993	Iraq	Military tool	No	To quell opposition to his government, Saddam Hussein reportedly poisons and drains the water supplies of southern Shiite Muslims.	Gleick 1993
1993	Yugoslavia	Military target and tool	Yes	Peruca Dam intentionally destroyed during war.	Gleick 1993
1995	Ecuador and Peru	Military/political goal	Yes	Armed skirmishes arise in part because of disagreement over the control of the headwaters of Cenepa River. Wolf argues that this is primarily a border dispute simply coinciding with location of a water resource.	Samson and Charrier 1997; Wolf 1997
1997	Singapore and Malaysia	Political tool	No	Malaysia supplies about half of Singapore's water and in 1997 threatened to cut off that supply in retribution for criticisms by Singapore of policy in Malaysia.	Zachary 1997

Notes: Conflicts may stem from the drive to possess or control another nation's water resources, thus making water systems and resources a *political* or *military goal. Inequitable distribution and use* of water resources, sometimes arising from a *water development,* may heighten the importance of water as a strategic goal or may lead to a degradation of another's source of water. Conflicts may also arise when water systems are used as instruments of war, either as *targets* or *tools.* These distinctions are described in detail in Gleick 1993.

Sources

Bingham, G., A. Wolf, and T. Wohlegenant. 1994. "Resolving water disputes: Conflict and cooperation in the United States, the Near East, and Asia." US Agency for International Development (USAID). Bureau for Asia and the Near East, Washington, D.C.

Burns, J.F. 1992. "Tactics of the Sarajevo siege: Cut off the power and water," *New York Times* (September 25), p. A1.

Butts, K. (ed.). 1997. *Environmental Change and Regional Security.* Asia-Pacific Center for Security Studies, Center for Strategic Leadership, US Army War College, Carlisle, Pennsylvania.

Drower, M.S. 1954, "Water-supply, irrigation, and agriculture," in C. Singer, E.J. Holmyard, and A.R. Hall (eds.) *A History of Technology.* Oxford University Press, New York.

Gleick, P.H. 1991. "Environment and security: The clear connections." *Bulletin of the Atomic Scientists,* April, pp. 17–21.

Gleick, P.H. 1993. "Water and conflict: Fresh water resources and international security." *International Security,* Vol. 18, No. 1, pp. 79–112.

Gleick, P.H. 1994. "Water, war, and peace in the Middle East." *Environment,* Vol. 36, No. 3, pp. 6–42. Heldref Publishers, Washington, D.C.

Gleick, P.H. 1995. "Water and Conflict: Critical Issues." Presented to the 45th Pugwash Conference on Science and World Affairs. Hiroshima, Japan: 23–29 July.

Green Cross International. *The Conflict Prevention Atlas. http://dns.gci.ch/water/atlas.*

Hillel, D. 1991. "Lash of the dragon." *Natural History* (August), pp. 28–37.

Honan, W.H. 1996. "Scholar sees Leonardo's influence on Machiavelli." *The New York Times* (December 8), p. 18.

Husarska, A. 1995. "Running dry in Sarajevo: Water fight." *The New Republic,* July 17 and 24.

Moorehead, A. 1960. *The White Nile.* Penguin Books, England.

Murphy, I.L., and J. E. Sabadell. 1986. "International river basins: A policy model for conflict resolution." *Resources Policy,* Vol. 12, No. 1, pp. 133–144. Butterworth and Co. Ltd., United Kingdom.

Reisner, M. 1986 (1993 rev ed.). *Cadillac Desert: The American West and its Disappearing Water.* Penguin Books, New York.

Samson, P., and B. Charrier. 1997. "International freshwater conflict: Issues and prevention strategies." Green Cross International. *http://www.dns.gci.ch/water/gcwater/study.html.*

Wallenstein, P., and A. Swain. 1997. "International freshwater resources—Conflict or cooperation?" Comprehensive Assessment of the Freshwater Resources of the World, Stockholm Environment Institute, Stockholm.

Wolf, A.T. 1995. *Hydropolitics along the Jordan River: Scarce Water and its Impact on the Arab–Israeli Conflict.* United Nations University Press, Tokyo.

Wolf, A. T. 1997. "'Water wars' and water reality: Conflict and cooperation along international waterways." NATO Advanced Research Workshop on Environmental Change, Adaptation, and Human Security. Budapest, Hungary. 9–12 October.

Yang Lang. 1989/1994. "High Dam: The Sword of Damocles," in Dai Qing (ed.), *Yangtze! Yangtze!* Probe International, Earthscan Publications, London, pp. 229–240.

Zachary G.P. 1997. "Water pressure: Nations scramble to defuse fights over supplies." *Wall Street Journal* (December 4), p. A17.

References

Arenstein, J. 1996. "A river barely runs through it." *Mail and Guardian*, African Eye News, South Africa (October 13).

Baker, J. 1989. Speech by Secretary of State James Baker, August 7, 1989, Mexico City, Mexico. Seventh U.S.–Mexico Binational Commission meeting.

Biswas, A.K. (ed.). 1994. *International Waters of the Middle East: From Euphrates–Tigris to Nile*. Oxford University Press, Bombay and New Delhi, India.

Biswas, A.K., J. Kolars, M. Murakami, J. Waterbury, and A. Wolf. 1997. *Core and Periphery: A Comprehensive Approach to Middle Eastern Water*. Water Resources Management Series, Vol. 5. Oxford University Press, Delhi, India.

Böge, V. 1993. "Das Sardar-Sarovar projekt an der Narmada in Indien—Gegenstand ökologischen Konflikts." Environment and Conflicts Project Occasional Paper No. 8. Center for Security Studies and Conflict Research/Swiss Peace Foundation Berne, Switzerland.

Breen, C., N. Quinn, and A. Deacon. 1994. "A description of the Kruger Parks Rivers Research Programme (Second phase)." Foundation for Research Development, Pretoria, South Africa.

Brooks, D.B. 1997. "Between the great rivers: Water in the heart of the Middle East," in A.K. Biswas (ed.), *Management of International Rivers*. *International Journal of Water Resources Development*, Vol. 13, No. 3, pp. 291–309.

Carnegie Commission. 1997. *Preventing Deadly Conflict*. Final report of the Carnegie Commission on Preventing Deadly Conflict. Carnegie Corporation of New York, New York.

Chira, S. 1986. "North Korea dam worries the South." *The New York Times* (November 30), p. 3.

Communiqué between Botswana and Namibia. 1996 (June 27). Okavango River communiqué, signed by D.N. Magang, Minister of Mineral Resources and Water Affairs, Botswana; and N. Mbumba, Minister of Agriculture, Water and Rural Development, Namibia.

Communiqué between Botswana and Namibia. 1996 (October 15). "Namibian Okavango River—Grootfontein Pipeline Link," signed by M. Sekwale for Botswana and R. Fry for Namibia.

Conley, A.H. 1996. "A synoptic view of water resources in southern Africa." Department of Water Affairs and Forestry, Pretoria, South Africa.

Crow, B. 1985. "The making and the breaking of agreement on the Ganges," in J. Lundqvist, U. Lohm, and M. Falkenmark (eds.), *Strategies for River Basin Management*. D. Reidel Publishing, Dordrecht, The Netherlands, pp. 255–264.

Crow, B., A. Lindquist, and D. Wilson. 1995. *Sharing the Ganges*. Sage Publishers, New Delhi, India.

Dabelko, G.D., and D.D. Dabelko. 1996. "Environmental security: Issues of conflict and redefinitions." *Environment and Security*, Vol. 1, pp. 23–49.

Dupont, A. 1998. *The Environment and Security in East Asia*. Adelphi Paper, forthcoming. Research School of Pacific and Asian Studies, Strategic and Defence Studies Centre, Australian National University, Canberra, Australia.

Engelman, R., and P. LeRoy. 1993. "Sustaining water: Population and the future of renewable water supplies." Population Action International, Washington, D.C.

Falkenmark, M. 1986. "Fresh waters as a factor in strategic policy and action," in A.H. Westing (ed.), *Global Resources and International Conflict: Environmental Factors in Strategic Policy and Action*. Oxford University Press, New York, pp. 85–113.

Financial Times Global Water Report (FTGWR). 1997. "Jordan grasps the nettle." Issue 35, pp. 1–3 (November 20).

Gleick, P.H. 1990. "Environment, resources, and international security and politics" in E.H. Arnett (ed.), *Science and International Security: Responding to a Changing World.* American Association for the Advancement of Science (AAAS), Washington, D.C., pp. 501–523.

Gleick, P.H. 1991. "Environment and security: The clear connection." *Bulletin of the Atomic Scientists* (April), pp. 17–21.

Gleick, P.H. 1993. "Water and conflict." *International Security, Vol. 18,* No. 1, pp. 79–112 (Summer 1993).

Gleick, P.H. 1994. "Water, war, and peace in the Middle East." *Environment, Vol. 36,* No. 3, pp. 6–42. Heldref Publishers, Washington, D.C.

Gleick, P.H. 1996. "Conflict and cooperation over freshwater in Southern Africa: Background to the issues." Working Paper of the Pacific Institute for Studies in Development, Environment, and Security (prepared for the Carnegie Commission for the Prevention of Violent Conflict).

Gleick, P.H. 1997. "Water and conflict in the twenty-first century: The Middle East and California," in D.D. Parker and Y. Tsur (eds.), *Decentralization and Coordination of Water Resource Management,* Kluwer Academic Publishers, Massachusetts, pp. 411–428.

Gorbachev, M. 1987. "Reality and guarantees for a secure world." *Moscow News,* No. 39, p. 3287.

Hazell, P., N. Perez, G. Siam, and I. Soliman. 1995. "Impact of the structural adjustment program on agricultural production and resource use in Egypt." EPTD Discussion Paper No. 10, International Food Policy Research Institute (IFPRI), Washington, D.C.

Heyns, P. 1995. "Existing and planned water development projects on international rivers within the SADC region." Presented at the Conference of SADC Ministers *Water Resources Management in Southern Africa: A Vision for the Future.* Pretoria, South Africa 23 November.

Hillel, D. 1994. *Rivers of Eden: The Struggle for Water and the Quest for Peace in the Middle East.* Oxford University Press, Oxford, United Kingdom.

Homer-Dixon, T. 1991. "On the threshold: Environmental changes as causes of acute conflict." *International Security,* Vol. 16, No. 2, pp. 76–116.

Homer-Dixon, T. 1994 "Environmental scarcities and violent conflict: Evidence from cases." *International Security,* Vol. 19, No.1, pp. 5–40.

Homer-Dixon, T., and V. Percival. 1996. "Environmental scarcity and violent conflict: Briefing book." The Project on Environment, Population, and Security. American Association for the Advancement of Science and the University of Toronto.

Honan, W.H. 1996. "Scholar sees Leonardo's influence on Machiavelli." *The New York Times* (December 8), p. 18.

Isaac, J., and H. Shuval (eds.). 1994. *Water and Peace in the Middle East.* Elsevier Publishing, Amsterdam, The Netherlands.

Jayasankaran, S., and M. Hiebert. 1997. "Snipe, snipe: Malaysia–Singapore spat reflects growing economic rivalry." *Far Eastern Economic Review (FEER),* June 5, p. 24.

James, D. 1996. "Water project strains relationship between Botswana and Namibia." *Republic of South Africa Star,* October 1.

Kelly, K., and T. Homer-Dixon. 1995. "Environmental scarcity and violent conflict: The case of Gaza." The Project on Environment, Population, and Security. American Association for the Advancement of Science and the University of Toronto.

Khaldi, N. 1992. "The emergence of barley in the Middle East and North Africa: The case of Syria." IFPRI, Washington, D.C.

Klötzli, S. 1993. "Der slowakisch-ungarische Konflikt um das Staustufenprojekt Gabcíkovo." Environment and Conflicts Project Occasional Paper No. 7. Center for Security Studies and Conflict Research/Swiss Peace Foundation Berne, Switzerland.

Klötzli, S. 1994. "The water and soil crisis in Central Asia—A source for future conflicts?" Environment and Conflicts Project Occasional Paper No. 11. Center for Security Studies and Conflict Research/Swiss Peace Foundation Berne, Switzerland.

Koch, N. 1987. "North Korean dam seen as potential 'water bomb'." *Washington Post/San Francisco Chronicle* (September 30), p. 3.

Kolars, J. 1992. "Water resources of the Middle East," in E.J. Schiller (ed.), *Sustainable Water Resources Management in Arid Countries,* Journal of Development Studies Special Issue, pp. 103–119.

Lee, J. 1997. "KL uses water as political tool." *The Jakarta Post,* July 7.

Leitch, J. 1997. "Okavango, a disputed river." *Hydroplus: International Water Review.,* Vol. 77, pp. 20–24.

Libiszewski, S. 1995. "Water disputes in the Jordan Basin region and their role in the resolution of the Arab–Israeli conflict." Environment and Conflicts Project Occasional Paper No. 13. Center for Security Studies and Conflict Research/Swiss Peace Foundation Berne, Switzerland.

Lipschutz, R.D. 1997. "Damming troubled waters: Conflict over the Danube, 1950–2000." Environment and Security Conference, Institute of War and Peace Studies, Columbia University, October 24.

Lonergan, S.C., and D.B. Brooks. 1994. *Watershed: The Role of Fresh Water in the Israeli–Palestinian Conflict.* International Development Research Centre, Ottawa, Canada.

Lowi, M. 1992. "West Bank water resources and the resolution of conflict in the Middle East." Project on Environmental Change and Acute Conflict, American Academy of Arts and Sciences and the University of Toronto. Occasional Paper No. 1, pp. 29–60.

McCaffrey, S. 1993. "International water law," in P.H. Gleick (ed.), *Water in Crisis: A Guide to the World's Fresh Water Resources,* Oxford University Press, New York, pp. 92–104.

Myers, N. 1989. "Environment and security." *Foreign Policy,* Vol. 74, pp. 23–41.

Postel, S. 1997. "Water for food production: Will there be enough in 2025?" Presentation at the International Water Resources Association IXth Congress, Montreal, Canada, 3–7 September.

Regional Centre for Strategic Studies (RCSS). 1997. "Indo-Bangladesh water sharing accord." *Newsletter of the RCSS.* Vol. 3, No. 1, p. 1 (January).

Reisner, M. 1986 (1993 revision). *Cadillac Desert: The American West and its Disappearing Water.* Penguin Books, New York.

Rogers, P., P. Lydon, D. Seckler, and G.T.K. Pitman. 1994. *Water and Development in Bangladesh: A Retrospective on the Flood Action Plan.* Irrigation Support Project for Asia and the Near East, U.S. Agency for International Development, Washington, D.C.

Samson, P., and B. Charrier. 1997. "International freshwater conflict: Issues and prevention strategies." Green Cross International. Via *http://www.gci.ch/water/gcwater/study.html.*

Spiegel, S.L., and D.J. Pervin (eds.). 1995. *Practical Peacemaking in the Middle East,* Volume 2. *The Environment, Water, Refugees, and Economic Cooperation.* Garland Publishing, Inc., New York.

Straits Times. 1997. "PM Goh's speech in Parliament: I shall work for a new era of cooperation." *The Straits Times,* June 6.

United Nations. 1978. *Registry of International Rivers.* Centre for Natural Resources, United Nations. Pergamon Press, New York.

United Nations. 1997a. "Convention on the Law of the Non-Navigational Uses of International Watercourses." UN General Assembly A/51/869 (April 11). United Nations Publications, New York.

United Nations. 1997b. *Comprehensive Assessment of the Freshwater Resources of the World.* Prepared for the Commission for Sustainable Development and the General Assembly. Stockholm Environment Institute and the United Nations, New York.

U.S. Department of State. 1997. "Environmental Diplomacy: The Environment and U.S. Foreign Policy." U.S. Department of State, Washington, D.C.

Venter, F.J., and A.R. Deacon. 1995. "Managing rivers for conservation and ecotourism in the Kruger National Park." *Water Science Technology,* Vol. 32, Nos. 5–6, pp. 227–233.

Wolf, A.T. 1995. *Hydropolitics along the Jordan River: Scarce Water and its Impact on the Arab–Israeli Conflict.* United Nations University Press, Tokyo, Japan.

Wolf, A. T. 1997. "'Water wars' and water reality: Conflict and cooperation along international waterways." NATO Advanced Research Workshop on Environmental Change, Adaptation, and Human Security. Budapest, Hungary. 9–12 October.

Zachary, G.P. 1997. "Water pressure: Nations scramble to defuse fights over supplies." *The Wall Street Journal* (December 4), p. A17.

Climate Change and
Water Resources:
What Does the Future Hold?

The world's leading climate scientists believe that we are now on the verge of changing our climate through human activities that produce trace gases, including the burning of fossil fuels, the destruction of forests, and a wide range of industrial and agricultural activities. Indeed, a growing number of scientists believe that some human-induced climatic changes are already beginning to occur or are unavoidable even if we act now to reduce our emissions of these gases. These climatic changes—the so-called "greenhouse effect"—will have widespread consequences for every aspect of life on earth. The climate determines where we live and how we live; the kinds of crops we grow and their success or failure; the location, size, and operation of dams and reservoirs; the kinds of structures we build along our coastlines; and even the clothes we buy.

One of the most important consequences of the greenhouse effect will be impacts on water resources, including both the natural hydrologic system and the complex water-management schemes we have built to alter and control that system. We are just beginning to learn about the ways in which these impacts will be felt around the globe. Water managers, policymakers, and the public must begin now to think about the implications of climatic change for long-term water planning and management. Changes may be necessary in the design of projects not yet built. Modifications to existing facilities may be required to permit them to continue to meet their design objectives. New projects may need to be built or old projects removed and new institutions may need to be created or old ones revamped in order to cope with possible changes.

The goal of this chapter is not to try to prove that climate change is a real problem. This has already been done in hundreds of research programs, papers, and field studies. Comprehensive reviews of the problem have been produced by a coalition of the world's leading climate scientists through two independent, peer-reviewed assessments produced by the Intergovernmental Panel on Climate Change (IPCC). The World Meteorological Organization (WMO) and the United Nations Environment Programme (UNEP) sponsor the IPCC, with the cooperation of over 120 nations (IPCC, 1990, 1996a,b). This chapter lays out the serious challenges that face water policymakers and identifies actions, policies, and programs that may make the risks of climate change effects on water resources smaller and more manageable.

What Do We Know?

The earth's climate is powered by solar radiation falling on the planet's surface. The distribution and quantity of sunlight reaching the earth depends on the shape of the earth's orbit and our location around the sun, the nature and extent of cloud cover, and other factors, particularly the composition of the atmosphere. The atmosphere naturally contains a variety of trace gases, some of which permit incoming solar radiation to pass through to the surface of the earth but trap outgoing infrared radiation that would otherwise escape back to space. These trace gases are often called "greenhouse gases" because they act like the heat-trapping walls of a greenhouse. The principal greenhouse gases are carbon dioxide, water vapor, nitrous oxide, and methane.

The "greenhouse effect" is neither new nor due solely to human activities. It is a natural component of the earth's climatic balance, and without it, the surface of the planet would be far colder than it is. In the period of industrialization after 1750, however, human activities have led to the emission of additional greenhouse gases at rates far faster than natural processes can remove them. Most of these gases come from fossil-fuel combustion, biomass burning, the use of nitrogen fertilizer, and actions that eliminate natural sinks for greenhouse gases, such as deforestation. Some of these gases are new synthetic gases capable of trapping tremendous amounts of heat: the chlorofluorocarbons. The atmospheric concentrations of carbon dioxide, methane, and nitrous oxide have grown by about 30 percent, 145 percent, and 15 percent, respectively, since preindustrial times, and these trends are largely attributable to fossil fuel combustion and changes in land use (IPCC 1996a; see Table 5.1).

There is substantial scientific evidence that increasing concentrations of greenhouse gases will lead, and may already be leading, to global climatic changes. As the concentration of these gases grows, the amount of heat trapped by the atmosphere also grows. If current trends continue, the cumulative warming effect will raise the average surface temperature of the earth and lead to a host of other climatic changes. The IPCC concluded that the global average temperature in 2100 is likely to be 2°C warmer than at present, assuming a best guess for climate sensitivity and future global emissions. A change in global temperature of a few degrees does not seem that significant to some, especially since it would be imposed on a natural system that routinely experiences larger temperature swings on fine regional and temporal scales. But a difference of only 1°C in average global temperature is all that separates today's climate from that of the Little Ice Age in the period from the fourteenth to the seventeenth century. An increase of 2°C would push average global temperatures beyond anything experienced in the past 10,000 years. An average warming of 5°C from the present level would make the earth warmer than it has been since three million years ago, when there was no ice cap, there were tropical and subtropical regions in Canada and England, and sea level was 75 meters higher than today's (Mintzer 1992).

There have already been some important systematic changes in the earth's climate. Global mean surface air temperature has increased by between 0.3° and 0.6°C since the late nineteenth century. The past decade has been the warmest decade in the period of instrumental record. Precipitation has increased over land in the high latitudes of the Northern Hemisphere, particularly during winter. Global sea level has risen by between

TABLE 5.1   SAMPLE GREENHOUSE GASES AFFECTED BY HUMAN ACTIVITIES[a]

	Carbon Dioxide	Methane	Nitrous Oxide	CFC-11
Pre-industrial concentration	280 ppmv	700 ppbv	275 ppbv	0
1994 concentration	358 ppmv	1,720 ppbv	312 ppbv	268 pptv
Atmospheric lifetime	50–200 years	12 years	120 years	50 years

Source: IPCC 1996a.

[a] ppmv: parts per million (by volume); ppbv: parts per billion (by volume); pptv: parts per trillion (by volume).

10 and 25 centimeters over the past 100 years (IPCC 1996a). And some scientists believe the increasing frequency and duration of El Niño warm episodes, responsible for dramatic changes in agricultural and fisheries production and regional water supplies, are at least partly due to global warming (Trenberth and Hoar 1996).

While the earth's climate is naturally variable, considerable effort has gone into trying to determine whether the observed changes are the result of natural or anthropogenic influences. Most studies of past climate have determined that recent changes are both statistically significant and "unlikely to be entirely natural in origin" (IPCC 1996a). The IPCC goes on to conclude, "The balance of evidence suggests a discernible human influence on global climate."

Many uncertainties remain. Future emissions of greenhouse gases and their longevity in the atmosphere depend on a wide range of economic, social, and geophysical factors, some of which are unknown or unknowable. The climate models used to generate estimates of future conditions of the climate are in need of refinement and improvement in many areas. Gaps in data and basic understanding of fundamental climatological processes hinder definitive assessments. And future unexpected large and rapid changes in climate, of a type known to have occurred in the past, are by their very nature difficult to predict. These "surprises," inherently unpredictable, arise from the nonlinear nature of the climate system. More effort needs to be expended exploring these important issues.

Hydrologic Effects of Climate Change

In addition to the uncertainties involved in trying to model the climate, additional difficulties hinder clear assessments of the impacts of climate changes on global or regional hydrology. Many important hydrologic processes occur at fairly small spatial scales that are not yet capable of being accurately modeled. Limitations in data availability and quality affect our ability to accurately validate models or verify results. And the complex human modifications of watersheds must be incorporated into any detailed analysis of impacts and adaptation.

There is little doubt that climatic change will alter the hydrologic cycle in a variety of ways, but there is little certainty about the form these changes will take, or when they will be unambiguously detected. As a result, while global climatic changes may already be appearing, we are as yet unable to accurately determine how such changes will affect water-supply systems or water demands.

The hydrologic system—an integrated component of the earth's geophysical system—both affects and is affected by climatic conditions. Changes in temperature affect evapotranspiration rates, cloud characteristics, soil moisture, storm intensity, and snowfall and snowmelt regimes. Changes in precipitation affect the timing and magnitude of floods and droughts, shift runoff regimes, and alter groundwater recharge characteristics. Synergistic effects will alter cloud formation and extent, vegetation patterns and growth rates, and the behavior of soil moisture.

Most of the scientific and media attention given to the impacts of climate change on society has focused on a very limited aspect of those changes—the increase in temperature. Even the colloquial names given to the problem—"global warming" and "the greenhouse effect"—reflect this bias. Yet some of the most severe impacts of climate change are likely to result not from the expected increases in temperature per se but from changes in precipitation, evapotranspiration, runoff, and soil moisture: in short, from changes in the most important variables for water planning and management.

This partially misdirected focus of public attention arises from limitations in the ability of global models, including the most complex representations, the general circulation models, to incorporate and reproduce important aspects of the hydrologic cycle. Many important hydrologic processes, such as the formation and distribution of clouds and rain-generating storms, occur on a spatial scale far smaller than most models are able to resolve. We thus know less about how the water cycle will change than is necessary to make informed and rational decisions about how to plan, manage, and operate water systems (Gleick 1992). Despite these limitations, we do know some things about how hydrology and water-management systems will be affected by climatic changes and how we might strive to adapt to those changes.

At the most general level, the IPCC (1996a) assessment stated,

> Warmer temperatures will lead to a more vigorous hydrological cycle; this translates into prospects for more severe droughts and/or floods in some places and less severe droughts and/or floods in other places. Several models indicate an increase in precipitation intensity, suggesting a possibility for more extreme rainfall events.

But many more impacts will occur. Among the expected impacts of climatic changes on water resources are increases in global average precipitation and evaporation; changes in the regional patterns of rainfall, snowfall, and snowmelt; changes in the intensity, severity, and timing of major storms; rising sea level and saltwater intrusion into coastal aquifers; and a wide range of other geophysical effects. These changes will also have many secondary impacts on freshwater resources, altering both the demand and supply of water, and changing its quality.

Changes in Precipitation

There is a very strong consensus that global climate change will result in a wetter world, on average, through changes in precipitation—rain and snow. Changes in precipitation will affect river flow, groundwater recharge, rainfed crop production, ecosystem viability, and many other things. Climate models consistently project an increase in global mean precipitation of between 3 and 15 percent for a temperature increase of 1.5 to 3.5°C (Schneider et al. 1990; IPCC 1996b). The global average, however, will hide significant differences in regional precipitation patterns, with some regions showing increases, some decreases, and considerable interannual variability. Increases in precipitation are expected to occur more consistently and intensely throughout the year at middle to high latitudes. In many model estimates, summer rainfall decreases slightly over much of the northern mid-latitude continents while winter precipitation increases. Other changes in the mid-latitudes remain highly variable and ambiguous. Less (and less consistent) information is available on changes in precipitation in subtropical arid regions. Estimates suggest that for eastern and southern Australia precipitation could drop by 10 to 20 percent in winter on average by 2070. The models show both increases and decreases during summer (CSIRO 1996). Even small changes in arid and semi-arid zones can have significant implications for ecological and human systems.

Evaporation and Transpiration

Evaporation of water into the atmosphere is a function of the availability of water and energy: as global average temperatures rise, the energy available for evaporation increases and the atmospheric demand for water from land and water surfaces increases. A warmer atmosphere can also hold more water, but actual changes in evaporation will depend on both the ability of the atmosphere to hold water (the humidity) and changes in the movements of air (wind patterns). Reviews of state-of-the-art climate models suggest that global average evaporation may increase by 3 to 15 percent for an equivalent doubling of atmospheric carbon dioxide concentration. The greater the warming, the larger these increases. Moreover, regional increases in potential evaporation could be as high as 40 percent in humid temperate regions (IPCC 1996b).

A second component to evaporation is transpiration: the loss of water into the atmosphere from plants. Transpiration is affected by a wide range of factors, including plant type and cover, root depth, stomatal behavior, and the concentration of carbon dioxide in the atmosphere. Laboratory and limited field studies have shown that certain plants will decrease water use when exposed to higher carbon dioxide levels. Other studies suggest that much of this improvement is lost because plants grow more and the increased leaf area offsets increased water-use efficiency. Some evidence also suggests that some plants acclimatize to increased CO_2 levels, limiting improvements in water-use efficiency, or that nutrients other than water are what limit growth. Real-world effects also make laboratory findings hard to reproduce in the field or inappropriate to generalize to large catchments. These issues continue to be major concerns for researchers in the climate and water area.

Changes in Soil Moisture

Precipitation that does not evaporate back into the atmosphere, transpire immediately from vegetation, get captured by humans for direct use, or run off into rivers, lakes, or the ocean infiltrates into the soil. Soil moisture is a critically important variable in both supporting agricultural production and defining natural vegetative type and extent. Any change in climate that alters precipitation patterns and the evapotranspiration regime will directly affect soil-moisture storage, runoff processes, and groundwater dynamics. In regions with precipitation decreases, soil moisture may be significantly reduced. Even in regions with precipitation increases, soil moisture may still drop if increases in evaporation owing to higher temperatures are even greater.

All climate models show increased soil moisture in the highest northern latitudes, where increases in precipitation greatly outpace increases in evapotranspiration. At the same time, most models also suggest large-scale drying of the earth's surface over northern mid-latitude continents in northern summer owing to higher temperatures and either insufficient precipitation increases or actual reductions in rainfall. Drying in these regions could have significant impacts, particularly on agricultural production and water demand. Others report that soil-moisture reductions could be extensive throughout south-central Asia and Latin America, where increases in evapotranspiration also exceed increases in precipitation (Binnie 1996).

One consequence of this is an increased incidence of droughts (measured by soil-moisture conditions), even in some regions where precipitation increases, because of the increased evaporation (Rind et al. 1990). This finding has also been seen in some of the detailed hydrologic modeling of specific river basins, such as the Colorado, where large increases in precipitation are necessary in order to simply maintain river runoff at present historical levels as temperatures and evaporative losses rise (Nash and Gleick 1991, 1993).

Changes in Snowfall and Snowmelt

In some regions, the most important hydrologic impacts of climatic change will be changes in snowfall and snowmelt. In watersheds with substantial snow, changes in temperature alone will lead to important changes in water availability and quality and complicate the management of reservoirs and irrigation systems. In such basins, temperature increases have three effects: they increase the ratio of rain to snow in cold months, they decrease the overall duration of the snow season, and they increase the rate and intensity of warm season snowmelt. As a result of these three effects: average winter runoff and average peak runoff increase; peak runoff occurs earlier in the year; spring runoff ends sooner; and there is a faster and more intense drying of soil moisture during the warm season (Gleick 1986; IPCC 1996a). Because of these effects, far more attention needs to be paid in some regions to the risk of floods rather than droughts. One of the greatest concerns about the effect of higher temperatures is, therefore, the increased probability *and* intensity of flood flows. Earlier snowmelt will also have implications for reservoir storage capacity and operation and for the availability of stored water for domestic and agricultural use later in the year.

Changes in Storm Frequency and Intensity

Many of the most severe societal impacts resulting from the hydrologic cycle occur due to climatic extremes and storm events. An important but unresolved question, therefore, is what global climatic changes may do to the variability of climatic conditions—that is, the frequency and intensity of extremes, such as cyclones, hurricanes, and more systemic impacts such as those caused by El Niño. Although little work has been done on this issue, there are some indications that the variability (interannual standard deviation) of the hydrologic cycle increases when mean precipitation increases and vice-versa. In one model study, the total area over which precipitation fell over the earth decreased, even though global mean precipitation increased (Noda and Tokioka 1989). This implies more intense local storms and runoff. Other changes in variability are also likely, though at this point we have little confidence that we can predict what they will be. There are some indications that day-to-day and interannual variability of storms in the mid-latitudes will decrease. At the same time, there is evidence from both model simulations and empirical considerations that the frequency, intensity, and area of tropical disturbances may increase. Far more modeling and analytical efforts are needed in this area.

Changes in Runoff: Floods and Droughts

River flow has the potential to integrate changes in hydrologic characteristics over a large area and thus may be particularly valuable as an indicator of climatic change. Many estimates of changes in runoff due to climatic change have been produced using detailed regional hydrologic models of specific river basins. By using anticipated, hypothetical, or historical changes in temperature and precipitation and models that include realistic small-scale hydrology, modelers suggest that some significant changes in the timing and magnitude of runoff are likely to result from quite plausible changes in climatic variables. At the same time, society and natural ecosystems are highly dependent on river flows, and any changes caused by the greenhouse effect would be cause for concern.

For example, in a series of studies of basins in the western United States, temperature increases of 2 to 4°C, with no change in precipitation, result in decreases in runoff of up to 20 percent (Stockton and Boggess 1979; Gleick 1986, 1987a,b; Flaschka et al. 1987; Schaake 1990; Nash and Gleick 1993). Increases or decreases of precipitation of 10 and 20 percent tend to change runoff by about the same amount. In basins where demands for water are close to the limit of reliable supplies, such changes will have enormous policy implications.

There is also a risk of increased flooding. The authors of the second IPCC report conclude that "the flood-related consequences of climate change may be as serious and widely distributed as the adverse impacts of droughts" and "there is more evidence now that flooding is likely to become a larger problem in many temperate regions, requiring adaptations not only to droughts and chronic water shortages, but also to floods and associated damages, raising concerns about dam and levee failure" (IPCC 1996b).

Ironically, some regions will be subjected to both increases in droughts and increases in floods if climate becomes more variable. Even without increases in variability, both problems may occur in the same region. In the western United States, for example, where winter precipitation falls largely as snow, higher temperatures will increase the amount of rain and decrease the amount of snow, contributing to high winter and spring runoff—the period of time when flood risk is already highest. At the same time, summer and dry-season runoff will decrease because of a decline in snowpack and accelerated spring melting.

Societal Impacts of Changes in Water Resources

The impacts of climate change on water-resources supply and availability will in turn lead to direct and indirect effects on a wide range of institutional, economic, and social factors. The nature of these effects is not well understood, nor is the ability of society to adapt to them. Yet arguments about "adaptation" have become central to arguments about the costs and benefits of trying to reduce greenhouse gas emissions. If water managers and planners can easily and cheaply adapt to any climatic disruptions that may occur, actions to prevent climate change will be less urgent. If we overestimate our ability to adapt, we may ignore inexpensive and successful actions that can reduce the impacts of climate change early.

Adaptation and innovative management will certainly be useful and necessary responses to climatic changes. Several factors, however, suggest that relying solely, or even principally, on adaptation may prove a dangerous policy. First, the impacts of climate change on the water sector will be very complicated and at least partly unpredictable. Second, many impacts may be nonlinear and chaotic, characterized by surprises and unusual events. Third, climatic changes will be imposed on water systems that will be increasingly stressed by other factors, including population growth, competition for financial resources from other sectors, and disputes over water allocations and priorities. Finally, some adaptive strategies may help mitigate some adverse consequences of climate change while simultaneously worsening others.

There is a rapidly growing literature about the impacts of climate change on water systems, reservoir operations, water quality, hydroelectric generation, navigation, and other water-management concerns. At the same time, this large literature has barely scratched the surface of the potential range of impacts, and far more research is needed. One consistent finding is that water-supply systems are sensitive to changes in inflows and demands. Nemec and Schaake (1982), in one of the earliest studies on climate impacts, showed that large changes in the reliability of water yields from reservoir systems result from small changes in reservoir inflows. This finding has now been reproduced in many studies from many different regions (Cole et al. 1991; McMahon et al. 1989; Mimikou et al. 1991; Nash and Gleick 1993).

In a comprehensive analysis of the Colorado River Basin, the highly integrated system of linked reservoirs was shown to be very sensitive to both the physical characteristics of the system and to the way it is managed and operated. Under current operating conditions and rules, even the extremely large volume of storage in the basin protects against modest climate-induced reductions in runoff for only a few years before the reservoirs are drawn down and hydroelectric generation drops (Nash and Gleick 1991, 1993). A 10 percent decrease in average natural flow in the Colorado River Basin—a plausible result given current climate projections—would result in a 30 percent decrease in reservoir storage, a 30 percent decrease in hydroelectricity generation, and a violation of salinity standards in the lower river in almost all years.

In the Mesohora reservoir in Greece, a 10 percent decrease in precipitation leads to a tripling of the risk that the hydroelectric facility there will be unable to produce its design power (Mimikou et al. 1991). Because of conflicts between flood-control functions and hydropower objectives, climatic changes in California may require more water to be released from California reservoirs in spring to avoid flooding. This would result in a reduction in hydropower generation and the economic value of that generation. The IPCC reports that an increase in fossil-fuel use of 11 percent would be required to meet the same energy demands in California at a cost of hundreds of millions of dollars and a worsening of greenhouse-gas emissions (IPCC 1996b).

In an unusual study of the impacts of climatic change on river temperatures, Miller et al. (1993) concluded that the Tennessee Valley Authority would be forced to reduce power generation and shut down fossil and nuclear plants more frequently to avoid violating temperature standards set for regional rivers. Plant efficiencies, which depend in part on the temperature of cooling water, would be reduced and cooling towers required more often.

These studies represent only a few of the many analyses that have been done in the past decade. But almost all of the work suggests two important conclusions: first, existing systems tend to be optimally designed for current climatic conditions and to be sensitive to any changes in those conditions; and second, changes in operating rules need to be closely examined to see whether they can reduce the risks associated with fixed infrastructure and designs.

Is the Hydrologic System Showing Signs of Change?

Is it possible to see any changes in the hydrologic cycle yet, and if so, are such changes due to human influence on the climate? Despite some serious gaps in data, inadequate and uneven climate monitoring, short record length, and biases in instrumental data, recent research has begun to show changes and variations in the hydrologic cycle of the earth. A number of these changes are statistically significant—they are sufficiently different from the past record to be the result of something other than just natural variability. Only time will tell whether these changes are directly related to intensification of the greenhouse effect or due to some other effect we do not fully understand.

The change that has received the most attention is the increase in average global temperature. Global surface temperatures are constantly measured in thousands of places around the world, and data from the late 1800s are available from a network of ground and ocean-based sites. More recently, satellites have been used to measure atmospheric temperatures at different altitudes. These data suggest that the average surface temperature of the earth has increased by nearly 1°C over the past century. The 12 warmest years this century have all occurred since 1980, and 1995 and 1997 were the warmest years on record. In April 1998, researchers announced that 1990, 1995, and 1997 were the warmest years in the past 500 in the Northern Hemisphere and that rising levels of greenhouse gasses are probably responsible (Mann et al. 1998). The higher latitudes have warmed more than the equatorial regions, in agreement with what the climate models project for greenhouse warming (IPCC 1996a; OSTP 1997).

Additional evidence supports the temperature data. Between 1981 and 1991, satellite imagery documented an increase in the length of the growing season in the northern high latitudes (between 45 and 70°N) by a total of up to 12 days. Vegetation bloomed up to eight days earlier in spring and summer and continued to photosynthesize an estimated

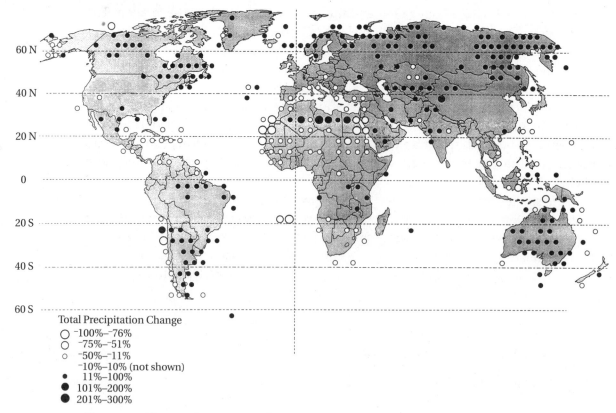

Total Precipitation Change
○ −100%–−76%
○ −75%–−51%
○ −50%–−11%
  −10%–10% (not shown)
● 11%–100%
● 101%–200%
● 201%–300%

MAP 5.1   GLOBAL PRECIPITATION CHANGES, 1901–1995

four days longer. Permafrost in the Alaskan and Siberian arctic is beginning to thaw, and mean sea level has risen between 10 and 25 centimeters since the 1890s (IPCC 1996b; OSTP 1997).

Precipitation patterns are also already showing some trends (IPCC 1990, 1996a; Dai et al. 1997; Karl and Knight 1998). By the late 1980s, observers had noticed a general increase in precipitation outside of the tropics, with a tendency for rainfall declines in the subtropics, particularly in the northern tropics of Africa (IPCC 1990). In recent analyses of data from 1900 to 1988, precipitation over land has increased by 2.4 millimeters (mm) per decade and global mean rainfall has risen by more than 2 percent. These results show that global rainfall now is about 22 mm per year higher than it was at the turn of the century (Dai et al. 1997). Map 5.1 shows global trends in precipitation over the past century. Precipitation has increased over land in the high latitudes of the Northern Hemisphere, particularly during winter. These trends have been supported by regional, national, and global studies, even correcting for known biases of precipitation measurements (Karl et al. 1995).

In another analysis, Karl and Knight (1998) show that not only has precipitation increased in the contiguous United States but this increase is also reflected primarily in an increase in heavy and extreme daily precipitation events. By analyzing long-term precipitation trends in the United States, they determined that:

• Precipitation over the contiguous United States has increased by about 10 percent since 1910;

- Increases in total precipitation are strongly affected by increases in both the frequency and the intensity of heavy and extreme events;

- The probability of precipitation on any given day has increased;

- The intensity of precipitation has only increased for very heavy and extreme precipitation days; and

- The proportion of total precipitation from heavy events has increased at the expense of moderate precipitation events.

These conclusions have enormous ramifications for water management, system operation, and water-related disasters.

A detailed study over the area of the former Soviet Union also shows increases in precipitation of around 10 percent over the past century, with larger increases in winter than summer (Groisman et al. 1994). Similar increases have been observed in southern Canada (Groisman and Easterling 1994). In Europe overall, the data suggest that precipitation has increased in northern Europe (north of 55°N) and decreased in southern Europe during this century (Nicholls et al. 1996). There is a strong seasonality associated with these changes, with most increases occurring during the autumn. No significant trends have been observed in central Europe.

No consistent continental-scale trends are evident in Central and South America, except in regions where the influence of the El Niño Southern Oscillation (ENSO) is clear (IPCC 1996a). In these regions, precipitation has decreased. A step-like decrease of precipitation has also been observed in the subtropics and tropics from northern Africa to Indonesia and Southeast Asia since 1960. Many of these areas also fall within the regions affected by El Niño, and the observed decreases probably reflect the influence of the recent increase in El Niño frequency. The same influence of ENSO events can be seen in the rainfall record in eastern Australia and New Zealand, which have experienced a decrease in annual rainfall over the past four decades. These connections seem so convincing that the indications in mid-1997 of the onset of another severe ENSO event led Australian farm analysts to significantly downgrade their estimates of crop production for the year (Global Change 1997).

Many parts of Africa typically experience highly variable precipitation, with significant regional variations and very high interannual variability. Rainfall in Sahelian West Africa in the last 35 years has been well below the amount received earlier in this century. While similar dry periods have been observed in the historical and recent geologic past, the recent period exhibits a tendency toward continental-scale dryness not seen in the past record (Nicholson 1994a,b, 1995). Far better monitoring of rainfall in arid regions is needed, however.

Snow cover over the Northern Hemisphere land surface has been consistently below the 21-year average (1974 to 1994) since 1988 (Robinson et al. 1993; Groisman et al. 1994), with an annual mean decrease in snow cover of about 10 percent over both North America and Asia. These changes are linked to increases in temperature. Other observed effects include earlier lake ice melting, earlier snowmelt-related floods in western Canada and the western United States, reduced duration of river ice in the former Soviet Union, and earlier warming of Northern Hemisphere land areas in the spring (Nicholls et al. 1996). At the same time that snow and ice cover seem to be decreasing and melting earlier, total annual snowfall in the far northern latitudes seems to be increasing, consistent with the observed increases in northern latitude precipitation.

River runoff is considered to be an excellent integrator of climatic factors, and some efforts have been made to look at long-term runoff records to see if any trends can be detected. One difficulty, however, is that although long records of runoff are essential to determining whether runoff is changing over time, very few rivers have reliable records any longer than several decades. Records longer than a century are extremely rare. Moreover, human interventions in the form of water withdrawals, the construction of dams and reservoirs, and land-use changes in watersheds have already caused significant changes in runoff regimes, greatly complicating the use of past runoff records to detect climate changes, or even trends in natural variability.

Nevertheless, there have been several recent efforts to identify secular trends in river flow, with mixed results. Studies in some regions have not found any clear trends. Chiew and McMahon (1993) looked at 30 Australian watersheds and found no apparent changes. Marengo (1995) looked at streamflow stations across South America and did not see any general change in flows. Some regional changes were observed related to ENSO signals. Trends and jumps in the hydrologic record in the Nordic region have been observed, but researchers could not directly link these observed changes with the greenhouse effect (Arnell 1996).

Other studies, however, have begun to see trends that cannot be explained by natural variability and that are consistent with modeling projections. Three studies published in 1994 all found evidence that certain rivers are exhibiting runoff trends consistent with the effects of global warming. Burn (1994) found a statistically significant trend toward earlier spring runoff in several rivers in western Canada—a finding predicted in model studies involving snowmelt done several years earlier for northern California (Gleick 1986, 1987b). Lins and Michaels (1994) also reported statistically significant increases in autumn and winter streamflow in North America between 1944 and 1988 and related these regional and seasonal increases to global warming. Lettenmaier et al. (1994) detected clear increases in winter and spring streamflow across much of the United States between 1948 and 1988. As Arnell (1996) states: "The evidence for global warming having a noticeable effect on hydrological behaviour is not yet convincing, but it does seem to be accumulating."

Similar studies are now being done looking at lake levels and soil moisture, which are also sensitive to both precipitation and temperature changes. For example, Lake Chad in northern Africa has shrunk from its greatest extent in the 1960s to about one-tenth that area in the 1980s because of decreased rainfall (Grove 1995). Similarly, Vinnikov et al. (1996) report that soil moisture over the past 30 years in the western part of the former Soviet Union has increased consistently with the observed increase in precipitation in this region. They also note that snowmelt and summer drying of soil moisture are beginning about a month earlier in grasslands there. Good, long-term data on soil moisture are not available in most places, complicating these assessments.

Recommendations and Conclusions

Impacts on water resources will be felt in virtually every aspect of natural resources management. All decisions about long-term water planning, the design and construction of new water-supply infrastructure, the type and acreage of crops to be grown, urban water allocations and rate structures, reservoir operation, and water-supply management depend on climatic conditions and what humans do to respond and adapt to those conditions.

In the past, water planning and management relied on the assumption that future climatic conditions would be the same as past conditions, and all our water-supply systems were designed with this assumption in mind. Dams are sized and built using available information on existing flows in rivers and the size and frequency of expected floods and droughts. Reservoirs are operated for multiple purposes using the past hydrologic record to guide decisions. Irrigation systems are designed using historical information on temperature, water availability, and soil water requirements.

This reliance on the past record now may lead us to make incorrect—and potentially fatal or expensive—decisions. Given that risk, it makes sense to both try to understand what the consequences of climate change will be for water resources and to begin planning for those changes. As the proverb says, "To see the future is good, to prepare for it is better."

The Second World Climate Conference concluded in 1991 that:

> The design of many costly structures to store and convey water, from large dams to small drainage facilities, is based on analyses of past records of climatic and hydrologic parameters. Some of these structures are designed to last 50 to 100 years or even longer. *Records of past climate and hydrological conditions may no longer be a reliable guide to the future. The design and management of both structural and non-structural water resource systems should allow for the possible effects of climate change.*" (Emphasis added) (Jager and Ferguson 1991)

A separate study focused on the implications of global climate changes for the water resources of the United States (Waggoner 1990). This study, a product of the Climate and Water Panel of the American Association for the Advancement of Science, chaired by Roger Revelle and Paul Waggoner, concluded:

> Among the climatic changes that governments and other public bodies are likely to encounter are rising temperatures, increasing evapotranspiration, earlier melting of snowpacks, new seasonal cycles of runoff, altered frequency of extreme events, and rising sea level. . . . *Governments at all levels should reevaluate legal, technical, and economic procedures for managing water resources in the light of climate changes that are highly likely.* [Italics in original]

Similarly, the IPCC (1996b) stated "freshwater resources in many regions of the world are likely to be significantly affected," and that many current freshwater problems will be made worse by the greenhouse effect. This report urges water managers to begin "a systematic reexamination of engineering design criteria, operating rules, contingency plans, and water allocation policies" and states with "high confidence" that "water demand management and institutional adaptation are the primary components for increasing system flexibility to meet uncertainties of climate change." This emphasis on demand management rather than construction of new facilities marks a change in traditional water-management approaches, which in the past have relied on the construction of large and expensive infrastructure.

Many uncertainties remain about the timing, direction, and extent of these climatic changes, as well as about their societal implications. Indeed, the most important effect of climatic change for water systems will be to greatly increase the overall uncertainty water managers face. These uncertainties greatly complicate rational water-resource planning for the future and have contributed to the ongoing debate over how to respond. But we cannot let these uncertainties paralyze us and stop rational and thoughtful actions from being taken. There is sufficient understanding of the current causes and future consequences of potential global climatic change to begin to rethink and reevaluate current policies, and to offer recommendations to water managers (AWWA 1997).

- While water management systems are often flexible, adaptation to new hydrologic conditions may come at substantial economic costs. Water agencies should begin now to reexamine engineering design assumptions, operating rules, system optimization, and contingency planning for existing and planned water-management systems under a wider range of climatic conditions than traditionally used.

- Water agencies and providers should explore the vulnerability of both structural and nonstructural water systems to plausible future climate changes, not just past climatic variability.

- Governments at all levels should reevaluate legal, technical, and economic approaches for managing water resources in the light of possible climate changes.

- Cooperation of water agencies with the leading scientific organizations can facilitate the exchange of information on the state-of-the-art thinking about climatic change and impacts on water resources. Among the organizations and activities relevant to these issues are the World Climate Research Programme of the World Meteorological Organization, the Global Energy and Water Cycle Experiment (GEWEX) of the International Council of Scientific Unions (ICSU), the American Geophysical Union, the American Association for the Advancement of Science, the various National Academy of Sciences organizations in leading countries, and many specific research activities such as the work of CSIRO in Australia, the U.S. Global Change Office, and others. In the United States, agencies such as NASA, the National Oceanic and Atmospheric Administration (NOAA), the U.S. Geological Survey, the U.S. Army Corps of Engineers, and the U.S. Environmental Protection Agency often play an important international role in research on these issues.

- The timely flow of information from the scientific global change community to the public and the water-management community would be valuable. Such lines of communication need to be developed and expanded.

REFERENCES

American Water Works Association. 1997. "Climate change and water resources." Committee Report of the Public Advisory Forum. *Journal of the American Water Works Association,* Vol. 89, Issue 11, pp. 107–110.

Arnell, N. 1996. *Global Warming, River Flows and Water Resources.* John Wiley and Sons, Chichester, United Kingdom.

Binnie, C. 1996. "Climate change and its potential effect on water resources." *Water Hong Kong '96,* pp. 421–429.

Burn, D.H. 1994. Hydrologic effects of climatic change in west-central Canada." *Journal of Hydrology,* Vol. 160, pp. 53–70.

Chiew, F.H.S., and T.A. McMahon. 1993. "Detection of trend or change in annual flow of Australian rivers." *International Journal of Climatology,* Vol. 13, pp. 643–653.

Cole, J.A., S. Slade, P.D. Jones, and J.M. Gregory. 1991. "Reliable yield of reservoirs and possible effects on climatic change." *Hydrological Sciences Journal,* Vol. 36, pp. 579–597.

CSIRO. 1996. *Climate Change Scenarios for the Australian Region.* Climate Impact Group, CSIRO Division of Atmospheric Research, Aspendale, Australia.

Dai, A., I.Y. Fung, and A.D. Del Genio. 1997. "Surface observed global land precipitation variations during 1900–1988." *Journal of Climate,* Vol. 10, pp. 2943–2962.

Flaschka, I.M., C.W. Stockton, and W.R. Boggess. 1987. "Climatic variation and surface water resources in the Great Basin region." *Water Resources Bulletin,* Vol. 23, No. 1, pp. 47–57.

Gleick, P.H. 1986. "Methods for evaluating the regional hydrologic impacts of global climatic changes." *Journal of Hydrology,* Vol. 88, pp. 97–116.

Gleick, P.H. 1987a. "The development and testing of a water-balance model for climate impact assessment: Modeling the Sacramento Basin." *Water Resources Research,* Vol. 23, No. 6, pp. 1049–1061.

Gleick, P.H. 1987b. "Regional hydrologic consequences of increases in atmospheric carbon dioxide and other trace gases." *Climatic Change,* Vol. 10, No. 2, pp. 137–161.

Gleick, P.H. 1992. "Effects of climate change on shared freshwater resources." In I.M. Mintzer (ed.), *Confronting Climate Change: Risks, Implications and Responses.* Cambridge University Press, Cambridge, United Kingdom, pp. 127–140.

Global Change. 1997. *Global Change: A Review of Climate Change and Ozone Depletion.* Electronic version via *http://www.globalchange.org.*

Groisman, P.Ya., and D.R. Easterling. 1994. "Variability and trends of precipitation and snowfall over the United States and Canada." *Journal of Climate,* Vol. 7, pp. 184–205.

Groisman, P.Ya., T.R. Karl, R.W. Knight, and G.L. Stenchikov. 1994. "Changes in snow cover, temperature, and the radiative heat balance over the Northern Hemisphere." *Journal of Climate,* Vol. 7, pp. 1633–1656.

Grove, A.T. 1995. "African river discharge and lake levels in the twentieth century." In T.C. Johnson and E. Odada (eds.), *The Limnology, Climatology and Paleoclimatology of the East African Lakes.* Gordon and Breach, London.

IPCC. 1990. *Climate Change. The IPCC Scientific Assessment.* J.T. Houghton, G.J. Jenkins, and J.J. Ephrauns (eds.). Cambridge University Press, Cambridge, United Kingdom.

IPCC. 1996a. *Climate Change 1995: The Science of Climate Change. Summary for Policymakers.* Contribution of Working Group I to the Second Assessment Report of the Intergovernmental Panel on Climate Change. Cambridge University Press, Cambridge, United Kingdom.

IPCC. 1996b. "Hydrology and freshwater ecology." In *Climate Change 1995: Impacts, Adaptations, and Mitigation of Climate Change.* Contribution of Working Group II to the Second Assessment Report of the Intergovernmental Panel on Climate Change. Cambridge University Press, Cambridge, United Kingdom.

Jager, J., and H. Ferguson. 1991. *Proceedings of the Second World Climate Conference,* Geneva, Switzerland.

Karl, T.R., and R.W. Knight. 1998. "Secular trends of precipitation amount, frequency, and intensity in the USA." Submitted to *Bulletin American Meteorological Society,* 10/27/97.

Karl, T.R., R.W. Knight, and N. Plummer. 1995. "Trends in high-frequency climate variability in the twentieth century." *Nature,* Vol. 377, pp. 217–220.

Lettenmaier, D. P., E.F. Wood, and J.R. Wallis. 1994. "Hydro-climatological trends in the continental United States 1948–1988." *Journal of Climate,* Vol. 7, pp. 586–607.

Lins, H.F., and P.J. Michaels. 1994. "Increased US streamflow linked to greenhouse forcing." *EOS,* Vol. 75, pp. 281–285.

Mann, M.E., R.S. Bradley, and M.K. Hughes. 1998. "Global-scale temperature patterns and climate forcing over the past six centuries." *Nature,* Vol. 393, No. 6678, pp. 779–800.

Marengo, J. 1995. "Variations and change in South American streamflow." *Climatic Change,* Vol. 31, pp. 99–117.

McMahon, T.A., R.J. Nathan, B.L. Finlayson, and A.T. Haines. 1989. "Reservoir system performance and climatic change." In G.C. Dandy and A.R. Simpson (eds.), *Proceedings of the National Workshop on Planning and Management of Water Resource Systems: Risk and Reliability.* Australian Government Publishing Service, Canberra, Australia, pp. 106–124.

Miller, B.A., V. Alavian, and M.D. Bender. 1993. "Impacts of changes in air and water temperature on thermal power generation." In R. Herrmann (ed.), *Managing Water Resources During Global Change.* Proceedings of the International Symposium of the American Water Resources Association, Bethesda, Maryland, pp. 439–448.

Mimikou, M., P.S. Hadjisavva, Y.S. Kouvopoulos, and H. Afrateos. 1991. "Regional climate change impacts: II. Impacts on water management works." *Hydrological Sciences Journal,* Vol. 36, pp. 259–270.

Mintzer, I.M. 1992. "Living in a warming world." In I.M. Minzter (ed.), *Confronting Climate Change: Risks, Implications and Responses.* Cambridge University Press, Cambridge, United Kingdom, pp. 1–14.

Nash, L.L., and P.H. Gleick. 1991. "The sensitivity of streamflow in the Colorado Basin to climatic changes." *Journal of Hydrology,* Vol. 125, pp. 221–241.

Nash, L.L., and P. H. Gleick. 1993. *The Colorado River Basin and Climatic Change: The Sensitivity of Streamflow and Water Supply to Variations in Temperature and Precipitation.* U.S. Environmental Protection Agency, EPA230-R-93-009, Washington, D.C., 121 pp.

Nemec, J., and J.C. Schaake. 1982. "Sensitivity of water resource systems to climate variation." *Hydrological Sciences Journal,* Vol. 27, pp. 327–343.

Nicholls, N., G.V. Gruza, J. Jouzel, T.R. Karl, L.A. Ogallo, and D.E. Parker. 1996. "Observed climate variability and change." *Climate Change 1995: The Science of Climate Change.* The Intergovernmental Panel on Climate Change, Working Group I Assessment. Cambridge University Press, pp. 133–192.

Nicholson, S.E. 1994a. "A review of climate dynamics and climate variability in eastern Africa." In T.C. Johnson and E. Odada (eds.), *The Limnology, Climatology and Paleoclimatology of the East African Lakes.* Gordon and Breach, London.

Nicholson, S.E. 1994b. "Recent rainfall fluctuations in Africa and their relationship to past conditions." *Holocene,* Vol. 4, pp. 121–131.

Nicholson, S.E. 1995. Variability of African rainfall on interannual and decadal time scales." In D. Martinson, K. Bryan, M. Ghil, T. Karl, E. Sarachik, S. Sorooshian, and L. Talley (eds.), *Natural Climate Variability on Decade-to-Century Time Scales.* United States National Academy Press, Washington, D.C.

Noda, A., and T. Tokioka. 1989. "The effect of doubling the CO_2 concentration on convective and non-convective precipitation in a general circulation model with a simple mixed layer ocean." *Journal of the Meteorological Society of Japan,* Vol. 67, pp. 1055–1067.

Office of Science and Technology Policy (OSTP). 1997. *Climate Change: State of Knowledge.* Executive Office of the President, Washington, D.C.

Rind, D., R. Goldberg, J. Hansen, C. Rosenzweig, and R. Ruedy. 1990. "Potential evapotranspiration and the likelihood of future drought." *Journal of Geophysical Review.* Vol. 95, pp. 9983–10004.

Robinson, D.A., K.F. Dewey, and R.R. Heim Jr. 1993. "Global snow cover monitoring: An update." *Bulletin of the American Meteorological Society,* Vol. 74, pp. 1689–1696.

Schaake, J. 1990. "From climate to flow," in P.E. Waggoner (ed.), *Climate Change and U.S. Water Resources.* John Wiley and Sons, Inc., New York, pp. 177–206.

Schneider, S.H., P.H. Gleick, and L.O. Mearns. 1990. "Prospects for climate change," in P.E. Waggoner (ed.), *Climate Change and U.S. Water Resources.* John Wiley and Sons, Inc. New York, pp. 41–73.

Stockton, C.W., and W.R. Boggess. 1979. "Geohydrological implications of climate change on water resource development." Report of the U.S. Army Coastal Engineering Research Center, Fort Belvoir, Virginia.

Trenberth, K., and T. Hoar. 1996. "The 1990–1995 El Niño-Southern Oscillation event: The longest on record." *Geophysical Research Letters,* Vol. 23, No. 1, pp. 57–60.

Vinnikov, K.Ya., A. Robock, N.A. Speranskaya, and C.A. Schlosser. 1996. "Scales of temporal and spatial variability of midlatitude soil moisture." *Journal of Geophysical Research,* Vol. 101, pp. 7163–7174.

Waggoner, P. (ed.) 1990. *Climate Change and U.S. Water Resources.* John Wiley and Sons, Inc., New York.

SIX

New Water Laws,
New Water Institutions

No water problems are purely hydrological. Fundamental to all questions of water quality, availability, allocation, and use is the question of what kinds of rules and institutions can best serve our needs. There are hundreds of organizations, legal rules, and principles at the international level that affect water planning, management, and development. There are even more such organizations and laws at national and regional levels. Given the failure of the world community to solve water problems by the end of the twentieth century, it is therefore natural—and healthy—to begin to question the effectiveness of these tools and to seek to improve the way they are designed, structured, and operated. But institutional change is difficult and slow.

In this light, the late 1990s have seen an unusual amount of institutional reevaluation and reform. There has been a move away from building or funding new large infrastructures. Many water providers are placing a growing emphasis on increasing the productivity of water use rather than finding new sources of supply. Two new international water organizations have been created, along with a commission to review concerns over large dams. Several countries are substantially rewriting their water laws. And broad legal principles for shared international watercourses have achieved acceptance by the international community.

This chapter reviews four major legal and institutional changes that have occurred in the past two years. In 1996 and 1997, South Africa promulgated new and comprehensive changes in water law and policy. These changes reflect the desire of the new government there to meet basic water needs for all citizens, to reallocate the inequitable water rights system put in place under apartheid, and to involve communities and local groups formerly excluded from water-policy decision making. During this same period, three new organizations and activities began at the international level: the Global Water Partnership, the World Water Council, and the World Commission on Dams. These groups are trying to find their place among existing water institutions and hope to address serious water problems. Each one, however, faces serious challenges in finding financing, developing an efficient organizational structure, and defining a clear role; it remains to be seen how successful they will be in accomplishing their professed missions.

Water Law and Policy in the New South Africa: A Move Toward Equity

> For us water is a basic human right, water is the origin of all things—the giver of life.
>
> —Antjie Krog

Water has long been a concern and focus of controversy in southern Africa, and particularly South Africa. The waters of the region are widely shared, and growing populations and economic development are putting increasing pressure on these resources. Even worse, there has been a long history of inequitable, undemocratic, and discriminatory resource allocation and use. Recently, however, remarkable changes have begun to alter the entire nature of national and international water policy and law in the region. As a result, South Africa is now taking the opportunity to become a world leader in rational water management and use. New attitudes toward international cooperation over shared rivers could successfully reduce the risks of water-related conflicts and promote broader regional cooperation. Efforts to manage domestic water resources to meet basic human needs for water, long ignored in the region, could reduce the gross inequities in water use. And an exciting debate over new water law and policy is redefining appropriate water rights and uses, with implications for people everywhere.

The Hydrology of Southern Africa

Southern Africa is a diverse region encompassing the 11 countries of Angola, Botswana, Lesotho, Malawi, Mozambique, Namibia, South Africa, Swaziland, Tanzania, Zambia, and Zimbabwe—all members of the Southern Africa Development Community (SADC). The region has an extremely variable climate and hydrologic regime: 22 percent of the area is arid and receives less than 400 millimeters of rain per year. Another 35 percent is considered semi-arid or dry "subhumid." South Africa itself is a good example of the hydrologic variability of the region—43 percent of the country's rain falls on 13 percent of its land with high annual variability. More than 60 percent of South Africa's land area receives less rain than dryland farming requires (Abrams 1997). Precipitation originates largely from the Indian Ocean and is highly seasonal, with most areas experiencing a five- to seven-month wet season during the October–April summer (Conley 1996). The variability of interannual precipitation is also high, resulting in unpredictable and often severe droughts. Droughts during the 1980s and 1990s have been particularly severe, leading to renewed interest in regional water management and planning.

The entire region depends largely on rainfall and river runoff for water supply, and every major perennial river in the region is shared by two or more nations. Tables 6.1 and 6.2 list the international rivers of the states of the SADC region. Map 6.1 shows the main river basins in the region. The Congo (formerly referred to by some as the Zaire) River dominates all other rivers on the continent, with nearly 30 percent of the total river flow of Africa. Despite the size of the Congo, its long distance from the demand centers in the south and the high cost of moving water make it unlikely that it will play much of a role in future water-supply considerations in the region.

TABLE 6.1   INTERNATIONAL RIVERS OF THE SADC REGION

Basin State	Number of International River Basins	River Basins
Angola	5	Cunene, Cuvelai, Okavango, Congo (Zaire), Zambezi
Botswana	5	Limpopo, Nata, Okavango, Orange, Zambezi
Lesotho	1	Orange
Malawi	2	Rovuma, Zambezi
Mozambique	9	Buzi, Incomati, Limpopo, Rovuma, Save, Maputo, Pungue, Umbeluzi, Zambezi
Namibia	5	Cunene, Cuvelai, Okavango, Orange, Zambezi
South Africa	4	Incomati, Limpopo, Maputo, Orange
Swaziland	3	Incomati, Maputo, Umbeluzi
Tanzania[a]	2	Rovuma, Zambezi
Zambia	2	Congo (Zaire), Zambezi
Zimbabwe	6	Buzi, Limpopo, Nata, Pungue, Save, Zambezi

Source: Ohlsson 1995.

[a]Part of Tanzania is in the Nile River basin, but the Nile is not considered a SADC basin.

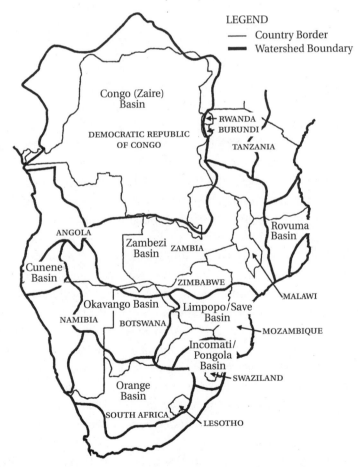

MAP 6.1   MAIN RIVER BASINS IN SOUTHERN AFRICA

TABLE 6.2   RIVER BASINS AND SADC BASIN STATES

River Basin	Basin States	Basin Area (km²)[a]
Buzi	Mozambique, Zimbabwe	30,000
Cunene	Angola, Namibia	110,000
Cuvelai	Angola, Namibia	125,000
Incomati	Mozambique, South Africa, Swaziland	54,000
Limpopo	Botswana, Mozambique, South Africa, Zimbabwe	385,000
Maputo	Mozambique, South Africa, Swaziland	34,000
Nata	Botswana, Zimbabwe	n.d.
Okavango	Angola, Botswana, Namibia	585,000
Orange	Botswana, Lesotho, Namibia, South Africa	950,000
Pungue	Mozambique, Zimbabwe	32,000
Rovuma	Malawi, Mozambique, Tanzania	167,000
Save	Mozambique, Zimbabwe	100,000
Umbeluzi	Mozambique, Zimbabwe	5,500
Zambezi	Angola, Botswana, Malawi, Mozambique, Namibia, Tanzania, Zambia, Zimbabwe	1,420,000
Congo (Zaire)[b]	Angola, Cameroon, Central African Republic, Congo, Burundi, Rwanda, Tanzania, Zaire, Zambia	3,800,000

Sources: UN 1978; Ohlsson 1995; Heyns 1995; Conley 1996.

[a] There is little agreement in the literature about actual river basin areas. These data come from several sources and are rounded off. New estimates of river basin areas are likely to become available in the next two years as new watershed maps are prepared from recently released digital elevation maps.

[b] The Congo (Zaire) is rarely considered relevant to SADC water discussions, given its distance from the major demand centers in the region.

South Africa has long dominated southern Africa politically, militarily, and economically. South Africa has a population of about 43 million and strong ecological and socio-economic diversity. It is one of the most unequal countries in terms of income distribution, with over 30 percent of the population unemployed. Until recently, official governmental policies—apartheid—supported the exclusion and marginalization of the black majority population. The water sector consequences of apartheid-era policies have been far reaching. Under apartheid, official state practice and law allocated water rights to those who owned land, automatically excluding the majority black populations of the country. As a result, when apartheid finally ended in the early part of this decade, between 10 and 20 million South Africans were without access to the most basic clean drinking water supplies and sanitation services, despite a sophisticated and well-developed water infrastructure. Even now thousands of children die annually from avoidable water-related diseases (Ohlsson 1995; SADWAF 1997a; Glazewski 1996; Abrams 1996, 1997). Lack of access to adequate water intensifies the severe poverty in parts of the country. Poverty, in turn, affects water resources by contributing to desertification and deforestation, loss of topsoil, widespread nonpoint source pollution, invasion of water-hungry invasive plant species, and other land-management practices that reduce groundwater recharge rates, increase siltation behind dams, and increase the likelihood and consequences of serious flooding.

The apartheid-era water law in South Africa was written in 1956 according to the traditional water conditions and laws of the well-watered countries of Europe. During apartheid, development efforts went to projects that increased the economic power and production of the minority population, ignoring for the most part the basic water requirements of millions of black South Africans. A fundamental part of the law was to preserve water rights for those with access to land.

Equally egregious was the implicit and explicit use of water deprivation as a means of controlling the rural and increasingly disaffected populations. In 1990, for example, the government of South Africa cut off basic water supplies as a weapon in its war against the growing power of opponents of apartheid. The black township of Wesselton, with a population of 50,000, had its water shut off following protests over miserable sanitation and living conditions (Pinder 1990).

In 1994, the apartheid government was thrown out and democratic elections were held. These elections offered an opportunity for fundamental political changes and a thorough review of policy in all sectors, including water. It is rare that such an opportunity presents itself. The new government set as a top priority efforts to address unmet needs for water. This government soon realized that meeting those needs would require a fundamental reallocation of water rights and water access. This priority is reflected in the new South African Constitution and the preparation of a new National Water Policy. The theme of all of the new water activities in South Africa is the concept of "Some, For All, Forever," which means access to a limited resource ("some") will be provided on an equitable basis ("for all"), now and in the future ("forever") (SADWAF 1997a).

Water in the New South African Constitution and Bill of Rights

The first democratic Constitution of South Africa was adopted on April 27, 1994. In this constitution, executive and legislative powers are subject to norms laid out in a Bill of Rights, including "fundamental" environmental rights. Section 27 of the Bill of Rights explicitly guarantees all South Africans "the right to have access to sufficient food and water" (Constitution of South Africa, Bill of Rights, Section 27 (1)(b)). This provision promises every person water of adequate quality for hygiene, survival, and basic needs. The government is instructed to "take reasonable legislative and other measures within its available resources to achieve the progressive realisation" of these rights. Section 29 further states that "every person shall have the right to an environment which is not detrimental to his or her health or well-being" (Section 29).

The new South African Constitution is binding on all law, courts, government departments and organizations, the government, and all South Africans. It sets out terms of rights, privileges, and benefits, as well as duties and responsibilities (Glazewski 1996). Among the rights related to access and allocation of water are:

Right to Equality: The right to equality "requires equitable access by all South Africans to, and benefit from, the nation's water resources, and an end to discrimination with regard to access to water on the basis of race, class, or gender."

Right to Dignity and Life: As stated in the background to the new Constitution, "The failure of the apartheid government to ensure the provision of sanitation and water for basic human needs such as washing, cooking, and drinking, for growing crops, and for economic development impacted significantly on both the right to dignity and the right to life amongst the black majority." Access to water is necessary to satisfy this right.

Environmental Rights: The right to an environment that is "not harmful to their health or well-being," requires government to make sure that water pollution is prevented, that there is sufficient water to maintain the ecological integrity of the water system, and that water conservation and sustainable "justifiable economic and social development" are promoted. This right requires that environmental, economic, and development goals be integrated.

The Right of Access to Sufficient Water: This provision of the constitution guarantees a basic water requirement to every person for hygiene, survival, and basic needs. The government is instructed to "take reasonable legislative and other measures within its available resources to achieve the progressive realisation" of these rights. This right permits the reallocation of existing water uses and makes the state the public guardian of water resources.

With the adoption of the new constitution, there was a call for all public institutions and laws related to water to be reassessed (SAMWAF 1996). As part of this, Professor Kader Asmal, Minister of Water Affairs and Forestry, initiated the Water Law review process and a Water Supply and Sanitation Policy White Paper was published in November 1994. In a parallel process, a National Policy White Paper laid out government policies regarding water-resources management and use.

The White Paper on Water Supply and Sanitation was the starting point in the review of water policy in South Africa. The principles around which the new national policy was based are:

1. Decisions about water use should be driven by demands and needs for community development;

2. Basic water services should be considered a human right;

3. The guiding principle of future water development should be "some for all" rather than "more for some";

4. The regional allocation of resources should be more equitable;

5. The economic values of water should be identified and used to make decisions;

6. Users should pay for water services;

7. Development of water management and supply systems should be integrated; and

8. The integrity of South Africa's environment should be protected and maintained (SADWAF 1997a).

These policy guidelines became part of the structure for a complete overhaul of the national water law. The need to review the old Water Act (Act 54 of 1956) arose from concerns about its inability to adequately address the modern complexities of the water situation in a semi-arid region with a growing industrial economy. This was an explicit acknowledgment that the 1956 Water Act had proven unable to prevent—and indeed had encouraged—extraordinarily unfair and inequitable water rights allocations and distributions.

The legal and policy review process was designed to encourage the involvement of a much broader set of the population than had traditionally been included in water planning and management in the past. The first phase was a document called "You and Your Water Rights," produced by the Ministry of Water Affairs and Forestry in early 1995, translated into the six major languages of South Africa, and widely distributed (SADWAF 1995). The ministry then established a Water Law Review panel to draft a set of principles on which a new water law could be based. The panel met in early September 1995 and many more times in the next few months, before releasing a set of principles for public review and comment. These principles were published on April 17, 1996, as a discussion document called the "Blue Book" (SADWAF 1996). This book offered to the public the first view of the standards on which a new water law would be based. Box 6.1 presents these principles.

Workshops were held around the country between May and July 1996 to present the principles to local organizations and to get informal and formal feedback. Special effort was made to bring disadvantaged communities into the review process. Traditional interest groups from agriculture, mining, forestry, industry, and local and provincial governments were also involved.

More than 1,500 comments, amendments, and suggestions were received during this period. There was strong support for the concept of managing water in an integrated fashion, for recognizing the economic value of water, and for protecting the fundamental resource base. Among the most contentious issues raised were the questions of water rights and whether existing private rights could be revoked and reallocated, or even whether the concept of private water rights was appropriate (SAMWAF 1996).

The single most controversial of the draft principles, not surprisingly, proved to be the one discussing allocation of water rights (see Principle F.1. in Box 6.1). In early drafts of the "fundamental principles," existing water rights were to be quantified, registered, and protected except where they conflicted with the protection of the "Reserve" of water for basic human and ecosystem needs. Any loss of existing rights was to be compensated if they were taken to meet a public interest (SAMWAF 1996). Serious opposition to this principle surfaced during the public review process. Among other things, it was noted that existing rights to use water were associated with land ownership, which under apartheid had resulted in inequitable and harmful consequences. The vast majority of existing rights rested with mining companies, large farmers, and industries. The idea that these rights would continue to be protected, and even compensated for if changed, caused enormous ire and distress among members of the new government, urban and rural community groups, and environmental advocates. At the October 1996 Water Law Review Conference in East London, Minister Asmal announced that this principle would be stricken. Final "Fundamental Principles and Objectives for a New Water Law" were produced at the Water Law Review National Consultative Conference in East London in October 1996 and approved by the cabinet in November 1996.

Two separate water bills were produced in 1997—a Water Services Bill and the National Water Bill. The Water Services Bill addresses provision of water services by local governments. The National Water Bill addresses the responsibility of the national government to establish a framework for all national norms and standards and introduces several measures that will have far-reaching impacts, not only in South Africa but in any other

BOX 6.1

South African Water Law Principles

Principle A. The Water Cycle

Principle A.1. In a relatively arid country such as South Africa, it is necessary to recognise the unity of the water cycle and the interdependency of its elements, where evaporation, clouds and rainfall are linked to underground water, rivers, lakes, wetlands, estuaries and the sea.

Principle A.2. The variable, uneven and unpredictable distribution of water in the water cycle should be acknowledged.

Principle B. Legal Aspects of Water

Principle B.1. All water, wherever it occurs in the water cycle, is a resource common to all, the use of which should be subject to national control. All water should have a consistent status in law, irrespective of where it occurs.

Principle B.2. There shall be no ownership of water but only a right to its use.

Principle B.3. The location of the water resource in relation to land should not in itself confer preferential rights to usage.

Principle C. Water Resources Management Priorities

Principle C.1. The objective of managing the quantity, quality and reliability of the nation's water resources is to achieve optimum long term social and economic benefit for society from their use, recognising the water allocations may have to change over time.

Principle C.2. The water required to meet peoples' basic domestic needs should be reserved.

Principle C.3. The quantity, quality and reliability of water required to maintain the ecological functions on which humans depend should be reserved so that the human use of water does not individually or cumulatively compromise the long term sustainability of aquatic and associated ecosystems.

Principle C.4. The water required to meet peoples' basic domestic needs and the needs of the environment should be identified as "the Reserve" and should enjoy priority of use.

Principle C.5. International water resources, specifically shared river systems, should be managed in a manner that will optimise the benefits for all parties in a spirit of mutual cooperation. Allocations agreed for downstream countries should be respected.

Principle D. Water Resource Management Approaches

Principle D.1. The national government is the custodian of the nation's water resources, as an indivisible national asset, and has ultimate responsibility for, and authority over, water resource management, the equitable allocation and usage of water, the transfer of water between catchments and international water matters.

Principle D.2. The development, apportionment and management of water resources should be carried out using the criteria of public interest, sustainability, equity and efficiency of use in a manner which reflects the value of water to society while ensuring that basic domestic needs, the requirement of the environment and international obligations are met.

Principle D.3. In as far as it is physically possible, water resources should be developed, apportioned and managed in such a manner as to enable all user sectors to gain equitable access to the desired quantity, quality and reliability of water, using conservation and other measures to manage demand where this is required.

Principle D.4. Water quality and quantity are interdependent and should be managed in an integrated manner, which is consistent with broader environmental management approaches.

Prinicple D.5. Water quality management options should include the use of economic incentives and penalties to reduce pollution; and the possibility of irretrievable environmental degradation as a result of pollution should be prevented.

Principle D.6. Water resource development and supply activities should be managed in a manner which is consistent with broader environmental management approaches.

Principle D.7. Since many land uses have a significant impact upon the water cycle, the regulation of land use should, where appropriate, be used as an instrument to manage water resources.

Principle D.8. Rights to the use of water should be allocated in good time and in a manner which is clear, secure and predictable in respect of the assurance of availability, extent and duration of use. The purpose for which the water may be used should not be arbitrarily restricted.

Principle D.9. The conditions subject to which water rights are allocated should take into consideration the investment made by the user in developing infrastructure to be able to use the water.

Principle D.10. The development and management of water resources should be carried out in a manner which limits to an acceptable level the danger to life and property due to natural or man-made disasters.

Principle E. Water Institutions

Principle E.1. The institutional framework for water management should, as far as possible, be simple, pragmatic and understandable. It should be self-driven, minimise the necessity for state intervention, and should provide for a right of appeal to or review by an independent tribunal in respect of any disputed decision made under the water law.

Principle E.2. Responsibility for the development, apportionment and management of available water resources should, where possible, be delegated to a catchment or regional level in such a manner as to enable interested parties to participate and reach consensus.

Principle E.3. Beneficiaries of the water management system should contribute to the cost of its establishment and maintenance.

Principle F. Existing Water Rights [Note: This draft provision was eliminated in late 1996.]

Principle F.1. Lawful existing water rights should be protected, subject to the public interest requirement to provide for the Reserve. Where existing rights are reduced or taken away, compensation should be paid wherever such compensation is necessary to strike an equitable balance between the interests of the affected person and the public interest. An existing right should not include a right which remains unquantified and unexercised at the time of the first publication of these principles.

Principle G. Water Services

Principle G.1. The right of all citizens to have access to basic water services (the provision of potable water supply and the removal and disposal of human excreta and waste water) necessary to afford them a healthy environment on an equitable and economically and environmentally sustainable basis should be supported.

Principle G.2. While the provision of water services is an activity distinct from the development and management of water resources, water services should be provided in a manner consistent with the goals of water resource management.

Principle G.3. Where water services are provided in a monopoly situation, the interests of the individual consumer and the wider public must be protected and the broad goals of public policy promoted.

Source: SADWAF 1996.

country considering redesign of national water policy. Among the recommendations elaborated in the White Paper are:

- The national government will act as custodian of the nation's water with its power to be exercised as a public trust.

- All water in the hydrologic cycle will be included in management.

- The former riparian system of allocation will be abolished.

- Water use allocations will no longer be permanent; they will be made tradable with government consent.

- Only that water required to meet basic human needs and maintain environmental sustainability will be guaranteed as a right: the "Reserve."

- In internationally shared basins, the government will ensure that legitimate requirements of neighboring countries can be met.

- Other water uses will be recognized if they are beneficial to the public interest and promote optimal use.

- In most cases, full financial costs will be charged users. Additional charges for "resource conservation" and environmental costs will also be levied. Subsidies for disadvantaged groups may be offered. Charges will also be waived for water to meet basic human needs.

The Fourth Draft of the new South African National Water Law was released on September 5, 1997, for public comment. This bill was prepared by a drafting team appointed by Minister Asmal, working under the guidance of the South African Water Law Review Policy and Strategy Team of the Department of Water Affairs and Forestry. In January 1998, the cabinet approved the act in preparation for taking it to the South African Parliament for final approval. How the principles will be codified into law remains to be seen, but much of the water world is eagerly watching.

One final aspect of South Africa's embrace of a new approach also deserves mention. Beginning around the same time as the review of South African water laws and policies, a small number of water experts, community activists, and forward-thinking individuals began aggressive efforts to develop and implement water conservation and demand management programs. These efforts have evolved into a sophisticated and successful set of activities under the umbrella of the National Water Conservation Campaign, whose patron is President Mandela with the strong support of Minister Asmal. The National Water Conservation Campaign is chaired by Tami Sokutu and managed by Dr. Guy Preston. The official goal of the campaign is to increase the efficiency of water use in South Africa in a manner that increases equitable allocation and reduces conflicts over scarce resources, but an additional benefit is showing the real advantages of water conservation activities for local communities, the natural environment, and overall water management in the nation.

Two major activities of the national campaign are the Working for Water Programme and the Greater Hermanus Water Conservation Programme. The Working for Water Programme combines water conservation activities with community job creation, occupational training, and the creation of secondary industries. A major focus of the program has been efforts to clear invasive plants from vital watersheds. By the end of 1997, nearly 40 separate watershed projects were under way. These projects train workers to identify and clear nonnative plants that are infesting streams and rivers and reducing water flows, replant native vegetation, clear debris that contributes to flooding, generate building materials and charcoal for local communities, and restore aquatic ecosystems. Working for Water began in 1995–1996 with a budget of R25 million (Rand) and in two years has grown to a budget of R265 million, employing 40,000 people, over 60 percent of whom are women (G. Preston, personal communication, 1998). This project has been so successful that the Worldwide Fund for Nature gave the Working for Water Programme its award for the "Best Conservation Project in South Africa in 1996" (SADWAF 1997b).

The other major project under way is the Greater Hermanus Water Conservation Programme. This effort is an ambitious long-term water conservation initative in an urban area. Hermanus is a town with a permanent population of about 19,000 and a holiday population exceeding 60,000. Water-use levels vary enormously from season to season and the region has a limited water supply. Under the conservation program, 12 separate activities are under way to reduce water use by 30 percent within three years, paid for entirely by revenue generated by water sales. Among the actions being implemented are increasing block rates with an 11-block escalating tariff, school water audits and educational activities, inspections for unmetered or illegal connections, retrofit of water fixtures in almost all homes and buildings to highly efficient devices, demonstration gardens showing low-water plants and efficient irrigation techniques, prepayment meters in some locations, and a variety of communications and education programs to teach people about conservation possibilities. In just the first four months of the project, a 32 percent savings in water was achieved, which exceeded the three-year goals. Water revenues increased 60 percent during this period and this money is to be reinvested in the local community conservation programs. Ongoing efforts will work to ensure that the water savings are real and permanent (SANWCC 1997; G. Preston, personal communication, 1998).

The Global Water Partnership

In August 1995, a new partnership to address international water problems was proposed at the Stockholm Water Symposium by the World Bank, the United Nations Development Programme (UNDP), and international water experts. The World Bank and the UNDP subsequently organized a meeting hosted by the Swedish International Development Agency (SIDA) in December 1995 in Stockholm to discuss ways to set up a new organization to help coordinate their overlapping and sometimes conflicting water programs and to ensure that limited financial resources are used efficiently. Seventy-five representatives from more than 50 water agencies and governments around the world discussed the slow progress being made in achieving the goals expressed at the Mar del Plata water conference in 1977, the Dublin water conference, and the Earth Summit in Rio in 1992. Several more meetings were held early in 1996 to identify needs and to define a mission. Out of these discussions, the Global Water Partnership (GWP) was formally created. The GWP was officially inaugurated at the Stockholm Water Symposium in August 1996.

The GWP describes itself as an action-oriented network of organizations interested in water issues with a mission to transform the "Dublin Principles" into practical tools for solving water problems at the local and regional level. The official objectives of the GWP are to:

- support integrated water-resources management programs through collaboration with governments and existing networks and by creating new collaborative arrangements;

- encourage governments, aid agencies, and other stakeholders to adopt consistent, mutually complementary policies and programs;

- build mechanisms for sharing information and experiences;

- develop innovative and effective solutions to problems common to integrated water-resources management;

- suggest practical policies and good practices based on those solutions; and

- help match needs to available resources (GWP 1997a).

The GWP is open to all parties involved in water-resources management, including governments, UN agencies, multilateral banks, professional associations, research organizations, the private sector, and NGOs. Its structure is still developing. By early 1998 it had four components: (1) a Consultative Group (CG), which is the organization's policy-making body, chaired by Dr. Ismail Serageldin, Vice President of the World Bank; (2) a Technical Advisory Committee (TAC), consisting of 12 water professionals and scientists, chaired by Dr. Torkil Jonch-Clausen of VKI Water Quality Institute, Denmark; (3) a Steering Committee; and (4) the GWP Secretariat, based in Stockholm and headed by Johan Holmberg. Regional TACs are in the process of being set up. Table 6.3 lists the members of the GWP TAC and Steering Committee as of early 1998. Box 6.2 shows the major GWP activities that have taken place since its inception.

The GWP hopes to translate recommendations for actions on water management into specific services for developing countries and to advance funding mechanisms to permit those services to be implemented. To achieve these goals, the GWP has been organizing regional meetings and organizational workshops and will establish specific areas of focus for future work. The first two, defined in 1996, are water for household use and water for food production. Two more were chosen in 1997—water for ecosystems and the environment, and integrated water resources management (*Newsflow* 1997). Other program areas may include water for hydropower, industrial use, and navigation. These subsector programs are intended to stimulate investment in their specific fields. In spring 1998, an analysis of the global water sector was finished so that GWP can present a framework for action useful for the activities of the United Nations and the Commission on Sustainable Development, which has chosen to take on the water issue during the 1998 session.

GWP is not a donor agency. It is a network of members sharing a vision of water management and development. Hence the GWP "represents its members' interests (e.g., government water departments, NGOs, academic institutions), and on an equal basis those international organizations supporting the Partnership financially." GWP activities are to be "based on sharing and developing knowledge and experience in line with the Dublin–Rio principles and intended to support national, regional and international cooperation and coordination of activities" (GWP 1997a). The Dublin Principles are shown in Box 6.3.

TABLE 6.3   COMPOSITION OF THE TECHNICAL ADVISORY COMMITTEE AND THE STEERING COMMITTEE OF THE GLOBAL WATER PARTNERSHIP (AS OF JANUARY 1998)

Technical Advisory Committee

Ramesh Batia, Resources and Environment Group, India

Ivan Chéret, Lyonnaise des Eaux, France

Malin Falkenmark, Natural Science Research Council, Sweden

Torkil Jønch-Clausen, VKI Water Quality Institute, Denmark

Mohammed Aït Kadi, Ministère de l'Agriculture et de la Mise en Valeur Agricole, Morocco

Janusz Kindler, Warsaw University of Technology, Poland

Judith Rees, London School of Economics, United Kingdom

Romana P. de los Reyes, Ateneo de Manila University, Philippines

Peter Rogers, Harvard University, United States

Miguel Solanes, Economic Commission for Latin America and the Caribbean, Chile

Fernando Gonzalez Villarreal, Universidad Nacional Autonomade México, Mexico

Albert Wright, The World Bank Resident Mission, Ghana

Steering Committee of the Global Water Partnership

Piet Heyns, Department of Water Affairs, Namibia

Guowei Yang, Water Resources Commission, China

Jerson Kelman, Rio de Janeiro State Agency for Rivers and Lakes, Brazil

Athanase O. Compaoré, Ministère de l'Environnement et de l'Eau, Burkina Faso

Armon Hartmann, Swiss Development Cooperation Agency, Switzerland

John Hodges, Overseas Development Administration, United Kingdom

Hans Wolter, FAO, Rome

Pierre Icard, Ministère de la Cooperation, France

Andras Szöllösi-Nagy, UNESCO, France

Aly Shady, CIDA, Canada

N'Dri Koffi, Union Africaine des Distributeurs d'Eau, Cote d'Ivoire

Jon Lane, WaterAid, United Kingdom

George Varughese, Development Alternatives, India

Mônica Porto, Associação Brasileira de Recursos Hídricos, Brasil

David Seckler, IIMI, Sri Lanka

Jacob Kijne, Rose Cottage, United Kingdom

Sue Milner, Natural Resources Institute, United Kingdom

Kusum Athukorala, Development Research Consultants, Sri Lanka

René Coulomb, Lyonnaise des Eaux, France

Enrique Aguilar, Aguiliar & Associates, Mexico

Kristina Ringwood, World Business Council for Sustainable Development, Switzerland

John Briscoe, The World Bank

Roberto Lenton, UNDP

Johan Holmberg, Executive Secretary GWP, Sweden

BOX 6.2

Major Global Water Partnership Activities, to March 1998

Activity	Date	Activity	Date
Initial idea broached, Stockholm	August 1995	Technical Advisory Committee meeting, Manila	June 8–12, 1997
Organizational meeting, Stockholm	December 1995	GWP hosts workshop on Women and Gender	August 12, 1997
Interim committee meeting, Stockholm	February 23, 1996	Technical Advisory Committee meeting, Stockholm	August 13, 1997
Interim committee meeting, Washington	May 9–10, 1996		
Interim Technical Advisory Committee meeting, Copenhagen	June 10–11, 1996	Second Annual GWP Consultative Group Meeting, Stockholm	August 14–15, 1997
GWP Inauguration, Stockholm Water Symposium, Stockholm	August 9, 1996	Preparatory TAC meetings in Vitria, Brazil, for November TAC meeting	October 13–15, 1997
Interim Technical Advisory Committee meeting, Windhoek	November 4–8, 1996	Technical Advisory Committee meeting, Vitria, Brazil	November 14–19, 1997
Consultative Group and Secretariat Meeting with SADC Water, Maseru	January 6–7, 1997	Global Water Sector Steering Committee Meeting, Rome	December 8–9, 1997
Technical Advisory Committee Meeting, Rome	Late February 1997	GWP Meeting with Government of India	December 11–12, 1997
Semi-Annual Consultative Group Meeting, Marrakech	March 20, 1997	Semi-Annual Consultative Group Meeting, Marseilles	March 17–18, 1998
Participation in SADC–EU Conference on Management of Shared River Basins, Maseru	May 20–21, 1997		

The Technical Advisory Committee

The principal operating arm of the GWP is its Technical Advisory Committee. A goal of the TAC is to elaborate a comprehensive framework for the GWP, develop a set of criteria for sustainable water management, and identify priorities for action. The first TAC was appointed on an interim basis through 1997. Members are listed in Table 6.3.

The TAC held an initial meeting in Copenhagen in June 1996 and convened its first regional meeting for the area of southern Africa, in Windhoek, Namibia, November 4–8, 1996. In Windhoek, the TAC reviewed irrigation and drainage issues with a view to defin-

---

BOX 6.3

The Dublin Principles

In January 1992, some 500 representatives from 100 countries and 80 international and nongovernmental organizations met in Dublin, Ireland, to prepare for the Earth Summit in Rio de Janeiro in June 1992. At the closing session of the Dublin Conference (the "International Conference on Water and the Environment" (ICWE)), the participants adopted the "Dublin Statement." That statement offered specific recommendations and activities based on four guiding "principles"—now called the Dublin Principles. The Dublin Principles state that:

For the full text of the Dublin Statement, see *http://www.wmo.ch/web/homs/icwedece.html.*

- Fresh water is a finite and vulnerable resource, essential to sustain life, development and the environment.
- Water development and management should be based on a participatory approach, involving users, planners and policy-makers at all levels.
- Women play a central part in the provision, management and safeguarding of water.
- Water has an economic value in all its competing uses and should be recognized as an economic good.

---

ing activities to be undertaken. Follow-up meetings were held in Maseru, Lesotho, in early January and Rome in late February with senior representatives of most of the major institutions involved with the GWP. At these meetings, it was decided that the TAC would initiate a comprehensive study of irrigation and drainage issues with a view to identifying the major gaps and defining appropriate activities for the GWP. Since those first two meetings, work in southern Africa has continued to mobilize donor support for activities designed to strengthen integrated water-resources management within the region. Nominations for a separate TAC for southern Africa ("SATAC") have been received and GWP participated actively in the SADC–EU conference, "Management of Shared River Basins," organized by the government of the Netherlands in Maseru on May 20–21, 1997 (GWP 1996, 1997c).

On June 8–16, 1997, the TAC held a two-part meeting in Manila, Philippines, hosted by the Asian Development Bank (ADB). Ten of the 12 TAC members, participants from ADB, and individuals from other regional and international organizations attended. The initial TAC meeting discussed the conceptual aspects of sustainable water-resources management, women and water, integrated water-resources management, and water for food security. The subsequent regional meeting, partly organized by the GWP, was attended by over 95 representatives of the Association of Southeast Asian Nations (ASEAN) countries, donor agencies, NGOs, and the private sector (GWP 1997a). The regional meeting focused on three topics: water management at the national level; water management at the river-basin level; and management approaches for water conservation and savings. Activities to be taken up by GWP in cooperation with ADB in Southeast Asia were also discussed.

Nine priority actions were recommended at the Manila meeting (GWP 1997b). These recommendations and commitments are, in abbreviated form:

1. Move toward a regional water partnership by establishing a GWP Regional Technical Advisory Committee (now called the Southeast Asian TAC or SEATAC) with ongoing secretariat support.

2. Facilitate awareness of water issues by convening a Southeast Asian caucus of countries at the GWP Consultative Group Meeting in Stockholm, August 1997.

3. Collect and disseminate data and information, including establishing an Internet-based information service in the water sector.

4. Develop knowledge, methodologies, and guidelines and define water sector needs in countries of the region by compiling national profiles of the water sector.

5. Collect and share experience among countries through comparative analysis, based on a set of regional and global case studies. These studies should initially focus on policy, practice, and experience in integrated water-resource management.

6. Build capacity, develop guidance material on key issues, and transfer knowledge/technology through workshops, seminars, and other means.

7. Mount a major regional conference on institution-building in the water sector to create and raise "water awareness," disseminate the findings of the comparative analyses, and provide guidance and impetus for national actions.

8. Convene a Ministerial Conference on effective water policy and investment.

9. Establish an annual consultation meeting among groups active in the water sector.

During the Stockholm Water Symposium in August 1997, the GWP's Technical Advisory Committee, Consultative Group, and Steering Committee all met. The TAC had a two-day meeting (August 12–13) to discuss its global agenda and plan work for the coming year. The Consultative Group met on August 14 and 15 to discuss GWP governance, the analytical framework of the water sector proposed by TAC, and the new possibilities offered by the Water Forum (see below) to exchange information electronically.

In November 1997, the TAC convened a meeting in Vitoria, Brazil (November 15–16), in partnership with the Brazilian Association of Water Resources (Associação Brasileira de Recursos Hídricos ABRH) and the Inter-American Development Bank (IDB). Main issues considered were water for large cities and public–private partnerships for water management. Additional meetings of the TAC will be held during 1998.

The Secretariat and the Water Forum

The GWP Secretariat is currently located at the Swedish International Development Agency (SIDA) in Stockholm with a full-time staff of six professionals. The Secretariat provides support to the TAC, administers the GWP governance structure, maintains contacts with organizations and individuals interested in the GWP, and organizes meetings. Updated information about the GWP and its activities comes from a newsletter (*News-Flow*) and two Internet homepages (*http://www.gwp.sida.se; http://www.gwpforum.org*) maintained by the Secretariat. The first of these web sites is devoted to general information about the GWP, its organization, and its activities. The second site is home of the GWP "Water Forum" and offers more comprehensive connections with the world water community. The forum offers a tool for discussion and exchanging information. Within the forum, users may set up information stands or "kiosks" with links to other Internet water resources. It is also designed to "enable full participation by those in developing countries and those with limited web capabilities" (*Newsflow* 1997). The site provides access to web links and enables the user to search and sort through them for information on water issues.

GWP Finance

The stated goal of the GWP is that it should be self-financing and sustained, at least principally, by regular contributions from its members. This idea was first broached at the Stockholm Water Symposium in August 1996. A decision on such contributions was deferred to the August 1997 CG meeting pending discussion of possible GWP services to member organizations and then was deferred again. As of early 1998, membership in GWP was free, though it intends at some time to implement a membership fee, with a fee structure related to organizational ability to pay. The GWP has made it clear that it will seek financial contributions from members when a set of services can be offered on a sustained basis. Until such contributions make up a large part of the GWP budget, it seems likely that major support will have to continue to come from the international aid organizations involved in its formation. According to the GWP, 1996 expenses were about US$657,000. Contributions in 1996 came from:

- Sweden (SIDA), US$391,300
- The World Bank, US$250,000
- Denmark, US$300,000
- UNDP, US$50,000
- Additional contributions by the UNDP/World Bank Water and Sanitation Programme and by the Swedish Ministry for Foreign Affairs (altogether US$65,400).

By mid-1997, the GWP received the following contributions from aid donors (GWP 1997a):

- The World Bank, US$1 million
- Denmark, US$300,000
- Switzerland, US$250,000
- Sweden (SIDA), SEK 3,000,000 (about US$400,000)
- France, one officer to the Secretariat

Issues and Problems

There remains a considerable interest in the ideas underlying GWP. Nevertheless, several issues and problems need to be addressed. In particular, the GWP is still developing and its basic concepts and structure are subject to question and debate. There is uncertainty and doubt in the world water community about how practical change or improvements in water problems will be helped by the GWP and about the legitimacy of the new organization (Holmberg 1997). This doubt has been fueled by the focus so far on expert meetings, talks, the creation of new regional committees, and the lack of concrete action. GWP is attempting to coordinate aid donors and policymakers in the area of water-resources development—a notion that is intrinsically positive and yet difficult to implement. GWP intends to work to try to improve coordination of the use of financial and intellectual resources, but it needs to develop clear and focused priorities. While GWP insists that it is not an aid organization, per se, many countries are looking to the GWP to either facilitate aid decisions or help international aid organizations better define new avenues and priorities for giving. This has not yet occurred, though the pending reports from the TAC may help resolve some concerns.

The World Water Council

The World Water Council (WWC) is a nonprofit independent forum to promote awareness on critical global water issues. Discussion about the need for a WWC began in the late 1980s as water-resource problems became increasingly important on the international environmental agenda. The concept of a formal organization grew out of ideas put forth at large international meetings as well as during private informal conversations. In 1990, senior members of the International Water Resources Association (IWRA) began more seriously considering the creation of a new organization. The idea of a World Water Council evolved and was refined at international meetings in New Delhi, Dublin, Rio, Noordwijk, Cairo, Montreal, and elsewhere.

In 1992, the Canadian delegation to the UN International Conference on Water in Dublin formally proposed the creation of a world water council. The Dublin Statement included the following recommendation: ". . . to involve private institutions, regional and non-governmental organizations along with all interested governments in the assessment and follow-up [to the Dublin Statement], the Conference proposes for consideration by UNCED, a world wide forum or council to which all such groups could adhere " (WMO 1997). Following the Dublin Statement (see Box 6.3), and in response to recommendations of the Ministerial Conference on water held in Noordwijk, the 1994 VIIIth IWRA Congress in Cairo charged a committee to carry out the preparatory work to create a WWC. This committee met in Montreal, Canada, and Bari, Italy, and developed a conceptual framework, a draft constitution, and the institutional structure necessary for formal operations. The site of the first meeting, Montreal, also hosted the interim secretariat for a year and a half. After a competition, the city of Marseilles successfully offered to serve as the host for the organization, and the council was formally incorporated in France in June 1996. Nine organizations are considered "founding members" because of their role in the formation of the WWC (see Table 6.4). In July 1996, in Granada, the interim Board of Governors debated how the World Water Council could meet its objectives in promoting awareness of global water issues and facilitating conservation, protection, development, planning, and management of water resources. Some specific suggestions were also put forward concerning a technical program for 1996 and 1997. In August 1996, Dr. Guy Le Moigne was appointed as the Council's first Executive Director.

The World Water Council has set its goals to be "the world's water-policy thinktank" (WWC 1997b, 1997c). The first General Assembly of the WWC was held at the IXth IWRA Congress in Montreal on September 3–4, 1997. The major tasks of the General Assembly were to approve the draft constitution and bylaws of the organization and to elect a

TABLE 6.4 FOUNDING MEMBERS OF THE WORLD WATER COUNCIL

International Water Resources Association (IWRA)

International Commission on Irrigation and Drainage (ICID)

Canadian International Development Agency (CIDA)

The World Bank (WB)

International Association on Water Quality (IAWQ)

International Water Supply Association (IWSA)

United Nations Development Programme (UNDP)

The World Conservation Union (IUCN)

The Water Supply and Sanitation Collaborative Council (WSSCC)

formal Board of Governors to replace the interim board that had served the organization to that point. Though 19 members of a new board were elected at that meeting, major issues were left unresolved about board representation, election procedures, budget and funding priorities, and priorities for action (WWC 1997a).

Like the GWP, the early activities of the WWC have focused on holding international meetings. In March 1997, the First World Water Forum was held in Marrakech, Morroco. The forum attracted participants from over 60 countries, including seven ministers, three heads of UN organizations, and senior governmental and nongovernmental officials from the World Bank, the African Development Bank, the Asian Development Bank, and private sector companies. The forum heard 30 papers on water issues (Aït-Kadi et al. 1997) and issued a declaration mandating the council to prepare a "Vision for Water, Life, and the Environment." The vision is to be prepared over the next few years, with the goal of presenting it in the year 2000 during the Second World Water Forum. The goal of the proposed "vision" is to analyze and integrate the various water-related activities that affect human and ecosystem well-being and to propose options to:

- ensure food security through aquaculture and rainfed and irrigated agriculture;

- provide adequate water supply and sanitation services;

- manage and develop water resources for economic functions, including the production of electricity; and

- protect the environment, including coastal areas and wetlands (WWC 1997a, 1997c; IWRA 1997).

In many ways, the WWC is an organization of organizations. WWC solicits members, at $1,000 per year dues. As a result, its structure and function leads it to work with existing organizations, including the public and private sector, nongovernmental organizations, the wide range of United Nations groups working on water issues, and the newly formed Global Water Partnership. As of September 1997, the WWC had about 152 member organizations representing 34 different countries and 23 international organizations (IWRA 1997). While there has been discussion about developing ways of encouraging the participation of nongovernmental organizations and individuals unable to pay the membership fee, current membership remains limited to organizations or individuals with considerable financial resources.

Governing Structure of the WWC

The WWC consists of the General Assembly of members, which elects and appoints a maximum of 38 members to a Board of Governors. The board then elects officers and sets up committees for specific operations. A chair for each committee is selected by the board. An Executive Committee, comprised of the executive director and the officers of the Board of Governors, manages the operation of the WWC between meetings of the Board of Governors (IWRA 1997). The entire General Assembly of members is to meet at least once every three years.

In Montreal in September 1997, the members present at the first General Assembly of the WWC elected a board, which then appointed the following officers:

President: Dr. Mahmoud Abu-Zeid

Vice Presidents: René Coulomb, France; and Aly M. Shady, Canada

Treasurer and Chair, Finance Committee: Leonard Bays, International Water Supply Association

Executive Director: Guy Le Moigne

Chair, World Fund for Water: Pierre-Frédéric Ténière-Buchot, France

Chair, Regional Centres Committee: Madhav Chitale, Intenrational Commission on Irrigation and Drainage

Chair, Publications and Information Committee: Andras Szöllösi-Nagy, UNESCO

Chair, By-Laws Committee: Jacques Lecornu, International Commission on Large Dams

Water Policy Journal

One of the early efforts of the WWC was a commitment to publish a new water-policy journal to provide information to decision makers in government, international and nongovernmental organizations, and the water industry. The journal, *Water Policy,* is due to be published in 1998 and will include "analyses, reviews and debates on all policy aspects of water resources" (Elsevier Science 1998). The editor is Dr. Jerry Delli Priscoli of the U.S. Army Corps of Engineers Institute for Water Resources and the publisher is Elsevier Science Publishers in Oxford, England.

World Fund for Water

The WWC is considering the creation of a World Fund for Water. The idea for such a fund was first suggested at a meeting in Granada, Spain, in July 1996. Discussions about its structure and goals continued at meetings of the WWC and other international water conferences in 1997, including the First World Water Forum in Marrakech in March and the IXth International Water Resources Association Congress in Montreal in September. The purpose of such a fund would be to provide grants to other water-dedicated institutions, to offer loans for technical water-development projects and other research and development activities, and to support communications campaigns, events, and public relations surrounding water issues (Tenniere-Buchot 1997). Revenues would be generated from grants, a possible endowment, high-level donors, smaller donors, or innovative debt for development swaps.

By late 1997, no decision had been made to formally constitute the fund. Studies were continuing about legal and financial issues, whether or not donors could be found, and how best to set up the board and institutional structure of such an organization. Concerns about conflicts with the activities of other funding organizations in the water area and with the Global Water Partnership were also unresolved.

The World Water Council and the Global Water Partnership

The creation of both the WWC and the GWP in the mid-1990s was cause for some concern among members of the world water community who felt there was a risk of overlap, inefficient use of limited time and resources, and competition for funds. As a result, serious efforts were made to try to define the characteristics of each, to identify separate missions and strengths, and to work together where possible. The WWC is self-described as working on raising awareness of issues related to water-resources management, while the GWP is concerned with bringing sustainable water-resources management closer to the users in developing countries. Nevertheless, they have related objectives (see Box 6.4) and were launched at about the same time. The WWC and GWP organized several meetings in parallel and several members of each group are shared. As an example, the Executive Committee of the WWC met during the September 1996 triennial congress of the ICID in Cairo, and Johan Holmberg, Executive Secretary of the GWP, was present as an observer. The GWP organized its Consultative Group Meeting in March 1997 in Marrakech, preceding the WWC's First World Water Forum. The executive director of the WWC and members of the interim Board of Governors participated in the GWP Steering Committee meeting at the 1997 Stockholm Water Conference in August and other parallel meetings have been held in Stockholm. A major GWP meeting was held in Marseilles in conjunction with the WWC around World Water Day in mid-March 1998. The Technical Advisory Committees of both groups also share some members in common, and members of both organizations serve on the working group preparing for the Programme of Action "Water 21" of the UN Commission for Sustainable Development in 1998. The WWC and GWP will have to continue to clearly define their mutual interests and activities in order to avoid overlap, redundancy, and inefficient use of one of the most scarce resources of all—time.

The World Commission on Dams

In September 1996, the World Bank released a review of 50 World Bank–funded large dams (World Bank 1996). In its review, an effort was made to determine whether or not the promised benefits of World Bank–funded dam projects had been delivered and major environmental and social impacts identified and avoided. The World Bank has supported around 10 percent of large dams in developing countries, and it has had a leading role in funding some of the largest facilities anywhere. The Bank has directly funded four

---

BOX 6.4
Stated Objectives of the GWP and the WWC

The Global Water Partnership "Objectives":
The GWP will:

- support integrated water resources management programmes by collaboration, at their request, with governments and existing networks and by forging new collaborative arrangements;

- encourage governments, aid agencies and other stakeholders to adopt consistent, mutually complementary policies and programmes;

- build mechanisms for sharing information and experiences;

- develop innovative and effective solutions to problems common to integrated water resources management;

- suggest practical policies and good practices based on those solutions;

- help match needs to available resources.

Source: http://www.gwp.sida.se

The World Water Council "Objectives" (from the Constitution of the WWC):

- to identify critical water issues of local, regional and global importance on the basis of ongoing assessments of the state of water;

- to raise awareness about critical water issues at all levels of decision making, from the highest authorities to the general public;

- to provide the forum to arrive at a common strategic vision on integrated water resources management on a sustainable basis, and to promote the implementation of effective policies and strategies worldwide;

- to provide advice and relevant information to institutions and decision-makers on the development and implementation of comprehensive policies and strategies for sustainable water resources management, with due respect for the environment and social and gender equity; and

- to contribute to the resolution of issues related to transboundary waters.

The World Water Council "Mission" statement reads:

The promotion of awareness about critical issues at all levels including the highest decision-making level; and the facilitation of efficient conservation, protection, development, planning, management and use of water in all its dimensions on an environmentally sustainable basis for the benefit of all life on the earth.

Source: http://www.worldwatercouncil.org

---

out of the five highest dams in developing countries outside China, three out of the five largest reservoirs in these countries, and three of the five largest hydroplants.

In the Bank's own review, the impacts of most of the dams reviewed were considered "acceptable" or "potentially acceptable." Thirteen of the projects in the review were regarded as acceptable, 24 as potentially acceptable, and 13 as unacceptable. The review concluded that in most of the cases reviewed, benefits far outweighed costs, including the costs of adequate resettlement programs, environmental safeguards, and other mitigation measures. The Bank also noted, however, that while 90 percent of the projects met the standards applicable when they were approved, only 25 percent were implemented in a way that complied with current World Bank policies (World Bank 1996).

In early 1997, a leading advocacy group working to stop large dams released a detailed critique of the review, accusing it of being "based on flawed methodology and inadequate data" (McCully 1997). They proposed that the World Bank impose a moratorium on the provision of loans, credits, guarantees, and other forms of support for large dams until a new independent review could be conducted. This is not the first call for a moratorium in World Bank support for large dams. Such a moratorium was one of the demands in the 1992 Manibeli Declaration, endorsed by more than 326 NGOs and coalitions in 44 countries. A similar demand was made in the March 1997 Curitiba Declaration approved at the First International Meeting of People Affected by Dams.

Partly in response to concerns about the World Bank review, the IUCN and the World Bank agreed to hold a workshop on large dams in early 1997. The workshop was intended to discuss the findings of this report and their implications "for the design, methodology and process of a proposed in-depth study (Phase II) on large dams (including comparative evidence from industrialized countries) to be undertaken by the World Bank in 1997/98." The workshop, entitled "Large Dams: Learning from the Past, Looking at the Future," was held on April 10–11, 1997 at IUCN headquarters in Gland, Switzerland. Participants included representatives from the World Bank, IUCN, the private dam industry, public dam-building agencies, academia, and NGOs active in the debate over large dams. The list of participants is in Table 6.5. A number of observers and journalists were also present (IUCN–World Bank 1997).

The workshop brought together leading experts and representatives from the different stakeholders for an open and transparent dialogue. The objectives were to review the 1996 study, compare the results to experience from other projects, develop a method for analyzing critical issues for future dam development, and identify appropriate actions necessary for developing new standards for designing, building, and operating large dams. During the workshop discussions, the participants identified social, environmental, economic, and engineering issues that need to be addressed in any effort to develop a new consensus on the role of large dams in economic development.

The most important outcome of the workshop was an agreement to establish, by November 1997, a World Commission on Dams (WCD) with a mandate to review the development effectiveness of large dams and develop standards, criteria, and guidelines to advise future decision making. As their initial step, the IUCN and the World Bank established an Interim Working Group to coordinate the establishment of the commission. A Secretariat, jointly chaired by IUCN and the World Bank, was formed to support the Working Group in Washington, D.C. In October 1997, Minister Kader Asmal from South Africa was chosen to be the chair of the commission.

In addition to the chair, the independent commission was set up to consist of five to eight members, an executive secretariat, and a consultative group. The members of the commission were originally to be chosen for their ability to be "objective, independent, and representative of the diversity of perspectives" on the issue (WCD 1997b, 1998). The products of the commission, which is to have a two-year duration, include a formal report with recommendations to the President of the World Bank, the Director General of IUCN, and the international community. The commission may produce recommendations for international performance standards for dam construction, criteria for land and water management, and guidelines for reparations for people affected by construction.

NGO participants present at the meeting believe that an independent international review commission "holds great potential for furthering the aims of dam critics" and could seriously restrict the building of destructive dams on the assumption that a review would show that they have failed to meet expected goals (McCully, personal

TABLE 6.5   LARGE DAMS WORKSHOP, GLAND, SWITZERLAND, APRIL 1997: PARTICIPANTS AND OBSERVERS

Mike Acreman, Institute of Hydrology, United Kingdom (rapporteur)

Sanjeev Ahluwalia, Tata Energy Research Institute, India

Ricardo Bayon, IUCN, Switzerland (observer)

Peter Bosshard, Berne Declaration, Switzerland

John Briscoe, World Bank

Wenmei Cai, Beijing University, China

Stuart Chape, IUCN, Laos

Edwardo de la Cruz Charry, ISAGEN, Colombia

Timothy Cullen, World Bank (observer)

Shripad Dharmadhikary, Narmada Bachao Andolan, India

Mbarack Diop, Tropica Environmental Consultants Ltd., Senegal

Tony Dorcey, University of British Colombia, Canada (facilitator)

Steve Fisher, Intermediate Technology Development Group, United Kingdom

Stephanie Flanders, *Financial Times*, United Kingdom (media observer)

Robert Goodland, World Bank

George Greene, IUCN, Switzerland

Daniel Hoffman, *Neue Zuercher Zeitung*, Switzerland (media observer)

David Iverach, Nam Theun 2 Electricity Consortium, Laos

E. A. K. Kalitsi, Volta River Authority, Ghana

Andres Liebenthal, World Bank

Gideon Lichfield, *The Economist*, United Kingdom (media observer)

Richard Meagher, Harza Engineering Company, United States

Patrick McCully, International Rivers Network, United States

Jeff McNeely, IUCN, Switzerland

Kathryn McPhail, World Bank

Reatile T. Mochebelele, Lesotho Highland Project, Lesotho

Ricardo Luis Montagner, Movimento dos Atingidos por Barragens, Brazil

Engelbertus Oud, Lahmeyer International, GMBH, Germany

Bikash Pandey, Alliance for Energy, Nepal

Thomas Philippe, Electricité de France, France

Robert Picciotto, World Bank

Jean Yves Pirot, IUCN, Switzerland

Martyn Riddle, International Finance Corporation, United States

Thayer Scudder, California Institute of Technology, United States

Aly Shady, International Commission on Irrigation and Drainage, Canada

Kalpana Sharma, *The Hindu*, India (media observer)

Andrew Steer, World Bank

Achim Steiner, IUCN, United States (coordinator)

Richard Stern, World Bank

Jan Strömblad, Asea Brown Boveri-ABB, Sweden

Theo Van Robbroeck, International Commission on Large Dams, South Africa

Pietro Veglio, World Bank

Tanlin Yuan, Ministry of Water Resources, China

Robert Zwahlen, Electrowatt Engineering Ltd., Switzerland

communication, 1997; IRN 1998). While NGOs have expressed some optimism about the outcome of such as review, they are also concerned that the process be transparent and independent of construction interests. The World Bank, on the other hand, hopes to improve its ability to avoid the worst projects, better understand concerns of NGOs, and reduce opposition to large dams by bringing opponents into the process.

In October 1997, the names of potential commission members had been assembled and a list of ten members proposed (WCD 1997a). This list drew immediate opposition from many sides and the process promptly bogged down in squabbling over membership. Some argued for equal representation of dam opponents and proponents. Others suggested that it was too heavy with dam proponents, or too light with representatives of people affected by dams. Still others suggested that the commission should consist of members with no apparent vested interests either way who would be capable of doing good independent assessments. The major roadblock came from some NGOs from the Gland meeting who felt the proposed list had "inadequate representation of dam-affected people and anti-dam NGOs" (IRN 1997, 1998). By late November, when the final list was to be announced, no resolution of the impasse had been reached.

In early 1998, a meeting was held in Cape Town, South Africa, to review candidates and propose a final list of commissioners (see Table 6.6). At that meeting, substantial agreement on membership was reached, increasing the total number of members from eight to at least twelve, a vice-chair proposed, and the launch rescheduled for February 1998. The Commission faces a difficult challenge ahead because of the complex and often rancorous politics surrounding dam development. Even if the WCD ultimately falls apart, there is a strong consensus that an independent assessment of dams and dam policy is needed. How to conduct such an assessment successfully in a field so fraught with controversy remains unanswered.

TABLE 6.6   WORLD COMMISSION ON DAMS COMMISSIONERS, FEBRUARY 1998

Professor Kader Asmal, Chair, Minister for Water Affairs and Forestry, South Africa.

Donald J. Blackmore, Chief Executive of the Murray-Darling Basin Commission, Australia.

Joji Cariño, Executive Secretary of the International Alliance of Indigenous-Tribal Peoples of the Tropical Forest.

José Goldemberg, Professor at the University of São Paolo, Brazil.

Judy Henderson, Chair of Oxfam International, Board Member of the Environmental Protection Agency of NSW, Australia.

Laxmi Chand Jain, Vice Chair, former Chairperson of the Industrial Development Services, India, Indian High Commissioner in South Africa.

Göran Lindahl, President and CEO of ABB Asea Brown Boveri, Ltd., Zurich.

Deborah Moore, Senior Scientist at the Environmental Defense Fund, United States.

Mehda Patkar, social scientist and founder of the Narmada Bachao Andolan (Struggle to Save the Narmada River) in India, founding member of the National Alliance of People's Movement.

Wolfgang Pircher, consultant in the field of hydropower, past President and current Honorary President of the International Commission on Large Dams.

Thayer Scudder, Professor of Anthropology at the California Institute of Technology.

Shen Guoyi, Director-General of the Department of International Cooperation in the Ministry of Water Resources, People's Republic of China.

Source: WCD 1998

REFERENCES

Abrams, L.J. 1996. "Policy development in the water sector—The South African experience." Paper presented at the Cranfield International Water Policy Conference, Cranfield University, Bedford, United Kingdom (September).

Abrams, L.J. 1997. "Management of water scarcity: National water policy reform in South Africa in relation to regional development co-operation—in southern Africa." Via *http://wn.apc.org/afwater.*

Aït-Kadi, M., A. Shady, and A. Szollosi-Nagy. 1997. *Water, The World's Common Heritage.* Proceedings of the First World Water Forum, Marrakesh, Morocco, 21–22 March. Elsevier Science, Oxford, United Kingdom.

Conley, A.H. 1996. "A synoptic view of water resources in southern Africa." Department of Water Affairs and Forestry, Pretoria, South Africa.

Elsevier Science. 1998. "Water policy: Information for authors." Elsevier Science Ltd. Oxford, United Kingdom.

Glazewski, J. 1996. "Environmental rights and the new South African Constitution," in A.E. Boyle and M.R. Anderson (ed.), *Human Rights Approaches to Environmental Protection.* Clarendon Press, Oxford, pp. 177–197.

Global Water Partnership (GWP). 1996. "Summary report: Technical Advisory Committee regional meeting, Namibia, 4–8 November 1996." Global Water Partnership, Stockholm, Sweden.

Global Water Partnership (GWP). 1997a. Global Water Partnership web site. Via *http://www.gwp.sida.se.*

Global Water Partnership (GWP). 1997b. "Summary of the Technical Advisory Committee (TAC) regional meeting, Manila, Philippines, June 8–12, 1997." Global Water Partnership, Stockholm, Sweden.

Global Water Partnership (GWP). 1997c. "Summary report: Maseru Workshop Maseru, Lesotho, 6–7 January 1997." Global Water Partnership, Stockholm, Sweden.

Heyns, P. 1995. "Existing and planned water development projects on international rivers within the SADC region." Presented at the Conference of SADC Ministers *Water resources management in southern Africa: A vision for the future.* Pretoria, South Africa 23 November.

Holmberg, J. 1997. "Lessons from Marrakesh." *NewsFlow,* No. 1/97 (April).

International Rivers Network. 1997. "NGOs condemn World Bank/IUCN dam-building commission." Press Release, November 25, Berkeley, California.

International Rivers Network. 1998. "Summary of status of World Commission on Dams." Email communication from Patrick McCully, January 22, Berkeley, California.

International Water Resources Association (IWRA). 1997. "The Ninth World Water Congress, Report of the Rapporteur-General." A.M. Kassem and A.M. Shady (eds.). Montreal, Canada.

IUCN–World Bank. 1997. *Large Dams: Learning from the Past, Looking at the Future.* T. Dorcey, A. Steiner, M. Acreman, and B. Orlando (eds.). Workshop Proceedings, Gland Switzerland, April 11–12, 1997. IUCN and the World Bank, Washington, D.C.

McCully, P. 1997. "A critique of 'the World Bank's Experience with Large Dams' ": A preliminary review of impacts." International Rivers Network, Berkeley, California.

Newsflow. 1997. The Newsletter of the Global Water Partnership, Vol. 1, Issue 2/97 (October).

Ohlsson, L. 1995. "Water and security in southern Africa." Department for Natural Resources and the Environment. Publication on Water Resources #1, Swedish International Development Cooperation Agency, Stockholm, Sweden.

Pinder, R. 1990. "50,000 blacks deprived of water." Reuters Press/ *San Francisco Chronicle* (October 24), p. A-11.

South African Department of Water Affairs and Forestry (SADWAF). 1995. "You and your water rights." Department of Water Affairs and Forestry, Republic of South Africa, Pretoria (March), 30 pp.

South African Department of Water Affairs and Forestry (SADWAF). 1996. "Water Law Principles Discussion Document." Department of Water Affairs and Forestry, Republic of South Africa, Pretoria, South Africa (April), 15 pp.

South African Department of Water Affairs and Forestry (SADWAF). 1997a. "White Paper on a National Water Policy for South Africa." Department of Water Affairs and Forestry, Republic of South Africa. Pretoria, South Africa (April), 37 pp.

South African Department of Water Affairs and Forestry (SADWAF). 1997b. "The Working for Water Programme: 1996/1997 Annual Report." Department of Water Affairs and Forestry, Republic of South Africa. Pretoria, South Africa (April), 16 pp.

South Africa Ministry of Water Affairs and Forestry (SAMWAF). 1996. "Fundamental Principles and Objectives for a New Water Law in South Africa." Report to the Minister of Water Affairs and Forestry of the Water Law Review Panel. Pretoria, South Africa (January), 33 pp.PP.

South African National Water Conservation Campaign (SANWCC). 1997. "The Greater Hermanus Water Conservation Programme." National Water Conservation Campaign and the Department of Water Affairs and Forestry, Cape Town, South Africa.

Tenniere-Buchot, P.F. 1997. "The World Water Fund." Paper presented at *Water Resources Outlook for the 21st Century: Conflicts and Opportunities*. The IXth World Water Congress, Montreal, Canada 1–6 September.

United Nations. 1978. *Registry of International Rivers*. Centre for Natural Resources, United Nations. Pergamon Press, New York.

World Bank. 1996. "The World Bank's experience with large dams: A preliminary review of impacts." (A. Liebenthal and staff). Operations Evaluation Department, The World Bank, Washington, D.C.

World Commission on Dams. 1997a. "Proposed list of Commission members." Letter to the Reference Group, October 22. Interim Secretariat, World Commission on Dams, Washington, D.C.

World Commission on Dams 1997b. "Background materials for the interim working group meeting." Cape Town, South Africa, January 24–25.

World Commission on Dams. 1998. "Minutes of Cape Town meeting: List of proposed Commission members." Letter to the Reference Group, January 27. Interim Secretariat, World Commission on Dams, Washington, D.C.

World Meteorological Organization (WMO). 1997. "The Dublin Statement on Water and Sustainable Development." Via *http://www.wmo.ch/web/homs/icwedece.html*.

World Water Council (WWC). 1997a. "Message from the President." *World Water Council Newsletter,* Vol. 1, No. 1. (September).

World Water Council (WWC). 1997b. *Constitution of the World Water Council*. Marseilles, France.

World Water Council (WWC). 1997c. WWC web site, available at *http://www.worldwater-council.org*.

SEVEN

Moving Toward a Sustainable Vision for the Earth's Fresh Water

Alice: "Would you tell me, please, which way I ought to go from here?"
Cheshire Cat: "That depends a good deal on where you want to get to."

Alice's Adventures in Wonderland, LEWIS CARROLL (1865)

Introduction

It seems inevitable that humans—barring some unforeseen catastrophe—will require a larger share of the earth's limited renewable fresh water in the future than we do today. As the new millennium approaches, almost six billion people use nearly 30 percent of the world's total accessible renewable supply of water to grow food, run industries and cities, cool power plants, and meet other needs for drinking, cleaning, cooking, and playing. As the world's population climbs to seven, eight, and nine billion, more and more water will be required to satisfy our basic needs and our social, cultural, and economic desires. This water will come at an increasing financial and ecological price.

People like to speculate about the future, to tell stories about alternative futures, about what tomorrow might look like. Perhaps the only certainty for anyone looking ahead is that the future is uncertain, unpredictable, and complex. But our present predicament—a consequence of our current policies, technologies, and institutions—is clear. Billions of people lack access to the most basic of water services. Millions of people die every year from a wide range of water-related diseases. Aquatic ecosystems and fisheries worldwide are being degraded and destroyed. Political disputes are flaring up over water that crosses political borders, and there is growing national and international competition for water in water-short regions. Despite increasing interest and concern over these problems, most of them have been getting worse over the past few decades, not better. As a result, while our water future can be only dimly seen, there is ample evidence that we won't like where we appear to be heading.

Water "experts" have been peering into the future for centuries, dreaming of taming wild rivers for power and wealth, planning to bring water to cities built in the desert, designing complex systems to let us grow food where natural conditions don't permit it. The goal has always been the same: to conquer the difficult challenges posed by acquiring and using finite and vulnerable resources to meet humanity's needs and desires. The result has been increasing wealth for a few, the satisfaction of basic human needs for some, and disappointment and misfortune for too many.

Because projections about the future are inherently uncertain, forecasters use scenarios to tell stories about the future, using past experience, history, expert judgment, and their imagination to articulate different visions. Present methods for projecting water demands typically assume that societal structures and desires, key driving forces, and management policies in the future will be virtually identical to those in place today. Resource, environmental, or even economic constraints are rarely considered. The aim of such "conventional development" or "business as usual" scenarios is to examine a future that looks in most ways like a continuation of the present. This approach has both advantages and limitations. Its principal advantage is that it gives us a glimpse of where society is heading. Perhaps its greatest disadvantage is that it routinely produces scenarios with irrational conclusions, such as water demands exceeding supply and water withdrawals unconstrained by environmental or ecological limits. Put most concisely, conventional development scenarios usually describe a future that we wouldn't want to visit if we had a choice.

We *do* have a choice. Decisions made every day about economic policies, technological choices, and institutional structures all affect how we use resources to accomplish desired goals. By explicitly elaborating on what society wants, we can intentionally set goals for water use that are both sustainable and achievable. By describing a vision of where we want to be, it becomes possible to craft the policies and institutions—and to apply the technologies and tools—that will take us there.

The following "Vision for 2050: Sustaining Our Waters" is a story about a *possible* attractive future. It is not a prediction or a projection or a prescription. It is a tool to help policymakers, managers, and the public explore how the world might turn out given a particular set of decisions and actions. Even if there were widespread agreement that such a future is both desirable and achievable, there is no guarantee that any part of it will come about. Indeed, given today's trends, policies, and planning, the future is much more likely to look like the many conventional development scenarios traditionally developed by water experts than the one described here. And this particular vision will not come about because of actions taken by any single large institution or organization; rather, it would result from small actions and decisions taken on local and regional levels.

Many different dreams and visions can be described. Without some positive vision, without some thought about what truly sustainable water use means, society risks continuing on a path that will take us further and further in the wrong direction. We can choose a different path and try to define and attain a different future. But we must make that choice soon.

A Vision for 2050: Sustaining Our Waters

It is the year 2050. After the ecological and human catastrophes of the late twentieth century and the early decades of the twenty-first century, major efforts have been made to restore the environmental balances necessary to support the continued existence of humankind at a decent quality of life. These efforts have been particularly extensive for fresh water, recognized as the most essential renewable resource, endowed with sacred meaning and essential for human and ecological health, the production of food and energy, and industrial production and transportation.

During the years of the twentieth century, the population of the world grew from 1,600 million to over 6,000 million people. World and regional wars as well as peaceful transitions reshaped the political landscape, while revolutions in transportation, technology, and communications led to unprecedented social and cultural changes. These changes also unleashed powerful forces that altered the earth's ecological landscape, from the local scale to the global scale. Nowhere was this more evident than in our modifications to the water cycle of the planet. By the year 2000, nearly half of the accessible renewable water of the earth was appropriated for human use. Water was being moved thousands of kilometers from water-rich to water-poor regions. Tens of thousands of major dams—many of them monuments of engineering wonder—had altered the hydrology and ecology of almost every river. Even the very rotation of the earth had been measurably affected by the mass of water stored in artificial reservoirs. Yet the same scientific and engineering skills that had supported the hydrologic revolution of the twentieth century had proved inadequate for meeting the needs of most of the world's people and had caused devastating environmental damage.

By the end of the century, a series of major international water conferences and meetings focused global attention on growing freshwater issues, particularly the enormous human suffering resulting from the lack of access to basic water supplies, inadequate availability of sanitation services, and the growing threats to global food security. At the same time, political conflicts over shared international rivers and watercourses raised the issue of environmental security to the highest political levels.

By the end of the 1990s, some progress had been made toward identifying a series of explicit goals to guide long-term water planning and management. Building on the principles enunciated at the Mar del Plata, Dublin, and Earth Summit conferences, these goals have been modified and refined in the intervening decades, but they marked the first explicit attempts to integrate water-resources supply, use, and management in a truly sustainable way.

Basic human needs for water were finally acknowledged as the top international priority and these needs are now largely met. *At the Earth Summit of 2007, "Universal Access" programs for basic human needs were adopted by all nations* *and access to clean water was made a top and permanent priority. Governments, international aid agencies, private corporations, and nongovernmental organizations joined forces in a new partnership to meet the goal of providing this basic need universally with a flexible and varied combination of technologies and institutions. By 2035, some 95 percent of the world's population had access to a basic water requirement for fundamental domestic needs, including drinking, sanitation, cleaning, and food preparation. The financial cost of meeting these basic needs proved to be far outweighed by vast savings in health costs, improvements in worker productivity, and the freeing of time for women and children for educational, commercial, and community activities.*

At the same time, domestic water use in the industrialized world has become much more efficient and equitably allocated. Improvements in efficiency begun in the late 1990s to cope with droughts and avoid the need for expensive and controversial new supply projects have been extended to all aspects of domestic life, and inexpensive water-efficiency equipment is widely available. Municipal water supplies in most places are supplemented by extensive use of reclaimed urban wastewater for nonpotable uses, and potable reuse is being implemented in seriously water-short urban areas.

The rush to privatize municipal water-supply systems in Europe and the Americas in the late 1990s and early 2000s slowed after the failure of some private companies to provide for long-term maintenance needs and equitable access to water. This, in turn, led to some local water quality and health disasters. A more effective balance of public and private ownership and operation is now the norm.

Large-scale, cost-effective seawater desalination systems are still not available, but there has been a significant expansion in the use of less-expensive brackish water desalination systems in arid and semi-arid regions to tap saline groundwater sources and to purify contaminated surface water sources. Some basic needs are being met by flexible small solar desalination systems. More widespread solar desalination is practiced in arid coastal countries where water-use efficiency is high, water availability low, economic capital available, and solar energy abundant, particularly in the Persian Gulf region and North Africa. But the water produced by these systems is no longer used for anything but the highest valued industrial and domestic uses.

Water-related diseases rampant at the end of the twentieth century are being conquered. *The revitalized international efforts to meet the basic water requirements of all people, combined with effective education about sanitation practices and wide improvements in access and quality of medical care, have greatly reduced both the prevalence and severity of human suffering from water-related diseases. Guinea worm was the first water-related disease to be eliminated, around the turn of the century. Attention then turned to trypanosomiasis and onchocerciasis, both of which were largely controlled by 2030. The incidence of childhood diarrhea was also drastically reduced as its sources were identified and attacked and as treatment became universally available. The seventh and eighth great cholera pandemics were brought under control. Malaria and typhoid have expanded in range and still plague certain regions, but biological controls of dis-*

ease vectors are making inroads on the drug-resistant strains prevalent during the early part of the century. The links between a variety of cancers and chemical contamination of water in the heavily industrialized nations continue to be discovered, but new methods for preventing such contamination and for cleaning up contaminated waters continue to reduce health risks.

The spread of cryptosporidium and new strains of bacteria throughout North America, Japan, and Europe was halted by 2020 through the wider application of a combination of watershed protection policies and effective large-scale filtration technology. At the height of the worst urban outbreaks, an enormous demand for bottled water and home water-purification systems developed in the richer markets. This market has now largely disappeared and many large cities now send samples of their drinking water to the annual taste-testing competition held every summer at the famous Stockholm Water Festival in an effort to win the prestigious prize for the best-tasting urban drinking water on each continent.

Agricultural water is now efficiently used and allocated. *One of the greatest concerns facing the world in the early part of the twenty-first century was the challenge of producing food for the world's billions. Shortfalls of grain began to appear in the first decade of the new millennium as major nations such as China began to make large purchases on the international markets. By 2012, China, India, Nigeria, Indonesia, and Bangladesh were competing in world markets for grain, along with most of the nations of the Middle East and Africa. At the same time, traditional exporters such as Argentina, Australia, Canada, and the United States had cut back on export volumes to meet internal and regional needs. Saudi Arabia, a major wheat exporter in the 1980s and 1990s, saw its agricultural exports collapse after groundwater overdraft depleted and permanently damaged its fossil aquifers.*

Between 2012 and 2018, the simultaneous Great North American, Chinese, and Indian droughts led to the reappearance of famines in Africa and Asia, as well as extremely high food prices in the United States and Canada. Food riots in the winter of 2017 in the United States and seven European countries forced a reevaluation of food and water policies, the reduction of water subsidies, widespread improvements in irrigation efficiency, and substantial shifts in cropping patterns.

By 2020, return of the rains and new policy changes began to reduce pressures and to increase the nutritional status of the world's poorest inhabitants. In particular, the enormous regional disparities in diet evident at the end of the twentieth century began to close as the large-scale consumption of meat in the industrialized world decreased. A rapid drop in beef and lamb consumption has occurred, driven by higher prices; public health disasters in Great Britain, the European Community, Japan, and the United States attributed to contaminated meat; better education about the adverse health effects of eating meat; and new policies on land management. This has freed up substantial quantities of land, irrigation capacity, and grain and cereal crops for direct human consumption.

At about the same time, new varieties of rainfed crops began to appear on the market. These genetically improved crops are less sensitive to the vagaries of natural rainfall and have substantially increased yields from rainfed lands, which remain the majority of all agricultural land. A renewed interest in traditional farming techniques in semi-arid regions combined with inexpensive high-tech water-monitoring equipment and new crop varieties has encouraged a rethinking of agricultural aid policies, improved production without new irrigation requirements, and resulted in great demand for farming advice from experts in developing countries.

On irrigated lands, overall water productivity has improved dramatically with the universal adoption of high-efficiency sprinklers and drip irrigation on appropriate crops and lands. Water-use efficiency has also improved due to advances in sensor and computer technology that permit farmers to monitor soil moisture inexpensively and accurately and to apply water only when needed. Many farmers are now tied directly into regional weather-forecasting centers that help avoid unnecessary irrigation prior to natural precipitation. The trend away from pesticide and herbicide use and toward integrated pest management and the use of innovative ground cover has further reduced overall irrigation requirements while maintaining high yields, soil fertility, and water quality.

In many arid countries, limited but efficient agricultural production is maintained with high-quality reclaimed urban wastewater. Middle East water experts, who pioneered this approach, are in high demand in many parts of Africa, Asia, and Latin America as countries work to maximize their use of this underutilized resource.

By 2030, great improvements in food distribution and storage permitted countries to rely more on international markets and reduced the impacts of severe droughts and other forms of climatic variability. Water trading among market sectors is common, reflecting a greater emphasis on economic mechanisms to meet water needs. Communities have a major say in water trading, however, and the price of water reflects community and environmental values, as well as purely market values.

A new focus on global food sufficiency has replaced the old nationalistic concept of national food security that led many countries in arid and semi-arid regions to overdraft fossil groundwater and invest in unsustainable irrigation projects during the late twentieth century. International development efforts have refocused on nonagricultural developments in water-short countries, such as industrial and commercial activities. These activities provide sufficient capital to permit food-buying nations to meet food shortfalls on the international market. The major food exporters have formally agreed to ban the use of food boycotts as a political weapon. No country in the world is completely food self-sufficient, yet the world as a whole maintains adequate food production and storage. Average populations in all regions now receive the minimum recommended number of calories.

Basic ecosystems water needs are being identified and met. *The mass extinctions of species in the Aral Sea and Lake Victoria in the 1980s and 1990s and in the Yangtze and Mekong rivers after 2000, combined with widespread extinctions in other aquatic systems, led to the adoption of national and international actions to protect ecosystems. Since 2025, the number and types of aquatic species formally* *listed as threatened and endangered have begun to diminish following implementation of comprehensive minimum environmental water commitments, the signing of international agreements on species protection and management, and the identification and protection of critical habitats around the world. All international aid and water-development projects now include explicit ecosystem protection and management components.*

Restoration efforts are well under way in coastal and inland wetlands around the world following the collapse of coastal fisheries in Asia, the North Atlantic, the Gulf of Mexico, and along the coastline of western Africa. International delta protection agreements are in place for the Mekong, the Nile, the Niger, the Zambezi, the Ganges/Brahmaputra, the Colorado, and dozens of other international rivers. The loss of wetlands has been stopped, and innovative management is now actually creating new wetlands at the mouth of many of the largest rivers in the world. The Mississippi River delta has begun to expand rapidly following a plan designed to give in to the river's natural inclinations to meander.

Regional monitoring programs are keeping exotic species invasions to a minimum, and international teams of ecologists are working to eliminate invasions that have successfully taken hold. The zebra mussel still clogs waterways in North America, but the water hyacinth is being defeated in Africa. The sea lamprey, accidentally introduced into the Great Lakes region of the United States and Canada, was unintentionally wiped out by commercial overfishing in 2010. It had been identified as a delicacy in the late 1990s and widely exported to Europe and Asia for over a decade.

Following the 2015 agreement among the nations sharing the Aral Sea basin, flows of the Amu Darya and Syr Darya rivers into the sea have reached their highest levels in half a century. The agreement instituted effective joint water management among the parties, a substantial reduction in cotton and rice production in the region, and vast improvements in irrigation efficiency. The surface area of the Aral Sea is now approximately 50,000 square kilometers, an increase of more than 25 percent since the 1990s but still more than 15,000 square kilometers below natural levels. The devastating health problems suffered by the region's inhabitants from the 1980s through the early part of the twenty-first century are abating, and work is under way to restore a fishery in the sea.

Similar efforts have protected the threatened bird and fish species at the mouth of the Colorado River in the Sea of Cortez. Local and international agreements during the twentieth century allocated all of the water of the Colorado to users and none to the natural ecosystems, leading to the ecological collapse of the system and severe impacts on the indigenous population dependent on fishing. After extensive negotiation with the many disparate parties along the river, including Mexico and all seven affected states in the United States, an agreement was signed guaranteeing minimum flows to the delta. The water for these flows came from a mix of agricultural and urban efficiency improvements, crop reallocations in the United States and Mexico, and innovative use of water reclaimed from cities and farms in both countries. The delta now supports one of the largest ecotourism and sports fishing industries in northern Mexico and is a major stop on the Pacific flyway for migratory birds. The success of these cases has provided the information and tools necessary to quantify and secure ecosystem water rights elsewhere and has helped avert other ecological disasters.

Serious water-related conflicts are now regularly resolved through formal negotiations. *Shared water resources are now seen as a major impetus for cooperation and international agreements, even though the early part of the twenty-first century saw a series of minor and major water-related conflicts. After major political conflicts and minor military skirmishes in southern Africa over shared rivers, the more serious intraregional conflicts in India and southern Asia, and instances of contamination of shared groundwater aquifers on the border of several countries, new international water tribunals were set up to hear and mediate disputes. By 2010, however, unresolved disputes in Asia over the development of regional rivers and the bombings of dams and water supply systems on major rivers in the Middle East led to a widely attended international diplomatic congress, at which binding principles of conflict resolution and negotiation were accepted.*

In the years following the congress, formal treaties and river basin commissions were put in place for nearly all of the world's major shared rivers. The New Nile River Treaty, for example, has been signed by all nations of the Nile Basin and includes provisions for sharing water, information, and experts. These treaties also include allocation agreements during droughts, provisions for formal negotiations of disputes, and a sharing of responsibility for environmental and ecological protections. Upon request, United Nation hydrologists and environmental scientists help to monitor water treaties using remote on-site survey equipment and the orbiting "Hydra" satellite system, which provides real-time observations of surface water conditions everywhere on the earth.

The Middle East—a region thought by many to be the most vulnerable to water-related conflicts—has turned out to be a model for regional cooperation and water sharing. Effective joint basin management commissions, first set up between Israel and Jordan in the treaty over the Jordan River, have now been established for the Tigris, Euphrates, and Orontes rivers. After serious disputes over dam projects on the Euphrates River, international negotiators helped the Turks, Syrians, Iraqis, and Kurds work out a sharing arrangement that equitably distributed both the benefits and the costs of river developments. A water-sharing agreement has been worked out between the Israelis and the Palestinians over groundwater aquifers in the West Bank.

Palestinians in Gaza are enjoying a flourishing economy that focuses on archeological tourism, small manufacturing, and moderate-cost Mediterranean holidays. Small solar desalination plants provide high-quality drinking water. Treated domestic wastewater meets the needs of important parts of the manufacturing sector. All water use is carefully monitored and managed by the Palestinian Water Authority. Some high-valued agricultural crops are produced using efficient drip systems and exported to markets throughout the Middle East and North Africa. While most of the Gaza coastal aquifer remains too contaminated to use, 30 years of active groundwater recharge and aquifer reconstruction has permitted modest groundwater withdrawals to resume for commercial use.

Among the unresolved problems of the twenty-first century is the question of how best to address the impacts of the developing greenhouse effect. *All efforts at rational water management continue to be complicated by the effects of global climate change. Global warming was recognized by the scientific community in the 1980s and 1990s, but it was not seriously acknowledged by politicians until well after 2000, at which point it was too late to prevent many major impacts.*

Climatic changes have had particularly severe effects on regional water-resources management. Rainfall patterns have changed, the frequency and intensity of storms have increased in many places, and reservoirs and municipal water-supply systems designed for one set of conditions have had to be redesigned or managed for very different conditions. One positive outcome has been the training of a whole new generation of water managers who rely on concepts of operational flexibility and resilient water management instead of using past trends in order to forecast future conditions.

The ability to predict El Niño/Southern Oscillation events a year or more in advance has greatly facilitated drought and flood management in regions with strong teleconnections. Australian farmers are now able to coordinate production with farmers in Africa and Asia quickly to shift crops and efforts to regions where weather forecasts suggest more favorable conditions. Water managers in the western United States routinely change operating regimes when El Niño events are expected.

Calving of ice in Greenland and Antarctica continues to be a problem for navigation. The Japanese have towed several large icebergs back to Japan to explore the feasibility of replacing their fleets of large plastic bags. These bags are currently used to tow water to Japan for high-valued industrial uses. The thriving "bag" business continues to serve temporary high-valued water needs in many parts of the world suffering from climatic disruptions, as well as to meet needs during natural disasters and other emergencies.

Finally, the wide sharing of water data and information on successful management strategies has led to a spirit of international cooperation throughout the world's water community. As high global populations now begin to drop, there is the feeling that the worst threats to global and regional stability from water problems are finally behind us and that sustainable water management will be a permanent fixture throughout the world.

Water Briefs

Welcome to the Water Briefs section of the book. This section contains updates to scientific discoveries related to fresh water, information on new technologies or trends, the text of important water agreements or laws, and the latest information on Internet resources devoted to freshwater issues. This edition introduces a major and unresolved scientific controversy over the origin of water on earth, news about the first successful tests and commercial applications of water bag technology, the full text of the new UN Convention on Non-Navigational Uses of International Watercourses, the 1996 India-Bangladesh treaty over the Farakka Barrage, and a comprehensive list of water-related web pages. The treaty texts and hot links to Internet water resources will also be made available on-line at *http://www.worldwater.org*. This site will be updated periodically with new information.

The Best and Worst of Science: Small Comets and the New Debate Over the Origin of Water on Earth

Science advances in different ways, from the small careful steps of meticulous research to dramatic abrupt shifts in fundamental principles and beliefs. Thomas Kuhn described the nature of these advances—or paradigm shifts—in his classic work, *The Structure of Scientific Revolutions,* published in 1962. Such a shift may now be occurring in the nature of our understanding of the origin of water on earth, accompanied by displays of the best and the worst of scientific behavior and debate.

In the past, most scientists—and most science textbooks—described the earth's water being present as the planet was formed, outgassed through intense volcanism in the planet's early days, or coming to earth in a small number of giant comets shortly after the planet was formed. In these theories, the earth's water has been essentially fixed in its present amount for over four billion years. This theory is now being challenged by new evidence that appears to suggest that a steady stream of water-bearing objects comparable to small comets is hitting the upper atmosphere of the planet.

The first evidence for this idea appeared in instrument readings taken in 1981 from an ultraviolet imager flown on NASA's *Dynamics Explorer 1* spacecraft. While most observers believed the readings represented instrument error, Professor Louis A. Frank and colleagues at the University of Iowa theorized in 1986 that dark spots seen by the satellite were atmospheric holes caused by the disintegration of small icy comets in the upper atmosphere. He suggested that the objects are made up mostly of water ice, with some sort of shell, perhaps of carbon, that keeps them together in outer space. The shell shatters from electrostatic stress as it approaches earth, allowing the ice to vaporize. Frank argued that the water vapor they produce momentarily absorbs ultraviolet solar radiation scattered from oxygen atoms in the upper atmosphere, preventing it from reaching the camera and resulting in the dark spots seen on the images captured by the satellite (Frank et al. 1986a,b). His theory introduced a new class of objects in the solar system and ignited a wide-ranging controversy. This early evidence was ambiguous and open to multiple interpretations. Most colleagues discounted the appearance of the holes as an instrument problem because of the poor resolution of the imager.

But Frank's hypothesis was more than just discounted; it was subject to ridicule of a kind often only reserved for ideas that challenge deeply entrenched scientific theories. While the public usually thinks that scientists have civilized discussions rather than messy disputes, challenges to deeply held beliefs rarely result in polite debates. This was seen when the theory of continental plate tectonics was put forth and when it was suggested that the great Cretaceous dinosaur extinctions were caused by the impact of a massive comet—two theories now considered true but subject to intense criticism when first proposed.

Frank's theory was, and continues to be, widely attacked. His proposal was released on April 1, leading some to question whether it was an April Fool's Day gag. One critic said, "This is the one theory I know of that has a thousand fatal flaws" (Kerr 1988). In August 1991, Alexander Dessler, an editor at the *Geophysical Research Letters*, a leading journal that published Frank's original report, published a lengthy critique of the idea. Dessler, who was criticized by colleagues for publishing the first report, has since become one of Frank's most vocal critics (*Science* 1992). "The small-comet hypothesis fails to agree with physical reality by factors that range from a thousand to a billion," said Dessler (Kerr 1997). Others have said that Frank "should know better" than to print such hard-to-swallow ideas. One physicist said, "I think a lot of people hate [Frank's] guts for publishing the thesis and hold Dessler 'responsible' for churning up a fruitless debate" (*Science* 1992).

While the theory may still ultimately be proven wrong, the scientific process is slowly churning forward and new tests, experiments, and evidence in favor of the hypothesis are accumulating. In 1988, a Jet Propulsion Laboratory scientist, Clayne Yeates, tried to test Frank's hypothesis using the Spacewatch Telescope at Kitt Peak in Arizona. Using data on the objects' number, speed, and direction of motion, Yeates decided to operate the telescope in an unusual fashion, leading his targets as if he were aiming a rifle at a skeet shoot, in order to hold the objects in view long enough to record their passing (Frank and Huyghe 1990; Sawyer 1997). After eliminating all the known objects (stars, satellites, cosmic rays, etc.) and false signals in 171 images, there were a few short, faint streaks from unidentified objects that Yeates and Frank considered a confirmation of the microcomets' existence. Additional observations to try to eliminate other possible causes for the streaks led Yeates to say, "Everything agreed precisely with the predictions. . . . This is a class of objects that agrees exactly with those proposed by Frank" (Kerr 1988). This evidence was not accepted as conclusive by much of the scientific community, however, many of whom continued to reject the results as possible background "noise."

In an effort to provide better evidence, Frank helped design two new cameras for NASA's *Polar* satellite, an ultraviolet imager (UVI) and a visible light imaging system (VIS). The VIS uses a filter to detect visible light emitted only by fragments of water molecules. In mid-1997, Frank announced that these new instruments had again detected objects streaking into the upper atmosphere, disintegrating at high altitudes, and depositing large clouds of water vapor that mix into the lower atmosphere and ultimately fall as rain. These results were reported at the 1997 spring meeting of the American Geophysical Union in Baltimore, Maryland (Sawyer 1997; Broad 1997).

The *Polar* cameras appear to have captured trails of light in both ultraviolet and visible wavelengths as the objects disintegrate 900 to 25,000 kilometers above the atmosphere. The comets are estimated to be 20- to 40-ton aggregates of ice with a velocity of 32,000 kilometers per hour (*USA Today* 1997). The new images from *Polar* also include observations of atmospheric holes in much greater detail than before. These holes have diameters estimated to be between 25 and 80 kilometers (NASA 1997a,b; EOS 1997a). The *Polar* data also seem to show that the holes change size when the satellite flies at different altitudes and that the frequency of the holes is altitude dependent, further reducing the chances that they are instrument errors. "Certainly, at a minimum, the argument has moved outside the detector," said Robert Hoffman, a NASA scientist working on an independent review of the data (EOS 1997a).

Frank's new observations are consistent with his 1986 theory and he believes the new data provide "irrefutable evidence that these atmospheric holes are a geophysical phenomenon" and not spurious instrumental readings (Frank and Sigwarth 1997a–d). The University of Iowa team concluded that the observed objects are depositing water in the atmosphere because water is the only common gaseous substance in the solar system that can efficiently absorb the dayglow along the line-of-sight of the *Polar* cameras in the proper wavelength and over the proper area.

The objects are coming in at a rate Frank estimates to be 5 to 30 small comets per minute, or thousands per day. If this rate has been constant over the history of the earth, it could account for the entire volume of water on earth. The objects would deliver enough moisture to earth to cover the entire planetary surface with 2.5 centimeters of water every 20,000 years, or about six kilometers over 4.5 billion years. Interestingly, the *Polar* satellite has also provided new evidence of how much water the earth is losing to space every year. In mid-1997 it was reported that polar winds are taking about four cubic meters of water out of the atmosphere each day in the form of ionized hydrogen and oxygen gases (EOS 1997b).

As more and more evidence accumulates, other experts are now beginning to reinterpret existing data or offer new suggestions for testing Frank's evidence and proving or disproving his theories, such as attempting to use satellite-borne lasers to measure the incoming water. In August 1997, another piece of supporting data was obtained. A satellite trailing the space shuttle *Discovery* observed hydroxyl in the atmosphere at the high northern latitudes (Dunn 1997; University of Iowa 1997b). During the shuttle flight one of the instruments aboard the satellite detected an unexpectedly large amount of hydroxyl, a product of the breakdown of water, above altitudes of 70 kilometers. One explanation for the existence of this water vapor could be the small comets proposed by Frank. Such a peak in hydroxyl has been detected before but was either ignored or interpreted differently. The first hint of unexpected amounts of water vapor at high attitudes actually came much earlier, in May 1987, and was reported by John Olivero and Dennis Adams at another American Geophysical Union meeting. Olivero and Adams initially set out to disprove Frank's theory by reanalyzing their collection of data on water-vapor concentrations in the mesosphere obtained with a microwave radiometer. Unexpectedly, they found temporary increases in the size and frequency of water vapor "bursts" one would expect if the small comets existed (*Science* 1992). Then, late in 1996, a team led by James M. Russell III of Hampton University in Virginia reported on its reanalysis of data gathered by the Halogen Occultation Experiment (HALOE) aboard the Upper Atmosphere Research Satellite (UARS) launched in September 1991. The data from the team's experiment, which used solar occultation to measure the presence of H_2O and other compounds in the mesosphere, also revealed a peak in water vapor at an altitude of about 70 kilometers (University of Iowa 1997a). These data were further confirmed by satellite observations taken in 1994 and published in 1996 (Conway et al. 1996).

The influx of small comets into earth's atmosphere can also be used to explain the composition and extreme height of noctilucent clouds—beautiful clouds only seen against a dark sky when illuminated by the setting sun. These clouds are seen over the polar regions during the summer months. They are thin clouds, wavy or banded, colored silver or bluish white that form at an altitude of nearly 90 kilometers in the coldest part of the upper atmosphere. No other cloud occurs so high in the sky. Atmospheric scientists have noted that these clouds require considerably more water vapor than can be expected from ocean evaporation. Rocket-borne experiments suggest that the clouds are composed of ice crystals formed around meteoric dust particles—a finding that supports the theory of small comets.

While the new evidence about the influx of small comets is increasingly convincing, it still remains enormously controversial. Among the difficulties with the theory is why the objects haven't been seen in other surveys that look for objects in near earth space. If there are very large numbers of these objects, they should be visible in observations of space. Observations of cratering on the moon also fail to support the theory, where the record includes estimates of craters down to the size of a few centimeters (Sawyer 1997). Another difficulty is that there seems to be no evidence from seismic detectors in place on the moon that such snowballs are hitting the surface there. Scientists such as the late Eugene Shoemaker of the U.S. Geological Survey, co-discoverer of the shattered comet Shoemaker-Levy 9, assert that at the high velocities observed the snowballs should deliver sufficient impact shock to be detected. Other scientists worry that if these comets are so common there should be more evidence of water on Venus, Mars, and elsewhere.

In response to the argument that there should be evidence from seismic instruments on the moon, Frank and his colleagues argue that these comets are more like "fluffy snowballs" than rocks and that they might not leave a permanent mark. They also argue that the lunar seismometers were not calibrated to detect the seismic signature of these kinds of objects. Low lunar gravity is used as a justification for the apparent lack of water on the moon. In this argument, the water vapor from the impact of small comets simply flies off, though some of the water molecules may condense in the crevices near the poles (University of Iowa 1997a). This was confirmed by data in March 1998 from the *Lunar Prospector* spacecraft. A possible explanation for the apparent discrepancy with the cratering record on the moon is that the observed shower may result from icy fragments of a single shattered comet traveling along roughly the same path around the sun as earth. If this had happened only within the last 100,000 years or so, it could account for the difference between current observations and moon crater record, though it would not support the theory that these comets represent the source of all water on earth.

In the past several years, there has been growing evidence that water is present in far larger quantities and far more places in space than previously expected. Water vapor has now been seen in the freezing atmospheres of Jupiter, Uranus, Neptune, and Saturn; there have been fleeting glimpses of water vapor in the sun; water has been discovered on the earth's moon, Saturn's moon Titan, and Jupiter's moon Europa (*Sky and Telescope* 1998). The presence of ice on our moon was first suggested by radar results from the *Clementine* spacecraft launched in 1994, but great uncertainties about those findings remained until March 1998 when data from the *Lunar Prospector* spacecraft confirmed that hundreds of millions of tons of water are locked up as ice in permanently shadowed regions of the lunar poles (Wilford 1998; *Sky and Telescope* 1998; NASA 1998).

In early April 1998, a team of U.S. astronomers announced the discovery of massive concentrations of water vapor being generated in a cloud of interstellar gas near the Orion nebula. The interstellar cloud is creating vast amounts of water—enough to fill the earth's oceans 60 times a day. Astronomers believe this discovery has implications for the origin of water in our own solar system. "It seems quite plausible that much of the water in the solar system was originally produced in a giant water vapor factory like the one we have observed in Orion," said Professor David Neufeld of the Physics and Astronomy Department at Johns Hopkins University—a member of the Infrared Space Observatory team that made the measurements (Brand 1998). These discoveries also help support the "growing belief that water on Earth may have been delivered by comets" (Radford 1998).

One of the reasons that the small comet theory has caused such a stir in the scientific community is that it suddenly may resolve a long-standing scientific debate about the origin of not only the earth's water but of the basic ingredients of life. One side of this debate has argued for nearly half a century that lightning strikes in earth's primitive seas and air produced the right kinds of chemicals for the emergence of life. The alternative theory suggests that comets bear the chemical precursors of life and that cometary impacts introduced these ingredients to earth. Astronomers recently noted that large comets like Hale-Bopp contain not just water but methanol, formaldehyde, carbon monoxide, hydrogen cyanide, hydrogen sulfide, and many carbon-based compounds—ideal for life. Proponents of the idea that the chemical precursors of life originated in the seas and atmosphere note that comets plunging through the atmosphere would generate heat sufficient to destroy most complex molecules. The small size of the newly discovered comets would make them ideal vehicles for bringing organic matter through the atmosphere since they would break up long before the onset of intense heating (University of Iowa 1997a; Broad 1997).

Many questions and uncertainties remain to be answered, and many scientists remain unconvinced. By late 1997, a number of scientific papers disagreeing with Frank's interpretation of the data were in preparation. At a packed meeting of the American

Geophysical Union (AGU) in December 1997 in San Francisco, George Parks (1997) presented a paper suggesting that the new observations from the *Polar* satellite are themselves instrument errors and not indicators of a real phenomena. His analysis, comparing images from the *Polar* visual imaging system (VIS) and the ultraviolet imager (UVI), concluded that the dark images were instrumental artifacts and noise, not real atmospheric holes. Frank (1997), in response, rejected that conclusion as "worthless" and reviewed evidence suggesting, among other problems, the UVI instrument was incapable of compensating for spacecraft motion adequately and hence did not have the sensitivity adequate for doing the kind of comparative analysis Parks was attempting. Frank also presented evidence that the data from *Polar* show a strong altitude dependence, which supports the conclusion that the atmospheric holes are real.

If the new hypothesis is true, it will essentially change fundamental ideas about the dynamics of growth of the oceans, the makeup of the atmosphere, the structure and composition of the solar system, and perhaps even the origin of life. Because of the importance of these questions, NASA has begun an interagency program to search for independent verification of the observations using a variety of space surveillance systems, including a focused radar fence in the southern United States, stereo-optical telescopes, a UHF radar system, and a 60-inch Schmidt telescope (Hoffman 1997). Results from some of these independent observations may be available by mid-1998, but regardless of the results, the debate over the origin of the earth's water is likely to continue for a long time as astrophysicists and others work to resolve uncertainties and unknowns. However, the initial ostracism and ridicule that followed Frank's first proposal may yet turn into the recognition and honors bold scientists deserve.

REFERENCES

Brand, D. 1998. "Astronomers discover a huge chemical 'factory' in interstellar space, suggesting origin of water in solar system." Press release, Cornell University, Ithaca, New York (April 9).

Broad, W.J. 1997. "Spotlight on comets in shaping of Earth." *The New York Times* (June 3) p. B7.

Conway, R.R., M.H. Stevens, J.G. Cardon, S.E. Zasadil, C.M. Brown, J.S. Morrill, and G.H. Mount. 1996. "Satellite measurements of hydroxyl in the mesosphere." *Geophysical Research Letters*, Vol. 23, No. 16, pp. 2093–2096 (August 1).

Dunn, M. 1997. "Discovery yields satellite surprise." Associated Press, August 11, 1997. Via *http://www.nacomm.org/news/1997/qtr3/atmosnow.htm*.

EOS. 1997a. "New 'small comet' images challenge researchers." *EOS: Transactions of the American Geophysical Union*, Vol. 78, No. 25, pp. 1–2.

EOS. 1997b. "In brief: Dehydration." *EOS: Transactions of the American Geophysical Union*, Vol. 78, No. 31, p. 2.

Frank, L.A. 1997. "The ultimate test for the reality of atmospheric holes." Presentation made December 9, 1997, at the American Geophysical Union fall conference, San Francisco, California.

Frank, L.A., and P. Huyghe. 1990. *The Big Splash*. Birch Lane Press, New York.

Frank, L.A., and J.B. Sigwarth. 1997a. "Transient decreases of Earth's far-ultraviolet dayglow." *Geophysical Research Letters*, Vol. 24, No. 19, pp. 2423–2426.

Frank, L.A., and J.B. Sigwarth. 1997b. "Simultaneous observations of transient decreases of Earth's far-ultraviolet dayglow with two cameras." *Geophysical Research Letters,* Vol. 24, No. 19, pp. 2427–2430.

Frank, L.A., and J.B. Sigwarth. 1997c. "Detection of atomic oxygen trails of small comets in the vicinity of Earth." *Geophysical Research Letters,* Vol. 24, No. 19, pp. 2431–2434.

Frank, L.A., and J.B. Sigwarth. 1997d. "Trails of OH emissions from small comets near Earth." *Geophysical Research Letters,* Vol. 24, No. 19, pp. 2435–2438.

Frank, L.A., J.B. Sigwarth, and J.D. Craven. 1986a. "On the influx of small comets into the earth's upper atmosphere: I. Observations." *Geophysical Research Letters,* Vol. 13, pp. 303–306.

Frank, L.A., J.B. Sigwarth, and J.D. Craven. 1986b. "On the influx of small comets into the earth's upper atmosphere: II. Interpretation." *Geophysical Research Letters,* Vol. 13, pp. 307–310.

Hoffman, R. 1997. "The search for small comets." Presentation made December 9, 1997, at the American Geophysical Union fall conference, San Francisco, California.

Kerr, R. 1988. "In search of elusive little comets." *Science,* Vol. 240, pp. 1403–1404 (June 10).

Kerr, R.A. 1997. "Spots confirmed, Tiny comets spurned." *Science,* Vol. 276, p. 1333 (May 30). Via *http://www.sciencemag.org.*

Kuhn, T. 1962. *The Structure of Scientific Revolutions.* The International Encyclopedia of Unified Science. Volume 2, No. 2, Chicago Press, Chicago, Illinois.

NASA. 1997a. Interplanetary ice comets. Thursday, July 24, 1997. Via *http://www.aufora. org/news/15.html.*

NASA. 1997b. "Polar spacecraft images support theory of interplanetary snowballs spraying Earth's upper atmosphere." Release 97-59. Via *ftp://pao.gsfc.nasa.gov/pub/ pao/releases/1997/97-59.htm*

NASA. 1998. "Ice on the moon." Via *http://lunar.arc.nasa.gov/science/results/ lunarice/index2.html.*

Parks, G. 1997. "Do ultraviolet and visible cameras on Polar detect cometesimals?" Presentation made December 9, 1997, at the American Geophysical Union fall conference, San Francisco, California.

Radford, T. 1998. "Masses of water discovered in space, scientists say." Scripps Howard, London, April 8, via *http://www.nando.net.*

Sawyer, K. 1997. "Evidence of cosmic snowballs starts scramble for explanations." *Washington Post* (June 1), p. A3. Via *http://wp2.washingtonpost.com/wp-srv/digest/ daily/june/02/cosmic.htm.*

Science. 1992. "Small comets/Big flap." *Science,* Vol. 257, pp. 622–623 (July 31).

Sky and Telescope. 1998. "Water on the moon; Europan ocean support deepens." *Sky and Telescope Weekly News Bulletin,* March 6, via *http://www.skypub.com/news/mar 0698.html.*

University of Iowa. 1997a. "The small comets Frequently Asked Questions list." Version 2.0, July 29, 1997. Via *http://smallcomets.physics.uiowa.edu/www/faq.htmlx.*

University of Iowa. 1997b. "Experiment aboard the space shuttle backs small comet discovery." Via *http://smallcomets.physics.uiowa.edu/www/crista.html.*

USA Today. 1997. "New evidence of space snowballs hitting atmosphere." October 23. Via *http://wsf2.usatoday.com/life/science/space/lss090.htm.*

Wilford, J.N. 1998. "Seeking water on moon in new NASA mission." *The New York Times* (January 4), p. 10.

Water Bag Technology

Beginning in the early 1980s, a few individuals and organizations began serious efforts to develop the technology to move large quantities of fresh water through the oceans in huge, sealed fabric bags. These efforts proceeded with little attention until the mid-1990s, when certain groups started testing pre-commercial versions of their technology and began the search for customers to both buy and sell water. By late 1997, the first commercial contracts had been signed and small-scale deliveries were being made. What had seemed like a far-fetched, eccentric idea in the early 1990s appeared headed for commercial application by early 1998.

Many people believe there is a market for water transferred in this manner, though no detailed study of the size of the market has yet been done. John Hayward, a World Bank water expert, was quoted in 1996 as saying, "One way or another water will be moved around the world as oil is now" (Lawrence 1996; Spragg 1997a). Worldwide, modified tankers already deliver water to regions willing to pay top dollar for small amounts of fresh water in emergency situations. Barges carry loads of fresh water to islands in the Bahamas as needed. Tankers deliver water to Japan, Taiwan, and Korea on an intermittent basis (Maritime Activity Reports 1996). Such fixed tanker systems are expensive and limited. This has led to the interest in the development of a way to move larger amounts of water more cheaply.

Economically, such transfers must be able to provide reliable water at a price below the marginal cost of alternatives, particularly desalination. This means a commercial system will have to be able to deliver water for under approximately $1.50 per cubic meter, and in many areas for well under $1.00 per cubic meter. Each of the bag manufacturers claims this is possible in the long run, but the economics of actual systems are just being revealed by actual deals.

There are also political constraints to such transfers. Regions and nations do not typically treat water resources as a standard commodity, capable of being bought and sold. Even within a single country, transfers from one region to another often generate significant opposition and hostility, even if that water can be considered "surplus." The political complications increase enormously when deliveries are to be made from one country to another.

Bag Technology

The idea of towing bags of fresh water through the ocean comes from the facts that large fixed pipelines have high capital costs, fresh water floats in salt water, and there has been a proven history of demand for water delivered in tankers. In theory, fresh water can be put into large buoyant fabric or plastic bags that can then be towed to almost any port. The technology for the movement of fresh water through the oceans in fabric bags appears on the verge of commercial viability. In its most basic form, the approach involves loading fresh water into a large sealed bag, towing it through the ocean to a buyer, and unloading it at the other end. While loading and unloading facilities can be constructed from existing technology, up until recently no large waterproof fabric bags capable of withstanding the strain of an ocean voyage had been fabricated.

A comprehensive water-delivery system would consist of a large array of facilities, including: (1) bag production facilities; (2) shoreside facilities to handle water from the source (i.e., pump stations, treatment facilities, water-storage tanks); (3) pipelines to a dock or offshore water-loading platform; (4) a water-loading platform to fill the bags; (5) a system to attach the bags to tugs for transport to delivery sites; (6) an off-loading facility to remove water from the bags; (7) facilities to handle empty bags and return them to the loading point; (8) mooring and bag handling facilities in the vicinity of the off-loading facility; and (9) ancillary facilities, such as water filtration plants, pump stations, and pipelines to the point of use.

One of the greatest difficulties with the large-scale transportation of water bags is making the size of the system large enough to permit economies of scale. Single or even several bags containing on the order of 1,000 cubic meters are unlikely to be commercially feasible. At the same time, making very large bags runs into problems with strain. Each of the groups manufacturing and testing bags has had to tackle the challenge of developing watertight fabrics sufficiently strong in large sizes to both move substantial amounts of water and hold up to the strain of ocean towing. A related problem is whether to develop single large bags or large numbers of smaller bags connected by some sort of coordinated towing system. Both alternatives are still under consideration.

The Players

Four groups are known to be actively working on water-bag technology. Detailed information on each groups' technology is limited because of concerns over patents, secrecy, and competition for customers, but the basic concepts are similar for each, with a few differences in approach and design.

Aquarius Water Trading and Transportation, Ltd.

In England, a company called Aquarius Water Trading and Transportation, Ltd., or Aquarius Holdings, Ltd., has been developing bags for bulk water transfers. Aquarius began in November 1989 as an independent venture with private funding and was formally set up in 1992. Aquarius now includes both private and corporate investors, including the Northumbrian Water Company, now part of Lyonnaise des Eaux, and some Norwegian companies (FTGWR 1997a). Aquarius Holdings offered public shares of the company through Karl Johan Fonds in Oslo starting in May 1997. Aquarius's commercial literature offers complete service, from bag supply and operation, water trading, construction and feasibility studies for loading and unloading systems, plus construction and operation of the infrastructure (Aquarius 1997).

Aquarius's bag design moved to the testing stage beginning in 1994 and 1995. Aquarius developed and modified its design through tank tests of small models at Southampton University in England in 1994 and 1995. It then constructed 20-cubic-meter scale models that were tested in the ocean off Scotland in 1995. In spring 1996 Aquarius produced two 12,000-cubic-meter bags. These bags were 180 meters in length and reportedly failed to perform adequately (Aquarius 1997). Aquarius then decided to build three smaller 750-cubic-meter bags and conducted sea trials in both English and Greek waters in fall 1996.

Aquarius currently has a small number of 750-cubic-meter bags available for use. These bags are only 2.1 meters deep and can easily be docked in small harbors. The bags were developed by Bruce Banks Sails in Southampton, United Kingdom, with technical assistance from the Institute of Marine Aerodynamics at Southampton University. They use a polyurethane fabric overlaid with a polymer developed for automobile tires (FTGWR 1997b), with an ultraviolet coating on the outside. A food-grade coating lines the inside. Aquarius plans to begin manufacturing and testing a 2,000-cubic-meter bag in 1998. Currently the bags cost about $100,000 to manufacture and have an expected life span of five years.

Aquarius has begun the first commercial deliveries of fresh water by bag. In mid-1997, Aquarius Holdings, Ltd., held letters of intent from various city councils on Angistri and Aegina islands in the Saronic Gulf. Many Greek islands have water systems that are subject to periods of shutdown during dry seasons. Some of these islands currently rely on water deliveries in tankers or on unsustainable groundwater pumping.

A contract calling for 290,000 cubic meters of water to be delivered by December 1997 was signed by Aquarius and authorities on the island Aegina in mid-1997, subject to the issuance of a certificate by the Greek Ministry of Health (Aquarius 1997). According to the Financial Times Global Water Report (1997a), Aquarius began commercial deliveries of water in May 1997 from Piraeus, Greece, to Aegina, a distance of only 20 kilometers. Aegina currently also receives water from commercial tankers, and Aquarius's deliveries supplement those supplies, at a cost of approximately $1.20 to $1.50 per cubic meter (FTGWR 1997a). The Greek operation includes a joint venture set up in Athens with the largest towing and salvage company in Greece, the Matzas Group. The Aquarius–Matzas 750-horsepower (hp) tug is capable of pulling one bag. A 1,200-hp tug is reportedly required to pull three of the 750-cubic-meter bags in a delta formation.

Aquarius believes the current market for transporting water to the Greek islands may exceed 200 million metric tons per year. An additional benefit of this market is that it doesn't require an international transfer of water from one country to another, with the associated political and legal complications (FTGWR 1997a). Aquarius also says it has had indications of interest from Spain and as far away as Fiji, though no contracts have been signed yet.

Nordic Water Supply Company

A second major player in this area is the Nordic Water Supply (NWS) Company in Oslo, Norway. NWS developed its bag technology through a research program involving the Norwegian Research Council and the Norwegian Industrial and Regional Development Fund. The Turkish Ministry of Energy and Natural Resources also reportedly helped finance the development costs. The Turkish Government is interested in using the system to deliver water from Turkey to the proclaimed Turkish Republic of Northern Cyprus (FTGWR 1997b). The NWS bags, which were built by Norway's Protan company, are designed so they can be rolled up and stored on the deck of the tug that delivered them (FTGWR 1997c). Testing has occurred over the past two years in the Sogne Fjord of Norway. In 1997, NWS tested 10,800-cubic-meter bags, which reportedly failed due to towing stresses (FTGWR 1997a).

In October 1997, a contract was signed between Nordic Water Supply and Turkey that calls for NWS to deliver water to northern Cyprus. Within two years at least seven million cubic meters of water are to be delivered annually at a cost of $4.1 million per year, with volumes growing over time (FTGWR 1997c). This is equivalent to a cost of $0.55 per cubic meter, well below the price of desalinated water.

The NWS project is ambitious. In addition to the 10,800-cubic-meter bag described above, NWS is building another 10,000-cubic-meter bag and has ordered three 20,000-cubic-meter bags (FTGWR 1997c). The Turkish contract ultimately will require a delivery of over 75,000 cubic meters per day for the 100-kilometer trip from the Turkish loading terminal at Soguksu to northern Cyprus. Expansion of the project to the larger volumes called for in the contract will require completion of the Manavgat project, in which water from the Manavgat River will be fed to an offshore terminal. Construction of the terminal is expected to be completed in 1998.

Spragg Waterbags

Inventor Terry Spragg, working out of southern California, has been developing water-bag technology for many years, mostly on his own, with the support of privately hired scientists and consultants. While Aquarius and Nordic Water Supply appear to have obtained the first commercial contracts, Spragg's work in this field appears to predate anyone else's. Spragg began work on the "bag" option in the late 1980s, after having evaluated and then rejected the idea of towing icebergs for freshwater supply in the 1970s. Spragg has focused his efforts on developing several innovative technological solutions to some of the difficult problems of ocean transport, particularly the idea of linking multiple bags together in a "train" to greatly increase the total volume moved at one time. For obvious reasons, Spragg has named (and trademarked) his bags as "Spragg Bags."

Spragg's first field test occurred in December 1990 on a 75-meter-long bag containing nearly 3,000 cubic meters of fresh water in Port Angeles harbor, Washington, using a conventional towing system and an early fabric design. This test demonstrated that the concept was sound, though this early fabric and the connection system eventually failed due to towing stresses. Mechanical redesign and testing were done with engineers from MIT after the 1990 field test. New towing and longitudinal stress tests were conducted on a scale model in July 1993. A redesigned proprietary fabric manufactured in Europe and a large zipper connection system designed by Gianfranco Germani of Milan, Italy, were produced in September 1993 (Maritime Activity Reports 1996).

In early 1996, two bags were built in Seattle, Washington. In order to address the problem of size and bag stress, Spragg has developed a way of linking large numbers of bags together in a train using this new zipper system, a fabric sleeve on the end of each bag, and a shroud to connect the bags to the towing tug. For the 1996 test, the two bags were each approximately 9 meters in diameter and 75 meters long. These bags were inflated with air and joined together using the zipper sleeve system. An open ocean test was conducted by Spragg in April 1996. Two bags connected in a string were towed by a small tug. Each bag held nearly 3,000 cubic meters of water (Gleaves 1996). During this test, one bag developed a leak, though the connection system performed successfully.

Spragg estimates that his bags could ultimately average about 14 meters in diameter and approach 150 meters in length, containing as much as 17,000 cubic meters of water. Dozens of these bags could be combined in a "train" and theoretically as many as 60 such bags could be towed by a single vessel, according to Spragg, though no sea trial of such a large system has yet been attempted. His published engineering estimates from tests by his engineering team suggest that the fabric he has developed and the connection system would be sufficiently strong to permit a 4,300-hp tug to pull this system (Spragg 1997). In full commercial operation, the bags would be manufactured in a variety of sizes depending on the loading and unloading requirements and the nature of the facilities available.

The focus of Spragg and Associates's efforts to date has been bag development and a search for a market where the technology could be applied commercially. Spragg has discussed sea trials of multiple bag tows from Washington State to California and from Turkey to Israel and Gaza in the Middle East, but no date has been set. One commercial possibility for Spragg is water transfers within the United States from the Pacific Northwest to water-short areas of California (Spragg 1997). Spragg has a proposal with the U.S. State Department for a demonstration voyage of his technology through the Middle East region.

Medusa Corporation

The Medusa Corporation of Calgary, Canada, has also promoted the idea of water bags. Headed by James Cran, Medusa believes single massive bags offer more economies of scale than trains of smaller bags. Medusa is reportedly seeking to develop bags capable of holding up to 1,750,000 cubic meters (FTGWR 1997d). Cran reports that a manufacturer of framework structures, Birdair Corporation in New York, says such gigantic bags are technically feasible, though no such bags have yet been tested (Lawrence 1996). Little public information is available on Medusa.

Likely Markets and Costs

While the ultimate goal of these projects is to offer long-term reliable freshwater delivery, the earliest uses of such technology are likely to be limited to two types of situations: extremely water-short coastal regions with a reliable demand for expensive water; and temporary deliveries in emergency situations, such as droughts, natural catastrophes, or health disasters.

The first commercial market opened in the Greek Islands in May 1997 and, as mentioned earlier, is being serviced by small deliveries from the Aquarius group. The second commercial contract, and the first for an "international" delivery of water, was the one signed on October 3, 1997, by Nordic Water Supply and Turkey for delivery of water from Turkey to Northern Cyprus. This contract drew immediate protests from the government of Cyprus in Nicosia, which doesn't recognize the "Turkish Republic of North Cyprus." The Norwegian government noted that there are no UN mandatory sanctions against the Turkish regime in northern Cyprus that would prohibit such a deal, but the political situation between Turkey and Greece, and between the government of Cyprus and the "Republic of North Cyprus," will complicate any effective long-term transport in this region.

Another source of water long considered for water bags is the controversial Manavgat Water Supply Project in Turkey. Turkey has been building a water transfer facility at the mouth of the Manavgat River that will have the capability of loading up to 250,000 cubic meters of water a day into tankers or waterbags through a set of pipelines connected to offshore mooring buoys. The Manavgat's annual average flow is about 44,000 million cubic meters a year, and there has been talk of transferring as much as 1,900 million cubic meters per year to Mediterranean and Red Sea markets (FTGWR 1997d). The price of this water could be as low as $0.30 per cubic meter, though sources say that Turkey is considering a fee of around $1.00 per cubic meter, which would make it extremely expensive when delivery costs are included.

Long-term commercial viability of this concept is further complicated by the low value of water compared to most other products. Fresh water just isn't worth much money, except in the bottled, "luxury" water market. At $1.00 per cubic meter of water, the delivery of 1,000 cubic meters will only gross $1,000. The delivery of 1,000 cubic meters of oil at $20 per barrel would gross over $125,000. Another part of the economic difficulty includes finding a reliable source of water, with guarantees for long-term deliveries and specified standards for water quality.

The cost of water at the delivery site depends on the cost of water at its source, number of bags towed at a time, towing costs, distance between source and market, depreciation costs and assumptions, and operations costs for loading and off-loading facilities. Most cost estimates from the companies remain proprietary, but some information on the existing contracts is available. The price reportedly paid for the 1997 Aquarius deliveries in the Greek Islands ranges from $1.20 to $1.50 per cubic meter.

As mentioned earlier the contract signed in October 1997 by Nordic Water Supply and Turkey works out to an equivalent cost of $0.55 per cubic meter, which is well below the price of desalinated water. The Cyprus government purchases water from privately constructed desalination plants for $1.05 per cubic meter (FTGWR 1997c).

Spragg and Associates have not yet signed a formal contract, but they claim to be willing to deliver water for around $0.80 per cubic meter for delivery distances up to 1,600 kilometers (Spragg, personal communication, 1997).

References

Aquarius. 1997. Information from Aquarius Holding, Ltd. October 1997. Via *http://www. aquariuswater.co.uk/news.htm.*

Financial Times Global Water Report. 1997a. "Aquarius: Water bags seize commercial niche." September 11, Issue 30, pp. 15–16.

Financial Times Global Water Report. 1997b. "Finding the market: North Sea and Levantine prospects." September 11, Issue 30, p. 15.

Financial Times Global Water Report. 1997c. "Water bags: Norway, Turkey, and the Cyprus imbroglio." October 9, p. 1.

Financial Times Global Water Report. 1997d. "Manavgat: Money bags and water bags." March 12, Issue 18, pp. 6–7.

Gleaves, K. 1996. "Spragg Bags cruise Puget Sound." In *West Coast Review* (August 1996).

Lawrence, R. 1996. "A water solution in the bag." *The Journal of Commerce*, Vol. 409, No. 28798 (August 29).

Maritime Activity Reports. 1996. Via *http://www.marinelink.com/aug96/mr0811.html.*

Spragg. 1997. Information from Spragg and Associates "Spragg Water Bags." Via *http://www.waterbag.com/tech.html.*

Treaty Between the Government of the Republic of India and the Government of the People's Republic of Bangladesh on Sharing of the Ganga/Ganges Waters at Farakka

On December 12, 1996, India and Bangladesh signed a formal treaty that moves toward resolving their long-standing dispute over the Farakka Barrage and water flows in the Ganges/Brahmaputra river system. This agreement, made possible in part by changes in both the Indian and Bangladeshi governments, covers the sharing of Ganges river flows between January 1 and May 31, when water flows are low and demands are high. In addition to specifying water allocations in normal and dry periods, both governments agree to conclude water-sharing treaties and agreements over more than 50 other shared rivers and to find solutions to augmenting the flow of the Ganges in the dry season.

The Farakka Barrage was built by India across the Ganges River in the late 1960s and early 1970s just upstream of the Bangladeshi border to divert water into the Hooghly River for irrigation and navigation. The dam was built without international agreement and particularly affected dry-season flows to Bangladesh, leading to serious political disputes between the two nations. Several interim agreements over the operation of the Farakka Barrage were signed, ignored, and voided over the years after construction of the dam (see Chapter 4). In the early 1990s, dry-season flow in the Ganges reaching Bangladesh fell to one-tenth of average dry-season flows, causing serious problems for downstream agriculture and leading to new regional interest in negotiating a solution, and ultimately, to the new agreement. There is widespread hope that the new treaty will help to permanently resolve problems between the two nations over the shared water resources. In particular, there is hope that the Joint Rivers Commission specified in the treaty will work to resolve disputes and to set an example for other international water-sharing agreements. The full text of the treaty is presented below.

Signed on December 12, 1996.

The Government of the Republic of India and the Government of the People's Republic of Bangladesh,

Determined to promote and strengthen their relations of friendship and good neighbourliness,

Inspired by the common desire of promoting the well-being of their people,

Being desirous of sharing by mutual agreement the waters of the international rivers flowing through the territories of the two countries and of making the optimum utilisation of the water resources of their region in the fields of flood management, irrigation, river basin development and generation of hydro-power for the mutual benefit of the peoples of the two countries,

Recognizing that the need for making an arrangement for sharing of the Ganga/Ganges waters at Farakka in a spirit of mutual accommodation and the need for a solution to the long-term problem of augmenting the flows of the Ganga/Ganges are in the mutual interests of the peoples of the two countries,

Being desirous of finding a fair and just solution without affecting the rights and entitlements of either country other than those covered by this Treaty, or establishing any general principles of law or precedent,

Have agreed as Follows:

Article –I

The quantum of waters agreed to be released by India to Bangladesh will be at Farakka.

Article –II:

(i) The sharing between India and Bangladesh of the Ganga/Ganges waters at Farakka by ten day periods from the 1st January to the 31st May every year will be with reference to the formula at Annexure I and an indicative schedule giving the implications of the sharing arrangement under Annexure I is at Annexure II.

(ii) The indicative schedule at Annexure II, as referred to in sub para (i) above, is based on 40 years (1949–1988) 10-day period average availability of water at Farakka. Every effort would be made by the upper riparian to protect flows of water at Farakka as in the 40-years average availability as mentioned above.

(iii) In the event flow at Farakka falls below 50,000 cusecs in any 10-day period, the two governments will enter into immediate consultations to make adjustments on an emergency basis, in accordance with the principles of equity, fair play and no harm to either party.

Article –III

The waters released to Bangladesh at Farakka under Article –I shall not be reduced below Farakka, except for reasonable uses of waters, not exceeding 200 cusecs, by India between Farakka and the point on the Ganga/Ganges where both its banks are in Bangladesh.

Article –IV

A Committee consisting of representatives appointed by the two Governments in equal numbers (hereinafter called the Joint Committee) shall be constituted following the signing of the Treaty. The Joint Committee shall set up suitable teams at Farakka and Hardinge Bridge to observe and record at Farakka the daily flow below Farakka barrage, in the Feeder canal, at the Navigation Lock, as well as at the Hardinge Bridge.

Article –V

The Joint Committee shall decide its own procedure and method of functioning.

Article –VI

The Joint Committee shall submit to the two Governments all data collected by it and shall also submit a yearly report to both the governments. Following submission of the reports the two Governments will meet at appropriate levels to decide upon such further actions as may be needed.

Article –VII

The Joint Committee shall be responsible for implementing the arrangements contained in this Treaty and examining any difficulty arising out of the implementation of the above arrangements and of the operation of the Farakka Barrage. Any difference or dispute arising in this regard, if not resolved by the Joint Committee, shall be referred to the Indo-Bangladesh Joint Rivers Commission. If the difference or dispute still remains unresolved, it shall be referred to the two governments which shall meet urgently at the appropriate level to resolve it by mutual discussion.

Article –VIII

The two Governments recognise the need to cooperate with each other in finding a solution to the long term problem of augmenting the flows of the Ganga/Ganges during the dry season.

Article –IX

Guided by the principles of equity, fairness and no harm to either party, both the Governments agree to conclude water sharing Treaties/Agreements with regard to other common rivers.

Article –X

The sharing arrangements under this Treaty shall be reviewed by the two Governments at five year intervals or earlier, as required by either party and needed adjustments, based on principles of equity, fairness and no harm to either party made thereto, if necessary. It would be open to either party to seek the first review after two years to assess the impact and working of the sharing arrangements as contained in this Treaty.

Article –XI

For the period of this Treaty, in the absence of mutual agreement on adjustments following review as mentioned in Article X, India shall release downstream of Farakka Barrage, water at a rate not less than 90% (ninety percent) of Bangladesh's share according to the formula referred to in Article II, until such time as mutually agreed flows are decided upon.

Article –XII

This Treaty shall enter into force upon signatures and shall remain in force for a period of thirty years and it shall be renewable on the basis of mutual consent.

In witness whereof the undersigned, being duly authorised thereto by the respective Governments, have signed this Treaty.

Done at New Delhi, 12th December, 1996, in Hindi, Bangla and English languages. In the event of any conflict between the texts, the English text shall prevail.

Signed: the Prime Minister of the Republic of India; the Prime Minister of the People's Republic of Bangladesh.

ANNEXURE—I

Availability at Farakka	Share of India	Share of Bangladesh
70,000 cusecs or less	50%	50%
70,000–75,000 cusecs	Balance of flow	35,000 cusecs
75,000 cusecs or more	40,000 cusecs	Balance of flow

Subject to the condition that India and Bangladesh each shall receive guaranteed 35,000 cusecs of water in alternative three 10-day periods during the period March 1 to May 10.

ANNEXURE—II

(Indicative schedule giving the implications of the sharing arrangement under Annexure—I for the period 1st January to 31st May). Figures in cusecs.

Period	Average of Actual Flow (1949–1988)	India's Share	Bangladesh's Share
January			
1–10	107,516	40,000	67,516
11–20	97,673	40,000	57,673
21–31	90,154	40,000	50,154
February			
1–10	86,323	40,000	46,323
11–20	82,839	40,000	42,839
21–28	79,106	40,000	39,106
March			
1–10	74,419	39,419	35,000
11–20	68,931	33,931	35,000
21–31	63,688	35,000	29,688
April			
1–10	63,180	28,180	35,000
11–20	62,633	35,000	27,633
21–30	60,992	25,992	35,000
May			
1–10	67,251	35,000	32,351
11–20	73,590	38,590	35,000
21–31	81,834	40,000	41,854

United Nations Convention on the Law of the Non-Navigational Uses of International Watercourses

On May 21, 1997, the United Nations General Assembly adopted the Convention on the Law of the Non-Navigational Uses of International Watercourses. The convention should now be considered the leading body of international law regarding the development and protection of transboundary watercourses. The International Law Commission developed the new Convention after 27 years of work and the final product reflects the difficult and complex political nature of shared international waters. The convention provides principles and rules to guide states in negotiating future agreements on shared watercourses. The General Assembly adopted it with 103 votes in favor to 3 against (Turkey, China, and Burundi), with 27 abstentions. Despite the important rejection of China and Turkey—two countries that dominate water politics in their regions—and some reservations about the rights of upstream riparians, this new agreement will help focus more serious attention to issues of international freshwater conflict and may offer guiding principles for reducing the risks of those conflicts (see Chapter 4).

The convention is an effort to codify customary principles of international water law developed through state practice and specific case law and to set out procedural requirements for notification and consultation among nations regarding the use of international watercourses. International rivers pose particularly difficult problems for international law. River systems are part of complex hydrological units and activities of upstream nations can have direct effects on the nature of the river downstream. Both natural events and human actions have the potential to change the quantity, quality, or use of the water in another part of the watershed. Extensive development of water resources in an upstream area can reduce the flow or quality of water to lower riparians. Thus, efforts to address multi-jurisdictional issues require joint agreement and coordinated actions and would benefit from consistent and agreed-upon standards and principles. This need helped stimulate the efforts to develop the new convention.

Among the most important portions of the convention are Article 7, which obliges states to take all appropriate measures to prevent harm to other states from their use of water; Article 8, which obliges watercourse states to cooperate on the basis of equality, integrity, mutual benefit, and good faith in order optimally to use and protect shared watercourses; and Article 33, which offers provisions for the peaceful settlement of disputes by negotiation, mediation, arbitration, or appeal to the International Court of Justice. Among the weaknesses of the convention are the inherent conflict between equitable uses and the obligation not to cause appreciable harm. The principles set forth in the convention are not binding and offer little concrete guidance to countries trying to allocate scarce water resources, but they may prove effective at encouraging states to negotiate water disputes and to implement joint management activities.

The convention is now open for signing until 20 May 2000 and will come into force on the 90th day following the deposit of the 35th instrument of ratification at the United Nations. The full text of the convention is presented on the following pages.

CONVENTION ON THE LAW OF THE NON-NAVIGATIONAL USES OF INTERNATIONAL WATERCOURSES

Report of the Sixth Committee convening as the

Working Group of the Whole

Chairman: Mr. Chusei YAMADA (Japan)

I. INTRODUCTION

1. Pursuant to paragraph 2 of General Assembly resolution 51/206 of 17 December 1996, the Working Group of the Whole of the Sixth Committee, which was convened under General Assembly resolution 49/52 of 9 December 1994, held its second session from 24 March to 4 April 1997 to elaborate a framework convention on the law of the non-navigational uses of international watercourses.[1]

2. The Working Group was chaired, as was the case in the first session, by Mr. Chusei Yamada (Japan) and the Drafting Committee was chaired, also as during the first session, by Mr. Hans Lammers (Netherlands) (A/C.6/51/L.3, para. 2). Mr. Robert Rosenstock, former Special Rapporteur of the International Law Commission on the topic, acted as Expert Consultant to the Working Group.

3. The Working Group held 12 meetings from 24 March to 4 April 1997. The views of the representatives who spoke during those meetings are reflected in the relevant summary records (A/C.6/51/SR.51 to 62).

4. The Drafting Committee held six meetings, from 24 to 27 March 1997. The Chairman of the Drafting Committee introduced the report of the Drafting Committee (A/C.6/51/NUW/L.1/Rev.1 and Add.1). The statements by the Chairman of the Drafting Committee introducing its report are reflected in the relevant summary records (A/C.6/51/SR.24 and 53).

5. In the course of the discussion of the text of the draft convention, the following were appointed coordinators of informal consultations: Mr. Jean François Pulvenis (Venezuela), on the preamble; Mr. Robert Harris (United States of America) on article 3, paragraphs 2 and 4; Mr. Attila Tanzi (Italy) on article 3, paragraph 3; Ms. Socorro Flores (Mexico) on articles 5 and 6; Mr. Tobias Nussbaum (Canada) on article 7; Mr. Rolf Welberts (Germany) on article 8; Mr. M. P. Vorster (South Africa) on article 10 (2); and Mr. A. K. H. Morshed (Bangladesh) on articles 20 and 22, as regards the term "ecosystems".

II. CONSIDERATION OF PROPOSALS

6. The Working Group and the Drafting Committee had before them the draft articles adopted by the International Law Commission on the topic[2] and the text of their previous reports, including the oral report of the Chairman of the Drafting Committee (A/C.6/51/L.3; A/C.6/51/NUW/WG/L.1/Rev.1 and A/C.6/51/SR.24), as well as a

preliminary draft preamble and final clauses prepared by the Secretariat (A/C.6/51/NUW/DC/CRP.2).

7. The Working Group and the Drafting Committee also had before them the following proposals submitted by States. <u>In the Drafting Committee</u>: proposal submitted by Finland, India and Romania for the preamble to the Convention (A/C.6/51/NUW/DC/CRP.3); proposals submitted by Ireland to amend provisions on final clauses (A/C.6/51/NUW/DC/CRP.4); proposal submitted by Finland for the preamble to the Convention (A/C.6/51/NUW/DC/CRP.7); proposal submitted by Jordan for article 7 (A/C.6/51/NUW/DC/CRP.8); proposal by Ethiopia with respect to article 2 (A/C.6/51/NUW/DC/CRP.9); proposal for article 33 by the Syrian Arab Republic and Switzerland (A/C.6/51/NUW/DC/CRP.10); and amendments proposed by Guatemala to the proposal for article 33 contained in document A/C.6/51/NUW/DC/CRP.10 (A/C.6/51/NUW/DC/CRP.11). <u>In the Working Group</u>: proposals by the United States of America concerning articles 1, 2, 3 and 29 (A/C.6/51/NUW/WG/CRP.1); proposal by Canada concerning article 1 (A/C.6/51/NUW/WG/CRP.2); proposal by Romania on the preamble (A/C.6/51/NUW/WG/CRP.3); proposal by Turkey for article 1 (A/C.6/51/NUW/WG/CRP.4); proposed amendments by the Swiss delegation (A/C.6/51/NUW/WG/CRP.5); proposal submitted by Romania for article 3 (A/C.6/51/NUW/WG/CRP.6); proposal submitted by India for article 3 (A/C.6/51/NUW/WG/CRP.7); proposals submitted by Israel for article 3 (A/C.6/51/NUW/WG/CRP.8); proposals submitted by Ethiopia (A/C.6/51/NUW/WG/CRP.9); proposal submitted by Italy for article 3 (A/C.6/51/NUW/WG/CRP.10); proposals submitted by the Netherlands for articles 5, 8 and 10 (A/C.6/51/NUW/WG/CRP.11); proposal submitted by Turkey for article 3 (A/C.6/51/NUW/WG/CRP.12); proposal submitted by Iraq for article 5 (A/C.6/51/NUW/WG/CRP.13); proposal submitted by South Africa for article 4 (A/C.6/51/NUW/WG/CRP.14); proposal submitted by France for articles 1 and 3 (A/C.6/51/NUW/WG/CRP.15); proposal submitted by the Netherlands for article 3 (A/C.6/51/NUW/WG/CRP.16); proposals submitted by the Czech Republic for article 6 (A/C.6/51/NUW/WG/CRP.17); proposals submitted by Finland for articles 6 and 7 (A/C.6/51/NUW/WG/CRP.18); proposal submitted by Israel for article 10 (A/C.6/51/NUW/WG/CRP.19); proposal submitted by Egypt for article 7 (A/C.6/51/NUW/WG/CRP.20); proposal submitted by China for article 3 (A/C.6/51/NUW/WG/CRP.21); proposals submitted by Iraq for articles 3, 4, 7 and 12 (A/C.6/51/NUW/WG/CRP.22); proposal submitted by Romania for article 7 (A/C.6/51/NUW/WG/CRP.23); proposal submitted by Turkey for article 7 (A/C.6/51/NUW/WG/CRP.24); proposals submitted by Finland for articles 1 and 9 (A/C.6/51/NUW/WG/CRP.25); proposal submitted by Canada for article 7 (A/C.6/51/NUW/WG/CRP.26); proposal submitted by Hungary and Romania for article 8 (A/C.6/51/NUW/WG/CRP.27); proposals submitted by India for articles 5 and 6 (A/C.6/51/NUW/WG/CRP.28); proposal submitted by Egypt for article 3 (A/C.6/n51/NUW/WG/CRP.29); proposal submitted by the Sudan for article 10 (A/C.6/51/NUW/WG/CRP.30); proposal submitted by Mexico for articles 7 and 9 (A/C.6/51/NUW/WG/CRP.31); proposals submitted by Israel for articles 2 and 3 (A/C.6/51/NUW/WG/CRP.32); proposals submitted by India for the name of the Convention and for article 9 (A/C.6/51/NUW/WG/CRP.33); proposals submitted by the Russian Federation for articles 8, 8 <u>bis</u> and 10 (A/C.6/51/NUW/WG/CRP.34); proposals submitted by Canada, Germany, Italy, Romania and the United States of America for articles 5 and 6

(A/C.6/51/NUW/WG/CRP.35); proposals submitted by the Russian Federation for articles 24 and 25 (A/C.6/51/NUW/WG/CRP.36); proposal by Turkey concerning part III of the draft articles (articles 11-19) (A/C.6/51/NUW/WG/CRP.37); proposals submitted by the Netherlands for articles 12 and 14 (A/C.6/51/NUW/WG/CRP.38); proposal submitted by Iraq for article 3 (A/C.6/51/NUW/WG/CRP.39); proposal submitted by South Africa for articles 6, 8 and 10 (A/C.6/51/NUW/WG/CRP.40); proposal submitted by the Syrian Arab Republic for article 5 (A/C.6/51/NUW/WG/CRP.41); proposals submitted by Canada for article 7 (based on informal attempts at coordination submitted by Canada to the Chairman of the Working Group) (A/C.6/51/NUW/WG/CRP.42); proposals submitted by the Syrian Arab Republic for articles 7 and 8 (A/C.6/51/NUW/WG/CRP.43); proposals submitted by Romania for article 14 (A/C.6/51/NUW/WG/CRP.44); proposals submitted by Finland for article 33 (A/C.6/51/NUW/WG/CRP.45); proposals submitted by Iraq for cluster III (articles 11-19) and article 33 (A/C.6/51/NUW/WG/CRP.46); proposal submitted by the Sudan for article 33 (A/C.6/51/NUW/WG/CRP.47); proposal submitted by Finland for article 21 (A/C.6/51/NUW/WG/CRP.48); proposal submitted by the Syrian Arab Republic for article 33 (A/C.6/51/NUW/WG/CRP.49); proposals submitted by the Netherlands for articles 18, 20, 21, 25 and 26 (A/C.6/51/NUW/WG/CRP.50); proposals submitted by Ethiopia for articles 6 and 7 (A/C.6/51/NUW/WG/CRP.51); proposals submitted by China for articles 20, 22 and 33 (A/C.6/51/NUW/WG/CRP.52); proposal submitted by Egypt for article 6 (A/C.6/51/NUW/WG/CRP.53); proposal submitted by the United Kingdom of Great Britain and Northern Ireland for article 6 (A/C.6/51/NUW/WG/CRP.54); proposals submitted by France concerning article 33 (A/C.6/51/NUW/WG/CRP.55); proposals submitted by Romania for articles 24 and 25 (A/C.6/51/NUW/WG/CRP.56); proposal submitted by Egypt for article 2, paragraph (b) (A/C.6/51/NUW/WG/CRP.57); proposal submitted by Romania for article 33 (A/C.6/51/NUW/WG/CRP.58); proposals submitted by South Africa for articles 11, 12 and 18 (A/C.6/51/NUW/WG/CRP.59); proposal submitted by Turkey concerning article 2 (A/C.6/51/NUW/WG/CRP.60); proposal submitted by the Russian Federation for article 32 (A/C.6/51/NUW/WG/CRP.61); proposals submitted by Guatemala for articles 32 and 33 (A/C.6/51/NUW/WG/CRP.62/Rev.1); proposal submitted by the Russian Federation concerning final clauses (A/C.6/51/NUW/WG/CRP.63/Rev.1); proposal submitted by the Syrian Arab Republic for article (2 or 5) (A/C.6/51/NUW/WG/CRP.64); proposal submitted by the Netherlands for article 17 (A/C.6/51/NUW/WG/CRP.65); proposal submitted by Jordan for article 18 (A/C.6/51/NUW/WG/CRP.66); proposals submitted by Ethiopia for articles 13 and 14 (A/C.6/51/NUW/WG/CRP.67); proposal submitted by Ambassador F. M. Hayes (Ireland) for article 7 (A/C.6/51/NUW/WG/CRP.68); proposal submitted by Guatemala for article 32 (A/C.6/51/NUW/WG/CRP.69); proposals submitted by Portugal, Venezuela, the United States of America and the Netherlands for article 21, paragraph 3 (A/C.6/51/NUW/WG/CRP.70); proposal submitted by Finland, Greece and Italy for article 33 (A/C.6/51/NUW/WG/CRP.71); revised text of article 7 proposed by Austria, Canada, Portugal, Switzerland and Venezuela (A/C.6/51/NUW/WG/CRP.72); proposed interpretation of article 18 of the draft Convention submitted by Poland (A/C.6/51/NUW/WG/CRP.73); proposal submitted by the United Kingdom of Great Britain and Northern Ireland for article 29 (A/C.6/51/NUW/WG/CRP.74); proposal submitted by Italy for article 3 (A/C.6/51/NUW/WG/CRP.75); proposals submitted by the United States of America concerning articles 2 and 3 (A/C.6/51/NUW/WG/CRP.76); proposal submitted by the Syrian Arab Republic for article 6 (A/C.6/51/NUW/WG/CRP.77); proposal submitted by Canada and Venezuela concerning the preamble (A/C.6/51/NUW/WG/CRP.78); proposal submitted by Romania (A/C.6/51/NUW/WG/CRP.79); proposals by the Russian Federation concerning articles 4 and 5 (A/C.6/51/NUW/WG/CRP.80); proposal submitted by the Netherlands on behalf of the European Community and the United States of America concerning article 2 (A/C.6/51/NUW/WG/CRP.81); proposal submitted by China for article 33 (A/C.6/51/NUW/WG/CRP.82); proposal from the Chairman of the Drafting Committee for article 33 (A/C.6/51/NUW/WG/CRP.83); proposal submitted by Argentina, Austria, Egypt, Germany, Greece, Hungary, the Islamic Republic of Iran, Italy, Jordan, Malaysia, Mali, Portugal, Romania, the Syrian Arab Republic, Tunisia, the United Kingdom of Great Britain and Northern Ireland, the United States of America, Venezuela and Viet Nam concerning article 8 (A/C.6/51/NUW/WG/CRP.84/Rev.1); proposal submitted by Italy for article 7 (A/C.6/51/NUW/WG/CRP.85); proposal submitted by the Russian Federation concerning article 25 (A/C.6/51/NUW/WG/CRP.86); proposal by the Chairman of the Drafting Committee (A/C.6/51/NUW/WG/CRP.87); restructured version submitted by Guatemala of the proposal from the Chairman of the Drafting Committee concerning article 33 as contained in document A/C.6/51/NUW/WG/CRP.83 (A/C.6/51/NUW/WG/CRP.88); report of the Coordinator of article 3 (A/C.6/51/NUW/WG/CRP.89); proposal submitted by the Russian Federation for article 2 on a new paragraph (d) (A/C.6/51/NUW/WG/CRP.90); amendment by China to the revised text of article 7 proposed by Austria, Canada, Portugal, Switzerland and Venezuela (A/C.6/51/NUW/WG/CRP.91); proposal submitted by the United States of America for article 2 (A/C.6/51/NUW/WG/CRP.92); proposal submitted by Egypt concerning the relationship between articles 5 and 7 (A/C.6/51/NUW/WG/CRP.93); Chairman's proposal for articles 5, 6 and 7 (A/C.6/51/NUW/WG/CRP.94); and report of the Coordinator for article 10 (2) (A/C.6/51/NUW/WG/CRP.95).

8. During the elaboration of the draft Convention on the Law of the Non-navigational Uses of International Watercourses, the Chairman of the Working Group of the Whole took note of the following statements of understanding pertaining to the texts of the draft Convention:

As regards <u>article 1</u>:
(a) The concept of "preservation" referred to in this article and the Convention includes also the concept of "conservation";
(b) The present Convention does not apply to the use of living resources that occur in international watercourses, except to the extent provided for in part IV and except insofar as other uses affect such resources.

As regards <u>article 2 (c)</u>:
The term "watercourse State" is used in this Convention as a term of art. Although this provision provides that States and regional economic integration organizations can both fall within this definition, it was recognized that nothing in this paragraph could be taken to imply that regional economic integration organizations have the status of States in international law.

As regards <u>article 3</u>:

(a) The present Convention will serve as a guideline for future watercourse agreements and, once such agreements are concluded, it will not alter the rights and obligations provided therein, unless such agreements provide otherwise;

(b) The term "significant" is not used in this article or elsewhere in the present Convention in the sense of "substantial". What is to be avoided are localized agreements, or agreements concerning a particular project, programme or use, which have a significant adverse effect upon third watercourse States. While such an effect must be capable of being established by objective evidence and not be trivial in nature, it need not rise to the level of being substantial.

As regards <u>article 6 (1) (e)</u>:

In order to determine whether a particular use is equitable and reasonable, the benefits as well as the negative consequences of a particular use should be taken into account.

As regards <u>article 7 (2)</u>:

In the event such steps as are required by article 7 (2) do not eliminate the harm, such steps as are required by article 7 (2) shall then be taken to mitigate the harm.

As regards <u>article 10</u>:

In determining "vital human needs", special attention is to be paid to providing sufficient water to sustain human life, including both drinking water and water required for production of food in order to prevent starvation.

As regards <u>articles 21, 22 and 23</u>:

As reflected in the commentary of the International Law Commission, these articles impose a due diligence standard on watercourse States.

As regards <u>article 28</u>:

The specific reference to "international organizations" is by no means intended to undermine the importance of cooperation, where appropriate, with competent international organizations on matters dealt with in other articles and, in particular, dealt with in the articles in part IV.

As regards <u>article 29</u>:

This article serves as a reminder that the principles and rules of international law applicable in international and non-international armed conflict contain important provisions concerning international watercourses and related works. The principles and rules of international law that are applicable in a particular case are those that are binding on the States concerned. Just as article 29 does not alter or amend existing law, it also does not purport to extend the applicability of any instrument to States not parties to that instrument.

* * *

Throughout the elaboration of the draft Convention, reference had been made to the commentaries to the draft articles prepared by the International Law Commission to clarify the contents of the articles.

9. At its 62nd meeting, on 4 April 1997, the Working Group adopted by vote the draft Convention reproduced in paragraph 10 below.

III. RECOMMENDATION OF THE WORKING GROUP OF THE WHOLE

10. The Working Group of the Whole recommends to the General Assembly the adoption of the following draft Convention:

Convention on the Law of the Non-navigational Uses of International Watercourses

The Parties to the present Convention,

Conscious of the importance of international watercourses and the non-navigational uses thereof in many regions of the world,

Having in mind Article 13, paragraph 1 (a), of the Charter of the United Nations, which provides that the General Assembly shall initiate studies and make recommendations for the purpose of encouraging the progressive development of international law and its codification,

Considering that successful codification and progressive development of rules of international law regarding non-navigational uses of international watercourses would assist in promoting and implementing the purposes and principles set forth in Articles 1 and 2 of the Charter of the United Nations,

Taking into account the problems affecting many international watercourses resulting from, among other things, increasing demands and pollution,

Expressing the conviction that a framework convention will ensure the utilization, development, conservation, management and protection of international watercourses and the promotion of the optimal and sustainable utilization thereof for present and future generations,

Affirming the importance of international cooperation and good-neighbourliness in this field,

Aware of the special situation and needs of developing countries,

Recalling the principles and recommendations adopted by the United Nations Conference on Environment and Development of 1992 in the Rio Declaration and Agenda 21,

Recalling also the existing bilateral and multilateral agreements regarding the non-navigational uses of international watercourses,

Mindful of the valuable contribution of international organizations, both governmental and non-governmental, to the codification and progressive development of international law in this field,

<u>Appreciative</u> of the work carried out by the International Law Commission on the law of the non-navigational uses of international watercourses,

<u>Bearing in mind</u> United Nations General Assembly resolution 49/52 of 9 December 1994,

<u>Have agreed as follows:</u>

PART I. INTRODUCTION

Article 1
Scope of the present Convention

1. The present Convention applies to uses of international watercourses and of their waters for purposes other than navigation and to measures of protection, preservation and management related to the uses of those watercourses and their waters.

2. The uses of international watercourses for navigation is not within the scope of the present Convention except insofar as other uses affect navigation or are affected by navigation.

Article 2
Use of terms

For the purposes of the present Convention:

(a) "Watercourse" means a system of surface waters and groundwaters constituting by virtue of their physical relationship a unitary whole and normally flowing into a common terminus;

(b) "International watercourse" means a watercourse, parts of which are situated in different States;

(c) "Watercourse State" means a State Party to the present Convention in whose territory part of an international watercourse is situated, or a Party that is a regional economic integration organization, in the territory of one or more of whose Member States part of an international watercourse is situated;

(d) "Regional economic integration organization" means an organization constituted by sovereign States of a given region, to which its member States have transferred competence in respect of matters governed by this Convention and which has been duly authorized in accordance with its internal procedures, to sign, ratify, accept, approve or accede to it.

Article 3
Watercourse agreements

1. In the absence of an agreement to the contrary, nothing in the present Convention shall affect the rights or obligations of a watercourse State arising from agreements in force for it on the date on which it became a party to the present Convention.

2. Notwithstanding the provisions of paragraph 1, parties to agreements referred to in paragraph 1 may, where necessary, consider harmonizing such agreements with the basic principles of the present Convention.

3. Watercourse States may enter into one or more agreements, hereinafter referred to as "watercourse agreements", which apply and adjust the provisions of the present Convention to the characteristics and uses of a particular international watercourse or part thereof.

4. Where a watercourse agreement is concluded between two or more watercourse States, it shall define the waters to which it applies. Such an agreement may be entered into with respect to an entire international watercourse or any part thereof or a particular project, programme or use except insofar as the agreement adversely affects, to a significant extent, the use by one or more other watercourse States of the waters of the watercourse, without their express consent.

5. Where a watercourse State considers that adjustment and application of the provisions of the present Convention is required because of the characteristics and uses of a particular international watercourse, watercourse States shall consult with a view to negotiating in good faith for the purpose of concluding a watercourse agreement or agreements.

6. Where some but not all watercourse States to a particular international watercourse are parties to an agreement, nothing in such agreement shall affect the rights or obligations under the present Convention of watercourse States that are not parties to such an agreement.

Article 4
Parties to watercourse agreements

1. Every watercourse State is entitled to participate in the negotiation of and to become a party to any watercourse agreement that applies to the entire international watercourse, as well as to participate in any relevant consultations.

2. A watercourse State whose use of an international watercourse may be affected to a significant extent by the implementation of a proposed watercourse agreement that applies only to a part of the watercourse or to a particular project, programme or use is entitled to participate in consultations on such an agreement and, where appropriate, in the negotiation thereof in good faith with a view to becoming a party thereto, to the extent that its use is thereby affected.

PART II. GENERAL PRINCIPLES

Article 5
Equitable and reasonable utilization and participation

1. Watercourse States shall in their respective territories utilize an international watercourse in an equitable and reasonable manner. In particular, an international watercourse shall be used and developed by watercourse States with a view to attaining optimal and sustainable utilization thereof and benefits therefrom, taking into account the

interests of the watercourse States concerned, consistent with adequate protection of the watercourse.

2. Watercourse States shall participate in the use, development and protection of an international watercourse in an equitable and reasonable manner. Such participation includes both the right to utilize the watercourse and the duty to cooperate in the protection and development thereof, as provided in the present Convention.

Article 6
Factors relevant to equitable and reasonable utilization

1. Utilization of an international watercourse in an equitable and reasonable manner within the meaning of article 5 requires taking into account all relevant factors and circumstances, including:

(a) Geographic, hydrographic, hydrological, climatic, ecological and other factors of a natural character;

(b) The social and economic needs of the watercourse States concerned;

(c) The population dependent on the watercourse in each watercourse State;

(d) The effects of the use or uses of the watercourses in one watercourse State on other watercourse States;

(e) Existing and potential uses of the watercourse;

(f) Conservation, protection, development and economy of use of the water resources of the watercourse and the costs of measures taken to that effect;

(g) The availability of alternatives, of comparable value, to a particular planned or existing use.

2. In the application of article 5 or paragraph 1 of this article, watercourse States concerned shall, when the need arises, enter into consultations in a spirit of cooperation.

3. The weight to be given to each factor is to be determined by its importance in comparison with that of other relevant factors. In determining what is a reasonable and equitable use, all relevant factors are to be considered together and a conclusion reached on the basis of the whole.

Article 7
Obligation not to cause significant harm

1. Watercourse States shall, in utilizing an international watercourse in their territories, take all appropriate measures to prevent the causing of significant harm to other watercourse States.

2. Where significant harm nevertheless is caused to another watercourse State, the States whose use causes such harm shall, in the absence of agreement to such use, take all appropriate measures, having due regard for the provisions of articles 5 and 6, in consultation with the affected State, to eliminate or mitigate such harm and, where appropriate, to discuss the question of compensation.

Article 8
General obligation to cooperate

1. Watercourse States shall cooperate on the basis of sovereign equality, territorial integrity, mutual benefit and good faith in order to attain optimal utilization and adequate protection of an international watercourse.

2. In determining the manner of such cooperation, watercourse States may consider the establishment of joint mechanisms or commissions, as deemed necessary by them, to facilitate cooperation on relevant measures and procedures in the light of experience gained through cooperation in existing joint mechanisms and commissions in various regions.

Article 9
Regular exchange of data and information

1. Pursuant to article 8, watercourse States shall on a regular basis exchange readily available data and information on the condition of the watercourse, in particular that of a hydrological, meteorological, hydrogeological and ecological nature and related to the water quality as well as related forecasts.

2. If a watercourse State is requested by another watercourse State to provide data or information that is not readily available, it shall employ its best efforts to comply with the request but may condition its compliance upon payment by the requesting State of the reasonable costs of collecting and, where appropriate, processing such data or information.

3. Watercourse States shall employ their best efforts to collect and, where appropriate, to process data and information in a manner which facilitates its utilization by the other watercourse States to which it is communicated.

Article 10
Relationship between different kinds of uses

1. In the absence of agreement or custom to the contrary, no use of an international watercourse enjoys inherent priority over other uses.

2. In the event of a conflict between uses of an international watercourse, it shall be resolved with reference to articles 5 to 7, with special regard being given to the requirements of vital human needs.

PART III. PLANNED MEASURES

Article 11
Information concerning planned measures

Watercourse States shall exchange information and consult each other and, if necessary, negotiate on the possible effects of planned measures on the condition of an international watercourse.

Article 12
Notification concerning planned measures with possible
adverse effects

Before a watercourse State implements or permits the implementation of planned measures which may have a significant adverse effect upon other watercourse States, it shall provide those States with timely notification thereof. Such notification shall be accompanied by available technical data and information, including the results of any environmental impact assessment, in order to enable the notified States to evaluate the possible effects of the planned measures.

Article 13
Period for reply to notification

Unless otherwise agreed:

(a) A watercourse State providing a notification under article 12 shall allow the notified States a period of six months within which to study and evaluate the possible effects of the planned measures and to communicate the findings to it;

(b) This period shall, at the request of a notified State for which the evaluation of the planned measures poses special difficulty, be extended for a period of six months.

Article 14
Obligations of the notifying State during the period for reply

During the period referred to in article 13, the notifying State:

(a) Shall cooperate with the notified States by providing them, on request, with any additional data and information that is available and necessary for an accurate evaluation; and

(b) Shall not implement or permit the implementation of the planned measures without the consent of the notified States.

Article 15
Reply to notification

The notified States shall communicate their findings to the notifying State as early as possible within the period applicable pursuant to article 13. If a notified State finds that implementation of the planned measures would be inconsistent with the provisions of articles 5 or 7, it shall attach to its finding a documented explanation setting forth the reasons for the finding.

Article 16
Absence of reply to notification

1. If, within the period applicable pursuant to article 13, the notifying State receives no communication under article 15, it may, subject to its obligations under articles 5 and 7, proceed with the implementation of the planned measures, in accordance with the notification and any other data and information provided to the notified States.

2. Any claim to compensation by a notified State which has failed to reply within the period applicable pursuant to article 13 may be offset by the costs incurred by the notifying State for action undertaken after the expiration of the time for a reply which would not have been undertaken if the notified State had objected within that period.

Article 17
Consultations and negotiations concerning planned measures

1. If a communication is made under article 15 that implementation of the planned measures would be inconsistent with the provisions of articles 5 or 7, the notifying State and

the State making the communication shall enter into consultations and, if necessary, negotiations with a view to arriving at an equitable resolution of the situation.

2. The consultations and negotiations shall be conducted on the basis that each State must in good faith pay reasonable regard to the rights and legitimate interests of the other State.

3. During the course of the consultations and negotiations, the notifying State shall, if so requested by the notified State at the time it makes the communication, refrain from implementing or permitting the implementation of the planned measures for a period of six months unless otherwise agreed.

Article 18
Procedures in the absence of notification

1. If a watercourse State has reasonable grounds to believe that another watercourse State is planning measures that may have a significant adverse effect upon it, the former State may request the latter to apply the provisions of article 12. The request shall be accompanied by a documented explanation setting forth its grounds.

2. In the event that the State planning the measures nevertheless finds that it is not under an obligation to provide a notification under article 12, it shall so inform the other State, providing a documented explanation setting forth the reasons for such finding. If this finding does not satisfy the other State, the two States shall, at the request of that other State, promptly enter into consultations and negotiations in the manner indicated in paragraphs 1 and 2 of article 17.

3. During the course of the consultations and negotiations, the State planning the measures shall, if so requested by the other State at the time it requests the initiation of consultations and negotiations, refrain from implementing or permitting the implementation of those measures for a period of six months unless otherwise agreed.

Article 19
Urgent implementation of planned measures

1. In the event that the implementation of planned measures is of the utmost urgency in order to protect public health, public safety or other equally important interests, the State planning the measures may, subject to articles 5 and 7, immediately proceed to implementation, notwithstanding the provisions of article 14 and paragraph 3 of article 17.

2. In such case, a formal declaration of the urgency of the measures shall be communicated without delay to the other watercourse States referred to in article 12 together with the relevant data and information.

3. The State planning the measures shall, at the request of any of the States referred to in paragraph 2, promptly enter into consultations and negotiations with it in the manner indicated in paragraphs 1 and 2 of article 17.

PART IV. PROTECTION, PRESERVATION AND MANAGEMENT

Article 20
Protection and preservation of ecosystems

Watercourse States shall, individually and, where appropriate, jointly, protect and preserve the ecosystems of international watercourses.

Article 21
Prevention, reduction and control of pollution

1. For the purpose of this article, "pollution of an international watercourse" means any detrimental alteration in the composition or quality of the waters of an international watercourse which results directly or indirectly from human conduct.

2. Watercourse States shall, individually and, where appropriate, jointly, prevent, reduce and control the pollution of an international watercourse that may cause significant harm to other watercourse States or to their environment, including harm to human health or safety, to the use of the waters for any beneficial purpose or to the living resources of the watercourse. Watercourse States shall take steps to harmonize their policies in this connection.

3. Watercourse States shall, at the request of any of them, consult with a view to arriving at mutually agreeable measures and methods to prevent, reduce and control pollution of an international watercourse, such as:
   (a) Setting joint water quality objectives and criteria;
   (b) Establishing techniques and practices to address pollution from point and non-point sources;
   (c) Establishing lists of substances the introduction of which into the waters of an international watercourse is to be prohibited, limited, investigated or monitored.

Article 22
Introduction of alien or new species

Watercourse States shall take all measures necessary to prevent the introduction of species, alien or new, into an international watercourse which may have effects detrimental to the ecosystem of the watercourse resulting in significant harm to other watercourse States.

Article 23
Protection and preservation of the marine environment

Watercourse States shall, individually and, where appropriate, in cooperation with other States, take all measures with respect to an international watercourse that are necessary to protect and preserve the marine environment, including estuaries, taking into account generally accepted international rules and standards.

Article 24
Management

1. Watercourse States shall, at the request of any of them, enter into consultations concerning the management of an international watercourse, which may include the establishment of a joint management mechanism.

2. For the purposes of this article, "management" refers, in particular, to:
(a) Planning the sustainable development of an international watercourse and providing for the implementation of any plans adopted; and
(b) Otherwise promoting the rational and optimal utilization, protection and control of the watercourse.

Article 25
Regulation

1. Watercourse States shall cooperate, where appropriate, to respond to needs or opportunities for regulation of the flow of the waters of an international watercourse.
2. Unless otherwise agreed, watercourse States shall participate on an equitable basis in the construction and maintenance or defrayal of the costs of such regulation works as they may have agreed to undertake.

3. For the purposes of this article, "regulation" means the use of hydraulic works or any other continuing measure to alter, vary or otherwise control the flow of the waters of an international watercourse.

Article 26
Installations

1. Watercourse States shall, within their respective territories, employ their best efforts to maintain and protect installations, facilities and other works related to an international watercourse.

2. Watercourse States shall, at the request of any of them which has reasonable grounds to believe that it may suffer significant adverse effects, enter into consultations with regard to:
(a) The safe operation and maintenance of installations, facilities or other works related to an international watercourse; and
(b) The protection of installations, facilities or other works from wilful or negligent acts or the forces of nature.

PART V. HARMFUL CONDITIONS AND EMERGENCY SITUATIONS

Article 27
Prevention and mitigation of harmful conditions

Watercourse States shall, individually and, where appropriate, jointly, take all appropriate measures to prevent or mitigate conditions related to an international watercourse that may be harmful to other watercourse States, whether resulting from natural causes or human conduct, such as flood or ice conditions, water-borne diseases, siltation, erosion, salt-water intrusion, drought or desertification.

Article 28
Emergency situations

1. For the purposes of this article, "emergency" means a situation that causes, or poses an imminent threat of causing, serious harm to watercourse States or other States and that

results suddenly from natural causes, such as floods, the breaking up of ice, landslides or earthquakes, or from human conduct, such as industrial accidents.

2. A watercourse State shall, without delay and by the most expeditious means available, notify other potentially affected States and competent international organizations of any emergency originating within its territory.

3. A watercourse State within whose territory an emergency originates shall, in cooperation with potentially affected States and, where appropriate, competent international organizations, immediately take all practicable measures necessitated by the circumstances to prevent, mitigate and eliminate harmful effects of the emergency.

4. When necessary, watercourse States shall jointly develop contingency plans for responding to emergencies, in cooperation, where appropriate, with other potentially affected States and competent international organizations.

PART VI. MISCELLANEOUS PROVISIONS

Article 29
International watercourses and installations in time of armed conflict

International watercourses and related installations, facilities and other works shall enjoy the protection accorded by the principles and rules of international law applicable in international and non-international armed conflict and shall not be used in violation of those principles and rules.

Article 30
Indirect procedures

In cases where there are serious obstacles to direct contacts between watercourse States, the States concerned shall fulfil their obligations of cooperation provided for in the present Convention, including exchange of data and information, notification, communication, consultations and negotiations, through any indirect procedure accepted by them.

Article 31
Data and information vital to national defence or security

Nothing in the present Convention obliges a watercourse State to provide data or information vital to its national defence or security. Nevertheless, that State shall cooperate in good faith with the other watercourse States with a view to providing as much information as possible under the circumstances.

Article 32
Non-discrimination

Unless the watercourse States concerned have agreed otherwise for the protection of the interests of persons, natural or juridical, who have suffered or are under a serious threat of suffering significant transboundary harm as a result of activities related to an international watercourse, a watercourse State shall not discriminate on the basis of nationality or residence or place where the injury occurred, in granting to such persons, in accordance with its legal system, access to judicial or other procedures, or a right to claim compensation or other relief in respect of significant harm caused by such activities carried on in its territory.

Article 33
Settlement of disputes

1. In the event of a dispute between two or more Parties concerning the interpretation or application of the present Convention, the Parties concerned shall, in the absence of an applicable agreement between them, seek a settlement of the dispute by peaceful means in accordance with the following provisions.

2. If the Parties concerned cannot reach agreement by negotiation requested by one of them, they may jointly seek the good offices of, or request mediation or conciliation by, a third party, or make use, as appropriate, of any joint watercourse institutions that may have been established by them or agree to submit the dispute to arbitration or to the International Court of Justice.

3. Subject to the operation of paragraph 10, if after six months from the time of the request for negotiations referred to in paragraph 2, the Parties concerned have not been able to settle their dispute through negotiation or any other means referred to in paragraph 2, the dispute shall be submitted, at the request of any of the parties to the dispute, to impartial fact-finding in accordance with paragraphs 4 to 9, unless the Parties otherwise agree.

4. A Fact-finding Commission shall be established, composed of one member nominated by each Party concerned and in addition a member not having the nationality of any of the Parties concerned chosen by the nominated members who shall serve as Chairman.

5. If the members nominated by the Parties are unable to agree on a Chairman within three months of the request for the establishment of the Commission, any Party concerned may request the Secretary-General of the United Nations to appoint the Chairman who shall not have the nationality of any of the parties to the dispute or of any riparian State of the watercourse concerned. If one of the Parties fails to nominate a member within three months of the initial request pursuant to paragraph 3, any other Party concerned may request the Secretary-General of the United Nations to appoint a person who shall not have the nationality of any of the parties to the dispute or of any riparian State of the watercourse concerned. The person so appointed shall constitute a single-member Commission.

6. The Commission shall determine its own procedure.

7. The Parties concerned have the obligation to provide the Commission with such information as it may require and, on request, to permit the Commission to have access to their respective territory and to inspect any facilities, plant, equipment, construction or natural feature relevant for the purpose of its inquiry.

8. The Commission shall adopt its report by a majority vote, unless it is a single-member Commission, and shall submit that report to the Parties concerned setting forth its findings and the reasons therefor and such recommendations as it deems appropriate for an

equitable solution of the dispute, which the Parties concerned shall consider in good faith.

9. The expenses of the Commission shall be borne equally by the Parties concerned.

10. When ratifying, accepting, approving or acceding to the present Convention, or at any time thereafter, a Party which is not a regional economic integration organization may declare in a written instrument submitted to the Depositary that, in respect of any dispute not resolved in accordance with paragraph 2, it recognizes as compulsory ipso facto and without special agreement in relation to any Party accepting the same obligation:
(a) Submission of the dispute to the International Court of Justice; and/or
(b) Arbitration by an arbitral tribunal established and operating, unless the parties to the dispute otherwise agreed, in accordance with the procedure laid down in the annex to the present Convention.

A Party which is a regional economic integration organization may make a declaration with like effect in relation to arbitration in accordance with subparagraph (b).

PART VII. FINAL CLAUSES

Article 34
Signature

The present Convention shall be open for signature by all States and by regional economic integration organizations from ... until ... at United Nations Headquarters in New York.

Article 35
Ratification, acceptance, approval or accession

1. The present Convention is subject to ratification, acceptance, approval or accession by States and by regional economic integration organizations. The instruments of ratification, acceptance, approval or accession shall be deposited with the Secretary-General of the United Nations.

2. Any regional economic integration organization which becomes a Party to this Convention without any of its member States being a Party shall be bound by all the obligations under the Convention. In the case of such organizations, one or more of whose member States is a Party to this Convention, the organization and its member States shall decide on their respective responsibilities for the performance of their obligations under the Convention. In such cases, the organization and the member States shall not be entitled to exercise rights under the Convention concurrently.

3. In their instruments of ratification, acceptance, approval or accession, the regional economic integration organizations shall declare the extent of their competence with respect to the matters governed by the Convention. These organizations shall also inform the Secretary-General of the United Nations of any substantial modification in the extent of their competence.

Article 36
Entry into force

1. The present Convention shall enter into force on the ninetieth day following the date of deposit of the thirty-fifth instrument of ratification, acceptance, approval or accession with the Secretary-General of the United Nations.

2. For each State or regional economic integration organization that ratifies, accepts or approves the Convention or accedes thereto after the deposit of the thirty-fifth instrument of ratification, acceptance, approval or accession, the Convention shall enter into force on the ninetieth day after the deposit by such State or regional economic integration organization of its instrument of ratification, acceptance, approval or accession.

3. For the purposes of paragraphs 1 and 2, any instrument deposited by a regional economic integration organization shall not be counted as additional to those deposited by States.

Article 37
Authentic texts

The original of the present Convention, of which the Arabic, Chinese, English, French, Russian and Spanish texts are equally authentic, shall be deposited with the Secretary-General of the United Nations.

IN WITNESS WHEREOF the undersigned plenipotentiaries, being duly authorized thereto, have signed this Convention.

DONE at New York, this _____ day of _____ one thousand nine hundred and ninety-seven.

ANNEX
ARBITRATION
Article 1

Unless the parties to the dispute otherwise agree, the arbitration pursuant to article 33 of the Convention shall take place in accordance with articles 2 to 14 of the present annex.

Article 2

The claimant party shall notify the respondent party that it is referring a dispute to arbitration pursuant to article 33 of the Convention. The notification shall state the subject matter of arbitration and include, in particular, the articles of the Convention, the interpretation or application of which are at issue. If the parties do not agree on the subject matter of the dispute, the arbitral tribunal shall determine the subject matter.

Article 3

1. In disputes between two parties, the arbitral tribunal shall consist of three members. Each of the parties to the dispute shall appoint an arbitrator and the two arbitrators so appointed shall designate by common agreement the third arbitrator, who shall be the

Chairman of the tribunal. The latter shall not be a national of one of the parties to the dispute or of any riparian State of the watercourse concerned, nor have his or her usual place of residence in the territory of one of these parties or such riparian State, nor have dealt with the case in any other capacity.

2. In disputes between more than two parties, parties in the same interest shall appoint one arbitrator jointly by agreement.

3. Any vacancy shall be filled in the manner prescribed for the initial appointment.

Article 4

1. If the Chairman of the arbitral tribunal has not been designated within two months of the appointment of the second arbitrator, the President of the International Court of Justice shall, at the request of a party, designate the Chairman within a further two-month period.

2. If one of the parties to the dispute does not appoint an arbitrator within two months of receipt of the request, the other party may inform the President of the International Court of Justice, who shall make the designation within a further two-month period.

Article 5

The arbitral tribunal shall render its decisions in accordance with the provisions of this Convention and international law.

Article 6

Unless the parties to the dispute otherwise agree, the arbitral tribunal shall determine its own rules of procedure.

Article 7

The arbitral tribunal may, at the request of one of the Parties, recommend essential interim measures of protection.

Article 8

1. The parties to the dispute shall facilitate the work of the arbitral tribunal and, in particular, using all means at their disposal, shall:
(a) Provide it with all relevant documents, information and facilities; and
(b) Enable it, when necessary, to call witnesses or experts and receive their evidence.

2. The parties and the arbitrators are under an obligation to protect the confidentiality of any information they receive in confidence during the proceedings of the arbitral tribunal.

Article 9

Unless the arbitral tribunal determines otherwise because of the particular circumstances of the case, the costs of the tribunal shall be borne by the parties to the dispute in equal shares. The tribunal shall keep a record of all its costs, and shall furnish a final statement thereof to the parties.

Article 10

Any Party that has an interest of a legal nature in the subject matter of the dispute which may be affected by the decision in the case, may intervene in the proceedings with the consent of the tribunal.

Article 11

The tribunal may hear and determine counterclaims arising directly out of the subject matter of the dispute.

Article 12

Decisions both on procedure and substance of the arbitral tribunal shall be taken by a majority vote of its members.

Article 13

If one of the parties to the dispute does not appear before the arbitral tribunal or fails to defend its case, the other party may request the tribunal to continue the proceedings and to make its award. Absence of a party or a failure of a party to defend its case shall not constitute a bar to the proceedings. Before rendering its final decision, the arbitral tribunal must satisfy itself that the claim is well founded in fact and law.

Article 14

1. The tribunal shall render its final decision within five months of the date on which it is fully constituted unless it finds it necessary to extend the time limit for a period which should not exceed five more months.

2. The final decision of the arbitral tribunal shall be confined to the subject matter of the dispute and shall state the reasons on which it is based. It shall contain the names of the members who have participated and the date of the final decision. Any member of the tribunal may attach a separate or dissenting opinion to the final decision.

3. The award shall be binding on the parties to the dispute. It shall be without appeal unless the parties to the dispute have agreed in advance to an appellate procedure.

4. Any controversy which may arise between the parties to the dispute as regards the interpretation or manner of implementation of the final decision may be submitted by either party for decision to the arbitral tribunal which rendered it.

Notes

1. For the report of the Sixth Committee on the work of the Working Group at its first session, held from 7 to 25 October 1996, see document A/51/624.

2. Official Records of the General Assembly, Forty-ninth Session, Supplement No. 10 (A/49/10), chap. III.D.

Water-Related Web Sites

The use of the Internet has exploded in recent years for entertainment, research, education, and commerce. For those interested in any aspect of water resources, the Internet has become a truly useful tool for finding references, data, and information about publications, organizations, and individuals, as well as for educators interested in developing and presenting useful information at all levels. User beware! ***Much that is useful is not on the Internet. And much that is on the Internet is not useful.*** Powerful search tools, however, now make it possible to explore almost any subject and to find valuable data and information.

Following is a list of web pages that the reader may find of use. By the time this book is published, some of these Internet addresses may have changed, and new resources will have become available. While this web directory will continue to be a part of future editions of *The World's Water,* the most up-to-date information will be found on the Internet itself. For that reason, the reader is encouraged to look on one of the comprehensive water pages listed below for current information and web addresses. In addition, links to current water pages will be maintained on the web site of the Pacific Institute for Studies in Development, Environment, and security at *www.worldwater.org.*

International Governmental and Nongovernmental Organizations

African Water Page	*http://wn.apc.org/afwater/index. htm*
Aquastat: UN FAO Water Information System	*http://www.fao.org/waicent/ faoinfo/agricult/agl/aglw/ aquastat/aquastat.htm*
Directory of Water Resources Organizations in the Americas	*http://www.oas.org/L/americas.htm*
European Water Environment Information Bulletin	*http://pantar.vub.ac.be/*
Global Environmental Monitoring System (GEMS)	*http://www.cciw.ca/gems*
Global Runoff Data Centre, Germany	*http://www.wmo.ch/web/homs/ grchome.html*
Global Water Partnership (GWP)	*http://www.gwp.sida.se/*
Global Water Partnership Forum	*http://www.gwpforum.org*
Groundwater Foundation	*http://www.groundwater.org*
Institute of Hydrology (IH)	*http://www.nwl.ac.uk/ih/*
Inter-American Water Resources Network	*http://iwrn.ces.fau.edu/*
International Association of Hydrological Sciences	*http://www.wlu.ca/~wwwiahs/ index.html*
International Association on Water Quality (IAWQ)	*http://www.iawq.org.uk/*
International Commission on Irrigation and Drainage	*http://www.ilri.nl/icid/ciid.html*
International Committee on Large Dams	*http://genepi.louis-jean.com/ cigb/index.html*
International Geosphere Biosphere Programme	*http://www.igbp.kvs.se*
International Irrigation Management Institute (IIMI)	*http://www.cgiar.org/iimi*
International Lake Environment Committee	*http://www.biwa.or.jp/ilec/top/ top.html*
International Rivers Network	*http://www.irn.org/*
IRC International Water and Sanitation Centre	*http://www.oneworld.org/ircwater/*

Middle East Water Information Network (MEWIN)	*http://www.ssc.upenn.edu/~mewin/*
New Zealand Water and Wastes Association	*http://www.nzwwa.org.nz*
Programa Hidrologico Internacional (PHI): UNESCO	*http://www.unesco.org.uy/phi*
Stockholm Environment Institute (SEI)	*http://nn.apc.org/sei/*
United Nations Development Programme (UNDP)	*http://www.undp.org*
United Nations Educational, Scientific, and Cultural Organization (UNESCO)	*http://www.unesco.org*
United Nations Environment Programme (UNEP)	*http://www.unep.org*
United Nations Food and Agriculture Organization (FAO)	*http://www.fao.org/*
University of London: Water Issues Group	*http://endjinn.soas.ac.uk/geography/ waterissues/*
Water Engineering Development Centre	*http://info.lut.ac.uk/departments/ cv/wedc/index/html*
Water Environment Federation	*http://www.wef.org/*
Water Quality Institute, Denmark (VKI)	*http://risul1.risoe.dk/dandokbas/ vkidata.html*
Water Research Commission, South Africa	*http://www.ccwr.ac.za/wrc/*
World Bank	*http://www.worldbank.org/*
World Bank Group: Global Water Unit	*http://www-esd.worldbank.org/ water/*
World Bank Water Policy Reform Program	*http://www.worldbank.org/ html/edi/edien.html*
World Conservation Monitoring Centre	*http://www.wcmc.org.uk*
World Conservation Union (IUCN)	*http://www.iucn.org*
World Health Organization (WHO)	*http://www.who.org*
World Meteorological Organization (WMO)	*http://www.wmo.ch*
World Water and Climate Atlas	*http://atlas.usu.edu/*
World Water Council (WWC)	*http://www.worldwatercouncil.org/*

National Sites and Organizations

American Water Resources Association (U.S.)	*http://www.uwin.siu.edu/~awra/*
American Water Works Association (U.S.)	*http://www.awwa.org/*
American Water Works Association Research Foundation (U.S.)	*http://www.awwarf.com/*
Australian CSIRO Division of Water Resources (Australia)	*http://www.dwr.csiro.au/*
British Geological Survey (United Kingdom)	*http://www.nkw.ac.uk/bgs/home. html*
British Hydrological Society (United Kingdom)	*http://www.salford.ac.uk/docs/ depts/civils/BHS/BHS_ Homepage. html*
Canadian Water Resources Association (Canada)	*http://www.cwra.org/cwra/*
Denver Water Company (U.S.)	*http://www.water.denver.co.gov/*
Environment Canada (Canada)	*http://www.doe.ca/envhome.html*
Great Lakes Information Network	*http://www.great-lakes.net/*
National Groundwater Association (U.S.)	*http://www.h2o-ngwa.org/*
South Africa Water Research Commission	*http://www-wrc.ccwr.ac.za/*
United States Committee on Large Dams (U.S.)	*http://www.uscold.org/~uscold/*
Water Education Foundation (U.S.)	*http://www.water-ed.org/*
WateReuse Association of California (U.S.)	*http://www.watereuse.org/*

Publishers, Data Sources, References, Libraries and Information Networks

Encyclopedia of Water Terms: Texas Environmental Center	*http://www.tec.org/tec/terms2.html*
Geraghty and Miller Publisher, Water	*http://www.gmgw.com/ catalogindex.cfm*
Global Environment Monitoring System (GEMS), Freshwater Quality Programme, UNEP	*http://www.cciw.ca/gems/*
Glossary of Terms: Water Quality Association	*http://www.wqa.org/WQIS/ Glossary/GlossHome.html*
International Lake Environment Committee Foundation (ILEC) Database	*http://www.biwa.or.jp/ilec/data base/database.html*
Island Press Publisher, Environment	*http://www.islandpress.org/*
U.S. Department of Agriculture: Water Quality Information Center: Water and Agriculture	*http://www.nal.usda.gov/wqic/*
U.S. Water News Online	*http://www.uswaternews.com/*
Universities Water Information Network	*http://www.uwin.siu.edu/*
Virtual Irrigation Library	*http://www.wiz.uni-kassel.de/kww/ projekte/irrig/irrig_i.html*
Virtual Library: Wastewater	*http://www.halcyon.com/cleanh2o/ ww/welcome.html*
Water Librarian's Home Page	*http://www.wco.com/~rteeter/ waterlib.html*
Water Resources Center Archives: University of California, Berkeley	*http://www.lib.berkeley.edu/WRCA/*
Water Resources Publications, LLC	*http://www.waterplus.com/wrp/ind ex.html*
Water Utility Homepages: List at AWWA	*http://www.awwa.org/utility.htm*
WaterWiser: Water Efficiency Clearinghouse	*http://www.waterwiser.org/*
World Hydrological Cycle Observing System (WHYCOS)	*http://www.wmo.ch/web/homs/ whycos.html*

Lists of Links to Other Internet Resources

Flood-Related Links	*http://maligne.civil.ualberta.ca/ water/misc/floodlinks.html*
Global Applied Research Network (GARNET)	*http://info.lut.ac.uk/departments/ cv/wedc/garnet/grntback.html*
Groundwater and the Internet	*http://gwrp.cciw.ca/internet/*
Hydrology Web	*http://terrassa.pnl.gov:2080/EESC/ resourcelist/hydrology.html*
Hydrology Web Directory	*http://www.webdirectory.com/ Science/Hydrology/*
Hydrology Web: interface to many other sites	*http://etd.pnl.gov:2080/ hydroweb.html*
World Wide Water	*http://pubweb.ucdavis.edu/ documents/gws/envissues/ george_fink/masterw.htm*

Government Organizations in the United States

California Department of Water Resources	*http://wwwdwr.water.ca.gov/*
U.S. Agency for International Development (USAID)	*http://www.info.usaid.gov/*

U.S. Bureau of Reclamation's Water Conservation Page (USBoR)	*http://ogee.hydlab.do.usbr.gov/rwc/rwc.html*
U.S. Environmental Protection Agency (USEPA)	*http://www.epa.gov/epahome/index.html*
U.S. Environmental Protection Agency Surf Your Watershed	*http://www.epa.gov/surf/*
U.S. Environmental Protection Agency: Office of Water	*http://www.epa.gov/OWOW/*
U.S. Geological Survey: Hydrology Primer (USGS)	*http://wwwdmorll.er.usgs.gov/~bjsmith/outreach/hydrology.primer.html*
U.S. Geological Survey: San Francisco Bay/Delta	*http://sfbay.wr.usgs.gov/*
U.S. Geological Survey: Water Resources of California	*http://water.wr.usgs.gov/*
U.S. Geological Survey: Water Resources of the United States	*http://h2o.usgs.gov/index.html*

Data Section

Table 1. Total Renewable Freshwater Supply by Country

Description

Average annual renewable freshwater resources are listed by country. These data are typically comprised of both renewable surface water and groundwater supplies, including surface inflows from neighboring countries. The UN FAO refers to this as total natural renewable water resources. Flows to other countries are not subtracted from these numbers. All quantities are in cubic kilometers per year (km³/yr). These data represent average freshwater resources in a country—actual annual renewable supply will vary from year to year.

Limitations

These detailed country data should be viewed, and used, with caution. The data come from different sources and were estimated over different periods. Many countries do not directly measure or report internal water resources data, so some of these entries were produced using indirect methods. For example, Margat compiles information from a wide variety of sources and notes that there is a wide variation in the reliability of the data. In the past few years, new assessments have begun to standardize definitions and assumptions.

Not all of the annual renewable water supply is available for use by the countries to which it is credited here—some flows are committed to downstream users. For example, the Sudan is listed as having 154 cubic kilometers per year, but treaty commitments require them to pass significant flows downstream to Egypt. Other countries such as Turkey, Syria, and France, to name only a few, also pass significant amounts of water to other users. The annual average figures hide large seasonal, interannual, and long-term variations.

Sources

a: Total natural renewable surface and groundwater. Typically includes flows from other countries. (FAO: "Natural total renewable water resources.")

b: Estimates from Belyaev, 1987, Institute of Geography, National Academy of Sciences, Moscow, USSR.

c: Estimates from Food and Agriculture Organization, 1995, *Water Resources of the African Countries: A Review*, Food and Agriculture Organization, United Nations, Rome.

d: World Resources Institute, 1994, *World Resources 1994–95*, in collaboration with the United Nations Environment Programme and the United Nations Development Programme, Oxford University Press, New York.

e: Margat, J., 1989, "The Sharing of Common Water Resources in the European Community (EEC)," *Water International* 14: 59–91, as cited in Gleick, P., editor, *Water in Crisis*, Oxford University Press, New York, Table A11.

f: Estimates from Shahin, M., 1989, "Review and Assessment of Water Resources in the Arab Region," *Water International* 14(4): 206–219, as cited in Gleick, P., editor, *Water in Crisis*, Oxford University Press, New York, Table A17.

g: Estimates from Goscomstat, USSR, 1989, *Protection of the Environment and Rational Utilization of Natural Resources in the USSR, Statistical Handbook*, Government Committee on Statistics, Moscow (in Russian), as cited in Gleick, P., editor, *Water in Crisis*, Oxford University Press, New York, Table A16.

h: World Resources Institute, 1996, *World Resources 1996–97*, A Joint Publication of the World Resources Institute, United Nations Environment Programme, United Nations Development Programme, and the World Bank, Oxford University Press, New York.

i: Economic Commission for Europe, 1992, *Environmental Statistical Database. The Environment in Europe and North America*, United Nations, New York.

j: Estimates from Food and Agriculture Organization, 1997, *Water Resources of the Near East Region: A Review*, Food and Agriculture Organization, United Nations, Rome.

k: Estimates from Food and Agriculture Organization, 1997, *Irrigation in the Countries of the Former Soviet Union in Figures.* Food and Agriculture Organization, United Nations, Rome, figures.

TABLE I. TOTAL RENEWABLE FRESHWATER SUPPLY BY COUNTRY

Region and Country	Annual Renewable Water Resources[a] (km³/yr)	Year of Estimate	Source of Estimate
Africa			
Algeria	14.3	1997	c,j
Angola	184.0	1987	b
Benin	25.8	1994	c
Botswana	14.7	1992	c
Burkina Faso	17.5	1992	c
Burundi	3.6	1987	b
Cameroon	268.0	1987	b
Cape Verde	0.3	1990	c
Central African Republic	141.0	1987	b
Chad	43.0	1987	b
Comoros	1.0	1987	b
Congo	832.0	1987	b
Congo, Democratic Republic (formerly Zaire)	1,019.0	1990	c
Côte d'Ivoire	77.7	1987	b
Djibouti	0.3	1997	j
Egypt	86.8	1997	j
Equatorial Guinea	30.0	1987	b
Eritrea	8.8	1990	c
Ethiopia	110.0	1987	b
Gabon	164.0	1987	b
Gambia	8.0	1982	c
Ghana	53.0	1970	c
Guinea	226.0	1987	b
Guinea-Bissau	27.0	1991	c
Kenya	30.2	1990	c
Lesotho	5.2	1987	b
Liberia	232.0	1987	b
Libya	0.6	1997	c,j
Madagascar	337.0	1984	c
Malawi	18.7	1994	c
Mali	67.0	1987	b
Mauritania	11.4	1997	c,j
Mauritius	2.2	1974	c
Morocco	30.0	1997	c,j
Mozambique	216.0	1992	c
Namibia	45.5	1991	c
Niger	32.5	1988	c
Nigeria	280.0	1987	b
Rwanda	6.3	1993	c
Senegal	39.4	1987	b
Sierra Leone	160.0	1987	b
Somalia	15.7	1997	j
South Africa	50.0	1990	c
Sudan	154.0	1997	c,j
Swaziland	4.5	1987	b
Tanzania	89.0	1994	c

continued

TABLE 1. CONTINUED

Region and Country	Annual Renewable Water Resources[a] (km³/yr)	Year of Estimate	Source of Estimate
Togo	11.5	1987	b
Tunisia	4.1	1997	j
Uganda	66.0	1970	c
Zambia	116.0	1994	c
Zimbabwe	20.0	1987	b
North and Central America			
Barbados	<1	1962	d
Belize	16.0	1987	b
Canada	2,901.0	1980	d
Costa Rica	95.0	1970	d
Cuba	34.5	1975	d
Dominican Republic	20.0	1987	b
El Salvador	18.9	1975	d
Guatemala	116.0	1970	d
Haiti	11.0	1987	b
Honduras	83.4	1992	h
Jamaica	8.3	1975	d
Mexico	357.4	1975	d
Nicaragua	175.0	1975	d
Panama	144.0	1975	d
Trinidad and Tobago	5.1	1975	d
United States of America	2,478.0	1985	d
South America			
Argentina	994.0	1976	d
Bolivia	300.0	1987	b
Brazil	6,950.0	1987	b
Chile	468.0	1975	d
Colombia	1,070.0	1987	b
Ecuador	314.0	1987	b
Guyana	241.0	1971	d
Paraguay	314.0	1987	b
Peru	40.0	1987	b
Suriname	200.0	1987	b
Uruguay	124.0	1965	d
Venezuela	1,317.0	1970	d
Asia			
Afghanistan	65.0	1997	j
Bahrain	0.1	1997	j
Bangladesh	2,357.0	1987	b
Bhutan	95.0	1987	b
Cambodia	498.1	1987	b
China	2,800.0	1980	d
Cyprus	0.9	1997	d,j
India	2,085.0	1975	d
Indonesia	2,530.0	1987	b
Iran	137.5	1997	j
Iraq	96.4	1997	j

continued

TABLE 1. CONTINUED

Region and Country	Annual Renewable Water Resources[a] (km³/yr)	Year of Estimate	Source of Estimate
Israel	2.1	1986	d
Japan	547.0	1980	d
Jordan	0.9	1997	j
Korea, Democratic People's Republic	67.0	1987	b
Korea, Republic of	66.0	1992	h
Kuwait	0.0	1997	j
Laos	270.0	1987	b
Lebanon	4.8	1997	j
Malaysia	456.0	1975	d
Mongolia	24.6	1987	b
Myanmar	1,082.0	1987	b
Nepal	170.0	1987	b
Oman	1.0	1997	j
Pakistan	429.4	1997	j
Philippines	323.0	1975	d
Qatar	0.1	1997	j
Saudi Arabia	2.4	1997	j
Singapore	0.6	1975	d
Sri Lanka	43.2	1970	d
Syria	46.1	1997	j
Thailand	179.0	1987	b
Turkey	200.7	1997	j
United Arab Emirates	0.1	1997	j
Vietnam	376.0	1987	b
Yemen	4.1	1997	j
Europe			
Albania	21.3	1970	d
Austria	90.3	1980	d
Belgium	12.5	1980	e
Bulgaria	205.0	1980	d
Czech Republic	58.2	1990	h
Denmark	13.0	1977	e
Finland	113.0	1980	d
France	198.0	1990	i
Germany	171.0	1991	i
Greece	58.7	1980	e
Hungary	120.0	1991	i
Iceland	170.0	1987	d
Ireland	50.0	1972	e
Italy	167.0	1990	h
Luxembourg	5.0	1976	e
Malta	0.0	1997	j
Netherlands	90.0	1980	e
Norway	392.0	1991	i
Poland	56.2	1980	e
Portugal	69.6	1990	h
Romania	208.0	1980	d

continued

TABLE 1. CONTINUED

Region and Country	Annual Renewable Water Resources[a] (km³/yr)	Year of Estimate	Source of Estimate
Slovakia	30.8	1990	h
Spain	111.3	1985	e
Sweden	180.0	1980	d
Switzerland	50.0	1985	d
United Kingdom	120.0	1980	e
Yugoslavia	265.0	1980	d
Russia	4,498.0	1997	g,k
Armenia	10.5	1997	k
Azerbaidzhan	30.3	1997	k
Belarus	58.0	1997	k
Estonia	12.8	1997	k
Georgia	63.3	1997	k
Kazakhstan	109.6	1997	k
Kyrgyzstan	20.6	1997	k
Latvia	35.4	1997	k
Lithuania	24.9	1997	k
Moldavia	11.7	1997	k
Tadjikistan	16.0	1997	k
Turkmenistan	24.7	1997	k
Ukraine	139.5	1997	k
Uzbekistan	50.4	1997	k
Oceania			
Australia	343.0	1975	d
Fiji	28.6	1987	b
New Zealand	327.0	1991	h
Papua New Guinea	801.0	1987	b
Solomon Islands	44.7	1987	b

Table 2. Freshwater Withdrawal by Country and Sector

Description

The use of water varies greatly from country to country and from region to region. Data on water use by regions and by different economic sectors are among the most sought after in the water resources area. Ironically, these data are often the least reliable and most inconsistent of all water-resources information. The following table presents the most up-to-date data available on total freshwater withdrawals by country in cubic kilometers per year and cubic meters per person per year, using national population estimates from approximately the year of withdrawal. The table also gives the breakdown of that water use by the domestic, agricultural, and industrial sectors, in both percentage of total water use and cubic meters per person per year.

"Withdrawal" refers to water taken from a water source for use. It does not refer to water "consumed" in that use. The domestic sector typically includes household and municipal uses as well as commercial and governmental water use. The industrial sector includes water used for power plant cooling and industrial production. The agricultural sector includes water for irrigation and livestock.

Limitations

Extreme care should be used when applying these data. They come from a wide variety of sources and are collected using a wide variety of approaches, with few formal standards. As a result, this table includes data that are actually measured, are estimated, are modeled using different assumptions, or are derived from other data. The data also come from different years, making direct intercomparisons difficult. For example, some water use data are over 20 years old. New data are included for the newly independent states of the former Soviet Union. Data are also reported separately for former East and West Germany.

Another major limitation of these data is that they do not include the use of rainfall in agriculture. Many countries use a significant fraction of the rain falling on their territory for agricultural production, but this water use is neither accurately measured nor reported in this set. In the past several years, the Food and Agriculture Organization has begun a systematic reassessment of water-use data, and there is reason to hope that over the next several years a more accurate picture of global and regional water use will emerge.

Sources

a: Food and Agriculture Organization, 1995, *Water Resources of African Countries*, Food and Agriculture Organization, United Nations, Rome.

b: Food and Agriculture Organization, 1997, *Water Resources of the Near East Region: A Review*, Food and Agriculture Organization, United Nations, Rome.

c: World Resources Institute, 1994 and 1998, *World Resources*, in collaboration with the United Nations Environment Programme and the United Nations Development Programme, Oxford University Press, New York.

d: Estimates from Food and Agriculture Organization, 1997, *Irrigation in the Countries of the Former Soviet Union in Figures*. Food and Agriculture Organization, United Nations, Rome.

TABLE 2. FRESHWATER WITHDRAWAL BY COUNTRY AND SECTOR

Region and Country	Year	Total Freshwater Withdrawal (km³/yr)	Per-Capita Withdrawal (km³/yr)	Domestic Use (%)	Industrial Use (%)	Agricultural Use (%)	Domestic Use (m³/p/yr)	Industrial Use (m³/p/yr)	Agricultural Use (m³/p/yr)	Source
Africa										
Algeria	1990	4.50	180	25	15	60	45	27	108	a
Angola	1987	0.48	48	14	10	76	7	5	36	a
Benin	1994	0.14	31	23	10	67	7	3	21	a
Botswana	1992	0.11	87	32	20	48	28	18	42	a
Burkina Faso	1992	0.38	42	19	0	81	8	0	34	a
Burundi	1987	0.10	18	36	0	64	7	0	12	a
Cameroon	1987	0.40	34	46	19	35	16	6	12	a
Cape Verde	1990	0.03	70	10	19	88	7	14	61	a
Central African Republic	1987	0.07	23	21	5	74	5	1	17	a
Chad	1987	0.18	32	16	2	82	5	1	26	a
Comoros	1990	0.01	18	0	0	0	0	0	0	c
Congo	1987	0.04	18	62	27	11	11	5	2	a
Congo, Democratic Republic (formerly Zaire)	1990	0.36	10	61	16	23	6	2	2	a
Côte d'Ivoire	1987	0.71	59	22	11	67	13	6	40	a
Djibouti	1973	0.01	24	13	0	87	3	0	21	a
Egypt	1993	55.10	1,013	6	8	86	61	81	871	a
Equatorial Guinea	1987	0.01	29	81	13	6	23	4	2	a
Ethiopia	1987	2.20	45	11	3	86	5	1	38	a
Gabon	1987	0.06	51	72	22	6	37	11	3	a
Gambia	1982	0.02	23	7	2	91	2	0	21	a
Ghana	1970	0.30	20	35	13	52	7	3	10	a
Guinea	1987	0.74	128	10	3	87	13	4	112	a
Guinea-Bissau	1991	0.02	17	60	4	36	10	1	6	a
Kenya	1990	2.05	85	20	4	76	17	3	65	a
Lesotho	1987	0.05	28	22	22	56	6	6	16	a
Liberia	1987	0.13	50	27	13	60	14	7	30	a
Libya	1994	4.60	1,011	11	2	87	110	22	879	a
Madagascar	1984	16.30	1,358	1	0	99	14	0	1,345	a
Malawi	1994	0.94	107	10	3	86	11	4	92	a
Mali	1987	1.36	148	2	1	97	3	2	143	a
Mauritania	1985	1.63	807	6	2	92	50	15	742	a
Mauritius	1974	0.36	333	16	7	77	54	23	256	a
Morocco	1991	11.04	441	5	3	92	22	13	406	a
Mozambique	1992	0.60	39	9	2	89	3	1	34	a
Namibia	1991	0.249	140	29	3	68	40	4	96	a
Niger	1988	0.50	65	16	2	82	10	1	53	a
Nigeria	1987	3.63	33	31	15	54	10	5	18	a
Rwanda	1993	0.77	106	5	2	94	5	2	99	a
Senegal	1987	1.36	186	5	3	92	9	6	171	a
Sierra Leone	1987	0.37	89	7	4	89	6	4	79	a
Somalia	1987	0.81	108	3	0	97	3	0	105	a
South Africa	1990	13.31	377	17	11	72	65	41	272	a
Sudan	1995	17.80	706	4	1	94	32	8	667	a
Swaziland	1980	0.66	830	2	2	96	13	20	797	a
Tanzania	1994	1.17	43	9	2	89	4	1	38	a
Togo	1987	0.09	25	62	13	26	16	3	7	a
Tunisia	1990	3.08	376	9	3	89	32	11	333	a
Uganda	1970	0.20	11	32	8	60	3	1	6	a
Zambia	1994	1.71	202	16	7	77	32	14	156	a
Zimbabwe	1987	1.22	126	14	7	79	18	9	99	a
North and Central America										
Barbados	1990	0.03	117	52	41	7	61	48	8	c
Belize	1990	0.02		10	0	90	0	0	0	c
Canada	1990	42.20	1,752	11	80	8	193	1,402	140	c

continued

TABLE 2. **CONTINUED**

Region and Country	Year	Total Freshwater Withdrawal (km³/yr)	Per-Capita Withdrawal (km³/yr)	Domestic Use (%)	Industrial Use (%)	Agricultural Use (%)	Domestic Use (m³/p/yr)	Industrial Use (m³/p/yr)	Agricultural Use (m³/p/yr)	Source
Costa Rica	1990	1.35	779	4	7	89	31	55	693	c
Cuba	1990	8.10	868	9	2	89	78	17	773	c
Dominican Republic	1990	2.97	453	5	6	89	23	27	403	c
El Salvador	1990	1.00	241	7	4	89	17	10	214	c
Guatemala	1990	0.73	139	9	17	74	13	24	103	c
Haiti	1990	0.04	46	24	8	68	11	4	31	c
Honduras	1990	1.34	508	4	5	91	20	25	462	c
Jamaica	1990	0.32	157	7	7	86	11	11	135	c
Mexico	1990	54.20	901	6	8	86	54	72	775	c
Nicaragua	1990	0.89	370	25	21	54	92	78	200	c
Panama	1990	1.30	744	12	11	77	89	82	573	c
Trinidad and Tobago	1990	0.15	149	27	38	35	40	57	52	c
United States of America	1990	467.00	2,162	12	46	42	259	995	908	c
South America										
Argentina	1990	27.60	1,059	9	18	73	95	191	773	c
Bolivia	1990	1.24	184	10	5	85	18	9	156	c
Brazil	1990	35.04	212	43	17	40	91	36	85	c
Chile	1990	16.80	1,625	6	5	89	98	81	1,446	c
Colombia	1990	5.34	179	41	16	43	73	29	77	c
Ecuador	1990	5.56	561	7	3	90	39	17	505	c
Guyana	1990	5.40	7,616	1	0	99	76	0	7,540	c
Paraguay	1990	0.43	111	15	7	78	17	8	87	c
Peru	1990	6.10	294	19	9	72	56	26	212	c
Suriname	1990	0.46	1,181	6	5	89	71	59	1,051	c
Uruguay	1990	0.65	241	6	3	91	14	7	219	c
Venezuela	1990	4.10	387	43	11	46	166	43	178	c
Asia										
Afghanistan	1987	26.11	1,436	1	0	99	14	0	1,422	b
Bahrain	1991	0.24	609	39	4	56	238	24	341	b
Bangladesh	1990	22.50	211	3	1	96	6	2	203	c
Bhutan	1990	0.02	15	36	10	54	5	2	8	c
Cambodia	1990	0.52	69	5	1	94	3	1	65	c
China	1990	460.00	462	6	7	87	28	32	402	c
Cyprus	1993	0.54	807	7	2	91	56	16	734	b
India	1990	380.00	612	3	4	93	18	24	569	c
Indonesia	1990	16.59	96	13	11	76	12	11	73	c
Iran	1993	70.03	1,362	6	2	92	82	27	1,253	b
Iraq	1990	42.80	4,575	3	5	92	137	229	4,209	b
Israel	1990	1.90	447	16	5	79	72	22	353	c
Japan	1990	107.80	923	17	33	50	157	305	462	c
Jordan	1993	0.98	173	22	3	75	38	5	130	b
Korea, Democratic People's Republic	1990	14.16	1,649	11	16	73	181	264	1,204	c
Korea, Republic of	1990	10.70	298	11	14	75	33	42	224	c
Kuwait	1994	0.54	238	37	2	60	88	5	143	b
Laos	1990	0.99	228	8	10	82	18	23	187	c
Lebanon	1994	1.29	271	28	4	68	76	11	184	b
Malaysia	1990	9.42	765	23	30	47	176	230	360	c
Mongolia	1990	0.55	272	11	27	62	30	73	169	c
Myanmar	1990	3.96	103	7	3	90	7	3	93	c
Nepal	1990	2.68	155	4	1	95	6	2	147	c
Oman	1991	1.22	325	5	2	94	16	6	306	b
Pakistan	1991	155.60	2,053	2	2	97	41	41	1,991	b
Philippines	1990	29.50	693	18	21	61	125	146	423	c
Qatar	1994	0.28	415	23	3	74	95	12	307	b
Saudi Arabia	1992	17.02	255	9	1	90	23	3	230	b
Singapore	1990	0.19	84	45	51	4	38	43	3	c

continued

TABLE 2. CONTINUED

Region and Country	Year	Total Freshwater Withdrawal (km³/yr)	Per-Capita Withdrawal (km³/yr)	Domestic Use (%)	Industrial Use (%)	Agricultural Use (%)	Domestic Use (m³/p/yr)	Industrial Use (m³/p/yr)	Agricultural Use (m³/p/yr)	Source
Sri Lanka	1990	6.30	503	2	2	96	10	10	483	c
Syria	1993	14.41	449	4	2	94	18	9	422	b
Thailand	1990	31.90	599	4	6	90	24	36	539	c
Turkey	1992	31.60	317	16	11	72	51	35	228	b
United Arab Emirates	1995	2.11	565	24	9	67	136	51	379	b
Vietnam	1990	5.07	81	13	9	78	11	7	63	c
Yemen	1990	2.93	251	7	1	92	18	3	231	c
Europe										
Albania	1990	0.20	94	6	18	76	6	17	71	c
Austria	1990	3.13	417	19	73	8	79	304	33	c
Belgium	1990	9.03	917	11	85	4	101	779	37	c
Bulgaria	1990	14.18	1,600	7	38	55	112	608	880	c
Czechoslovakia	1990	5.80	379	23	68	9	87	258	34	c
Denmark	1990	1.46	289	30	27	43	87	78	124	c
Finland	1990	3.70	774	12	85	3	93	658	23	c
France	1990	40.00	728	16	69	15	116	502	109	c
German Democratic Republic	1990	9.13	545	14	68	18	76	371	98	c
Germany, Federal Republic of	1990	41.22	688	10	70	20	69	482	138	c
Greece	1990	6.95	721	8	29	63	58	209	454	c
Hungary	1990	5.38	502	9	55	36	45	276	181	c
Iceland	1990	0.09	349	31	63	6	108	220	21	c
Ireland	1990	0.79	267	16	74	10	43	198	27	c
Italy	1990	56.20	983	14	27	59	138	265	580	c
Luxembourg	1990	0.04	119	42	45	13	50	54	15	c
Malta	1995	0.06	68	87	1	12	59	1	8	b
Netherlands	1990	14.47	1,023	5	61	34	51	624	348	c
Norway	1990	2.00	489	20	72	8	98	352	39	c
Poland	1990	16.80	472	16	60	24	76	283	113	c
Portugal	1990	10.50	1,062	15	37	48	159	393	510	c
Romania	1990	25.40	1,144	8	33	59	92	378	675	c
Spain	1990	45.25	1,174	12	26	62	141	305	728	c
Sweden	1990	3.98	479	36	55	9	172	263	43	c
Switzerland	1990	3.20	502	23	73	4	115	366	20	c
United Kingdom	1990	28.35	507	20	77	3	101	390	15	c
Yugoslavia	1990	8.77	393	16	72	12	63	283	47	c
Former Soviet Union										
Armenia	1994	3	804	30	4	66	241	32	531	d
Azerbaijan	1995	17	2,177	5	25	70	109	544	1,524	d
Belarus	1990	3	264	22	43	35	58	114	92	d
Estonia	1995	0	107	56	39	5	60	42	5	d
Georgia	1990	3	637	21	20	59	134	127	376	d
Kazakhstan	1993	34	2,002	2	17	81	40	340	1,622	d
Kyrgyz Republic	1994	10	2,257	3	3	94	68	68	2,122	d
Latvia	1994	0	114	55	32	13	63	36	15	d
Lithuania	1995	0	68	81	16	3	55	11	2	d
Moldova	1992	3	667	9	65	26	60	434	173	d
Russian Federation	1994	77	521	19	62	20	99	323	104	d
Tajikistan	1994	12	2,001	3	4	92	60	80	1,841	d
Turkmenistan	1994	24	5,723	1	1	98	57	57	5,609	d
Ukraine	1992	26	504	18	52	30	91	262	151	d
Uzbekistan	1994	58	2,501	4	2	94	100	50	2,351	d
Oceania										
Australia	1985	14.60	1,306	65	2	33	849	26	431	c
Fiji	1987	0.03	37	20	20	60	7	7	22	c
New Zealand	1991	2.0	379	46	10	44	174	38	167	c
Papua New Guinea	1987	0.10	25	29	22	49	7	6	12	c
Solomon Islands	1987	0.00	18	40	20	40	7	4	7	c

Note: Figures may not add to totals due to independent rounding.

Table 3. Summary of Estimated Water Use in the United States, 1900 to 1995

Description

Water use in the United States, including both withdrawals and consumption, is given here for the years 1900 to 1995, in cubic kilometers per year. Data are provided for withdrawals of water by sector. Also provided are data on per-capita withdrawals and consumptive use in cubic meters per person per year. Trends in total withdrawals and per-capita withdrawals can be seen more clearly in Figures 1.3, 1.4, and 1.5.

Limitations

The size of the United States has changed over time. The original "total withdrawal" data from 1900 to 1980 are rounded to two significant figures; data for 1985, 1990, and 1995 are rounded to three significant figures, as in the original sources.

Sources

U.S. Geological Survey, 1993, *Estimated Use of Water in the United States in 1990*, U.S. Geological Survey Circular 1081. U.S. Dept. of the Interior. Washington, D.C.

Council on Environmental Quality, 1991, *Environmental Quality: 21st Annual Report*. Washington, D.C.

Data for 1995 from H. Perlman, USGS, via FTP from 144.47.32.102.

TABLE 3. SUMMARY OF ESTIMATED WATER USE IN THE UNITED STATES, 1900 TO 1995 (VALUES IN CUBIC KILOMETERS PER YEAR UNLESS OTHERWISE STATED)

	1900	1910	1920	1930	1940	1945	1950	1955	1960	1965	1970	1975	1980	1985	1990	1995
Population (millions)	76	92	105	125	132	140	151	164	179	194	206	216	230	242	252	267
Offstream use																
Total withdrawals	56	93	125	152	188	224	250	330	370	430	510	580	610	551	564	554
Public supply	4	7	8	11	14	17	19	24	29	33	37	40	47	50	53	55.6
Rural domestic/livestock	3	3	3	4	4	5	5	5	5	6	6	7	8	11	11	12
Irrigation	28	54	77	83	98	110	120	150	150	170	180	190	210	189	189	185
Industrial																
Thermoelectric	7	10	12	25	32	44	55	100	140	180	240	280	290	257	269	262
Other industrial	14	19	25	29	40	48	51	54	53	64	65	62	62	42	41	39
Consumptive use									84	110	120	130	140	128	130	138
Per-capita withdrawal (m³/p/yr)	737	1,011	1,190	1,216	1,424	1,600	1,659	2,012	2,064	2,219	2,477	2,680	2,657	2,273	2,235	2,071
Per-capita freshwater withdrawals (m³/p/yr)							1,573	1,921	1,836	1,925	2,108	2,170	2,265	1,927	1,854	1,757
Per-capita consumptive use (m³/p/yr)									471	568	585	601	610	528	515	517
Source of water																
Groundwater																
Fresh							47	65	69	83	94	110	120	101	110	105
Saline							0.8	0.6	0.7	1.4	1.4	1.2	0.90	1.7	1.6	
Surface water																
Fresh							190	250	260	290	340	360	400	366	358	364
Saline							14	25	43	59	73	95	98	82.3	94.2	82.5
Total freshwater withdrawals							237	315	329	373	434	470	520	467	468	469
Total saltwater withdrawals							14	26	43	60	75	97	99	83	96	84

Table 4. Total and Urban Population by Country, 1975 to 1995

Description

Total and urban population data by country for the years 1975 to 1995 are presented as a complement to Tables 5 and 6. Note that the world's population, particularly in Africa and Asia, is becoming increasingly urban. This has consequences for water planning and management around the globe.

Limitations

Data are not available for every country, and some countries do not have data for all years. All data originated in the United Nations Population Division but were gathered from numerous secondary sources.

Sources

a: United Nations data, cited by the World Bank, 1993, *World Tables 1993*, World Bank, Johns Hopkins University Press, Baltimore.

b: United Nations data, cited by the World Resources Institute, 1986, *World Resources 1986*, World Resources Institute and the International Institute for Environment and Development, Basic Books, New York.

c: United Nations data, cited by the World Resources Institute, 1996, *World Resources 1996–97*, A Joint Publication of the World Resources Institute, United Nations Environment Programme, United Nations Development Programme, and the World Bank, Oxford University Press, New York.

d: World Health Organization, 1996, *Water Supply and Sanitation Sector Monitoring Report: 1996 (Sector Status as of 1994)*, in collaboration with the Water Supply and Sanitation Collaborative Council and the United Nations Children's Fund, New York.

e: United Nations, 1995, *World Population Prospects, the 1994 Revision*, Department for Economic and Social Information and Policy Analysis, Population Division, United Nations, New York.

f: United Nations Environment Programme, 1993–94, *Environmental Data Report*, GEMS Monitoring and Assessment Research Centre in cooperation with the World Resources Institute and the UK Department of the Environment, Basil Blackwell, Oxford.

g: United Nations data, cited by the World Resources Institute, 1988, *World Resources 1988–89*, A Report by the World Resources Institute and the International Institute for Environment and Development in collaboration with the United Nations Environment Programme, Basic Books, New York.

TABLE 4. TOTAL AND URBAN POPULATION BY COUNTRY, 1975 TO 1995

Region and Country	Population (thousands)						Percentage Urban					
	1975	1980	1985	1990	1994	1995	1970	1975	1980	1985	1990	1995
World			4,842,048	5,284,832		5,716,426	37	38	40	41	43	45
Africa			553,210	632,669	724,254			25			32	34
Algeria	16,018	18,669	21,993	24,935		27,939	40	40	41	43	52	56
Angola			8,754	9,194	10,674	11,072	15	18	21	25	28	32
Benin	3,029	3,464	4,005	4,633	5,096	5,409	16	20	28	35	38	31
Botswana	755	902	1,079	1,276		1,487	8	12	15	19	25	28
Burkina Faso	3,202	6,962	6,939	8,987	9,772	10,319	6	6	7	8	15	27
Burundi	3,680	4,130	4,631	5,503	5,847	6,393	2	3	4	6	5	8
Cameroon	7,439	8,701	9,714	11,526		13,233	20	27	35	42	40	45
Cape Verde	278	289	321		350	392	6		5	5	29	
Central African Republic	2,034	2,320	2,567	2,927	3,001	3,315	30	34	38	42	47	39
Chad	4,030	4,477	5,018	5,533	6,183	6,361	11	16	21	27	32	21
Comoros	298	333	457			653	11	35	23	25	28	
Congo	1,380	1,630	1,740	2,232		2,590	35	35	37	40	41	59
Congo, Democratic Republic	23,251	27,009	33,052	37,436	39,939	43,901	30	30	34	37	28	29
Côte d'Ivoire	6,755	8,194	9,797	11,974	13,780	14,253	27	32	37	42	40	44
Djibouti			293		557	577	62		74	78	81	
Egypt	36,289	40,875	46,800	56,312	60,319	62,931	44	43	45	46	44	45
Equatorial Guinea	313	341	392	352	389	400	39	27	54	60	29	42
Eritrea				3,082				12				17
Ethiopia	32,954	37,717	36,454	47,423		55,053	9	10	11	12	12	13
Gabon	637	797	1,166	1,146		1,320	26	31	36	41	46	50
Gambia	533	634	643	923	1,042	1,118	15	17	18	20	23	26
Ghana	9,835	10,740	13,478	15,020	16,944	17,453	29	30	31	32	34	36
Guinea	4,149	4,461	5,429	5,755	6,501	6,700	14	16	19	22	26	30
Guinea-Bissau	627	809	889	964	1,050	1,073	18	16	24	27	20	22
Kenya	13,741	16,632	20,600	23,613	26,391	28,261	10	13	16	20	24	28
Lesotho	1,187	1,339	1,520	1,792	1,996	2,050	9	11	14	17	19	23
Liberia	1,609	1,879	2,191	2,575	2,941	3,039	26	30	35	40	45	45
Libya	2,446	3,043	3,604	4,545		5,407	36	61	57	65	82	86
Madagascar	7,604	8,714	10,012	12,571	14,303	14,763	14	16	19	22	24	27
Malawi	5,244	6,138	7,016	9,367	10,843	11,129	6	8	10	12	12	14
Mali	5,905	6,590	8,053	9,212	10,135	10,795	14	16	17	18	24	27
Mauritania	1,371	1,551	1,881	2,003	2,106	2,274	14	20	27	35	47	54
Mauritius	883	966	1,050	1,057	1,057	1,117	42	43	43	42	41	41
Morocco	17,305	19,382	23,602	24,334	26,945	27,028	35	38	41	45	46	48
Mozambique	10,606	12,103	14,085	14,187	15,527	16,004	6	9	13	19	27	34
Namibia	923	1,066	1,235	1,349	1,385	1,540		21			28	37
Niger	4,704	5,515	6,115	7,731	8,846	9,151	9	11	13	16	20	17
Nigeria	61,241	71,148	95,198	96,154	105,264	111,721	16	23	20	23	35	39
Reunion						653					64	
Rwanda	4,384	5,163	6,115	6,986		7,952	3	4	5	6	6	6
Sao Tome and Principe	81	94	103	115		133					42	
Senegal	4,806	5,538	6,520	7,327	7,902	8,312	33	34	35	36	40	42
Seychelles	59	63	65	68		73					59	
Sierra Leone	2,931	3,263	3,602	3,999	4,402	4,509	18	21	25	28	32	36
Somalia	4,967	5,746	5,552	8,677		9,250	23	21	30	34	24	26
South Africa	25,842	29,529	32,392	37,066	40,555	41,465	48	48	53	56	49	51
Sudan	16,550	19,152	21,550	24,585	26,641	28,098	16	19	20	21	23	25
Swaziland	482	565	649	744	809	855	10	14	20	26	26	31
Tanzania	15,379	18,098	22,499	25,600		29,685	7	10	17	22	21	24
Togo	2,282	2,615	2,923	3,531	3,763	4,138	13	16	19	22	29	31
Tunisia	5,611	6,384	7,209	8,080	8,407	8,896	44	50	52	57	56	57
Uganda	11,228	12,807	15,697	17,949	50,621	21,297	8	8	9	10	11	13
Zambia	4,841	5,647	6,666	8,150	9,198	9,456	30	35	43	50	42	43
Zimbabwe	6,065	7,009	8,767	9,903		11,261	17	20	22	25	29	32
North and Central America and Caribbean			400,802					57				68
Anguilla						8						
Antigua and Barbuda	71	75	76	79		66					32	
Aruba						70						
Bahamas	189	210	232	255		276					64	
Barbados	246	249	253	257		262	37		40	42	45	
Belize	129	144	166	189	210	215		50			51	47
British Virgin Islands						19						
Canada	22,697	24,043	25,605	27,791		76	76	76	76	76	77	77

continued

TABLE 4. CONTINUED

Region and Country	Population (thousands)						Percentage Urban					
	1975	1980	1985	1990	1994	1995	1970	1975	1980	1985	1990	1995
Cayman Islands						31					100	
Costa Rica	1,968	2,284	2,600	3,035	3,192	3,424	40	41	46	50	47	50
Cuba			7,029	10,598		11,041	60	64	68	72	74	76
Dominican Republic	5,048	5,697	6,416	7,110	7,543	7,823	40	45	51	56	60	65
Dominica						71						
El Salvador	4,085	4,525	4,739	5,172	5,517	5,768	39	40	39	39	44	45
Grenada		87	94	91		92						
Guadeloupe						428					49	
Guatemala	6,023	6,917	7,963	9,197		10,621	36	37	39	40	39	41
Haiti	4,920	5,353	5,865	6,486	7,035	7,180	20	22	25	27	29	32
Honduras	3,081	3,662	4,383	4,879	5,493	5,654	29	32	36	40	44	44
Jamaica	2,013	2,133	2,260	2,366		2,447	42	44	50	54	52	54
Martinique						379					75	
Mexico	58,876	67,046	74,766	84,511	90,027	93,674	59	63	66	70	73	75
Montserrat						11					12	
Netherlands Antilles						199						
Nicaragua	2,426	2,802	3,229	3,676	4,255	4,433	47	50	53	57	60	63
Panama	1,748	1,956	2,180	2,398	2,491	2,631	48	49	51	52	53	53
Puerto Rico						3,674					74	
St. Kitts	44	44	43	40		41					49	
St. Lucia	112	124	137	150		142					44	
St. Vincent	93	98	102	107		112					20	
Trinidad/Tobago	1,012	1,082	1,160	1,236		1,306	39	63	57	64	65	72
Turks/Caicos Islands						14					51	
United States of America	216,000	228,000	239,000	249,924			74	74	74	74	75	76
United States Virgin Islands						105					47	
South America			268,825					64			75	78
Argentina	26,052	28,237	30,564	32,547		34,587	78	81	83	85	86	88
Bolivia	4,894	5,581	6,371	6,573	6,893	7,414	41	42	44	48	51	61
Brazil	108,000	121,000	135,564	148,477	146,825	161,790	56	61	68	73	75	78
Chile	10,350	11,145	12,074	13,154	13,600	14,262	75	78	81	84	85	84
Colombia	23,776	26,525	28,714	32,300	33,985	35,101	57	61	64	67	70	73
Ecuador	7,035	8,123	9,380	10,264	10,980	11,460	40	42	47	52	56	58
Falkland Islands (Malvinas)						2						
French Guiana						147					75	
Guyana	730	760	790	796	808	835	29	30	31	32	33	36
Paraguay	2,682	3,147	3,681	4,317		4,960	37	39	42	44	48	53
Peru	15,161	17,295	19,698	21,588	22,886	23,780	57	61	65	67	70	72
Suriname	365	356	398	400		423	46	45	45	46	48	50
Uruguay	2,829	2,914	3,012	3,094		3,186	82	83	84	85	89	90
Venezuela	12,665	14,871	18,386	19,502	19,502	21,844	76	78	84	87	91	93
Asia			2,824,008									
Afghanistan			14,636	15,045	18,870	20,141	11	13	16	19	18	20
Armenia				3,352		3,599		63				69
Azerbaijan				7,117		7,558		52				56
Bahrain	262	334	431			564	78		81	82	83	
Bangladesh	76,582	86,700	101,147	108,118	117,787	120,433	8	9	10	12	16	18
Bhutan	1,047	1,165	1,417	1,544	600	1,638	3	3	4	5	5	6
Brunei Darus						285					58	
Cambodia			7,393	8,841		10,251	23	10	10	11	12	21
China	916,000	981,000	1,063,105	1,155,305	1,196,360	1,221,462	12	17	20	21	26	30
Cyprus	609	629	667	702		742	20		46	50	53	
East Timor						814					13	
Gaza Strip						793	20				94	
Georgia				5,418		5,457	17	50				58
Hong Kong	4,360	5,039	5,456	5,705		5,865	41				94	
India	613,000	684,000	761,175	850,638	918,570	935,744	56	21	23	26	26	27
Indonesia	133,000	148,000	164,887	182,812	191,671	197,588	84	19	22	25	29	35
Iran	33,206	39,124	46,374	58,946	62,507	67,283	71	46	49	52	57	59
Iraq			15,676	18,078	19,925	20,449	51	61	66	71	72	75
Israel	3,455	3,878	4,298	4,660		5,629	50	87	89	90	92	91
Japan	112,000	117,000	120,072	123,537		125,095	41	76	76	77	77	78
Jordan			3,509	4,259	4,443	5,439	78	55	60	64	68	71
Kazakhstan				16,670		17,111	10	52				60
Korea, Democratic People's Republic			20,082	21,774		23,917	41	56	60	64	60	61

continued

TABLE 4. CONTINUED

Region and Country	Population (thousands)						Percentage Urban					
	1975	1980	1985	1990	1994	1995	1970	1975	1980	1985	1990	1995
Korea, Republic of	35,281	38,124	40,872	42,869		44,995	38	48	57	65	72	81
Kuwait	1,007	1,375	1,785	2,143		1,547	59	84	90	94	96	97
Kyrgyzstan				4,362		4,745	27	38				39
Laos	3,024	3,205	3,594	4,202	4,605	4,882		11	13	16	19	22
Lebanon			2,688	2,555	3,700	3,009	45	67	75	80	84	87
Macau						410	4			*	99	
Malaysia			15,551	17,891		20,140	5	38	34	38	43	54
Maldives					238	254	25				29	
Mongolia	1,447	1,663	1,900	2,177		2,410	33	49	51	51	58	61
Myanmar (Burma)			39,487	41,813	44,596	46,527	80	24	24	24	25	26
Nepal	12,841	14,640	16,482	19,253	19,755	21,918	49	5	6	8	11	14
Oman	766	988	1,228	1,751	1,909	2,163	100	6	7	9	11	13
Pakistan	71,033	82,851	101,696	121,933	136,645	140,497	22	26	28	30	32	35
Philippines	43,103	48,323	54,709	60,779	63,427	67,581	43	36	37	40	43	54
Qatar			301			551	13		86	88	90	
Saudi Arabia	7,251	9,372	11,240	16,048		17,880	42	59	66	72	77	80
Singapore	2,037	2,282	2,572	2,705		2,848	18	100	100	100	100	100
Sri Lanka	13,496	14,738	16,404	17,225	17,671	18,354	8	22	22	21	21	22
Syria	7,438	8,704	10,581	12,348	13,696	14,661	32	45	47	50	50	52
Tajikistan				5,287		6,101		36				32
Thailand	41,359	46,700	51,571	55,583		58,791		15	17	20	22	20
Turkey	40,078	44,438	49,974	56,098		61,945	34	42	44	46	61	69
Turkmenistan				3,657		4,099	52	48				45
United Arab Emirates	505	1,015	1,312	1,671		1,904	94	65	81	78	81	84
Uzbekistan				20,420		22,843	52	39				41
Vietnam			59,451	66,689	72,931	74,545	55	19	19	20	20	21
Yemen Arab Republic	6,991	8,219	9,670	11,311		14,501	80	16	15	20	29	34
Yemen, People's Democratic Republic			2,124						37	40		
Europe			492,009	721,734		726,999		67			73	74
Albania			3,050	3,289		3,441	34	33	33	34	36	37
Andorra						68					63	
Austria	7,556	7,553	7,555	7,705		7,968	52	53	55	56	58	56
Belarus				10,212		10,141		50				71
Belgium	9,795	9,847	9,880	9,951		10,113	94	95	95	96	96	97
Bosnia and Herzegovina				4,308		3,459		31				49
Bulgaria	8,721	8,862	8,941	8,991		8,769	52	58	63	67		71
Channel Islands			15,648			148					31	
Croatia				4,517		4,495		45				64
Czechoslovakia/Czech Republic	14,802	15,262	15,500	10,306		10,296	55	58	62	65	77	65
Denmark	5,060	5,123	5,144	5,140		5,181	80	82	84	86	85	85
Estonia				1,575		1,530		68			72	73
Faeroe Islands						47					31	
Finland	4,711	4,780	4,875	4,986		5,107	50	58	60	64	60	63
France	52,699	53,880	54,608	56,718		57,981	71	73	73	73	73	73
German Democratic Republic			16,642				74		76	77		
Germany, Federal Republic of			61,106				81		84	86	85	
Germany	78,679	78,303	77,698	79,365		81,591		81				87
Gilbraltar						28					100	
Greece	9,047	9,643	9,932	10,238		10,451	53	55	58	60	63	65
Holy See						1						
Hungary	10,532	10,710	10,797	10,365		10,115	46	53	54	56	64	65
Iceland	218	228	241	255		269	85	87	88	89	91	92
Ireland	3,177	3,401	3,595	3,503		3,553	52	54	55	57	57	58
Isle of Man						74					74	
Italy	55,441	56,434	56,874	57,023		57,187	64	66	67	67	69	67
Latvia				2,671		2,557		65			71	73
Liechtenstein						31					20	
Lithuania				3,711		3,700		56			69	72
Luxembourg	361	365	367	382		406	68		78	81	84	
Macedonia, TFYR				2,046		2,163		51				60
Malta	328	364	382	344		366	78		83	85	87	
Moldova, Repbulic of				4,362		4,432		36				52
Monaco						32					100	
Netherlands	13,666	14,150	14,506	14,952		15,503	46	88	88	88	89	89
Norway	4,007	4,091	4,150	4,241		4,337	65	68	71	73	75	73

continued

TABLE 4. CONTINUED

Region and Country	Population (thousands)						Percentage Urban					
	1975	1980	1985	1990	1994	1995	1970	1975	1980	1985	1990	1995
Poland	34,022	35,578	37,556	38,119		38,388	52	55	58	61	62	65
Portugal	9,093	9,766	10,077	9,868		9,823	26	28	30	31	34	36
Romania	21,245	22,201	23,065	23,207		22,835	42	46	48	49	54	55
Russian Federation				147,913		147,000		66				76
San Marino						25					92	
Slovakia				5,256		5,353		46				59
Slovenia				1,918		1,946		42				64
Spain	35,937	37,386	39,019	39,272		39,621	66	70	73	76	78	76
Sweden	8,193	8,310	8,350	8,559		8,780	81	83	83	83	84	83
Switzerland	6,405	6,319	6,470	6,834		7,202	55	56	57	58	62	61
Ukraine				51,637		51,380		58				70
United Kingdom	56,226	56,330	56,618	57,411		58,258	89	89	91	92	89	89
Yugoslavia	21,365	22,304	23,191	10,156		10,849	35	43	42	46	56	57
USSR			278,373					63	66	66		
Oceania			24,820			28,549		72			71	70
American Samoa						54					47	
Australia	13,893	14,692	15,714	16,888		18,088		86	86	86	85	85
Cook Islands						19					25	
Fiji	576	634	684	726		784		37	39	41	39	41
French Polynesia						220					65	
Guam						150					53	
Kiribati						79					36	
Marshall Islands						54						
Micronesia						124						
Nauru						11						
New Caledonia						181					60	
New Zealand	3,087	3,113	3,291	3,360		3,575		83	83	84	84	86
Niue					2	2					23	
Northern Mariana Islands						47						
Palau						17					97	
Papua New Guinea	2,729	3,070	3,696	3,839	4,110	4,302		12	13	14	16	16
Pitcairn						0						
Samoa						171					22	
Solomon Islands	194	233	273	320		378		9	9	10	15	17
Tokelau					2	2						
Tonga	90	94	95	99	98	98					35	
Tuvalu					9	10						
Vanuatu	96	116	131	147		169					19	
Wallis and Futuna Islands						14						
Western Samoa	151	155	157	160								
Sources	a	a	b	c	d	e	f,g	c	f,g	f,g	g	c

Table 5. Percentage of Population with Access to Safe Drinking Water by Country, 1970 to 1994

Description

Access to safe drinking water is a cornerstone of any healthy population. Data are given here for urban, rural, and total populations in percentage of population with access to a safe water supply for 1970, 1975, 1980, 1985, 1990, and 1994—the most recent year for which data were available. Data for 1990 are based on original, unrevised World Health Organization (WHO) data from the end of the decade and do not reflect recent revisions by the Joint Monitoring Programme (JMP) of WHO (see descriptions of Tables 7 and 8).

The provision of safe drinking water for all was the goal of the United Nations' International Drinking Water Supply and Sanitation Decade (Decade) from 1981 to 1990. Most of the data presented here were assembled by WHO during the decade and were drawn from responses by national governments to WHO questionnaires. The JMP of WHO, the United Nations Children's Fund, and the Water Supply and Sanitation Collaborative Council have continued sector monitoring in the post-Decade years and aim to support and strengthen the monitoring efforts of individual countries.

Limitations

The definition of "access" is inconsistent from country to country and from year to year within the same country. Two-thirds of the countries reporting to the JMP in 1996 indicated how they defined "access." This definition most commonly centered on walking distance or time from household to a "safe" water source, such as a public standpipe, which varied from 50 to 2,000 meters and 5 to 30 minutes. Definitions sometimes included considerations of quantity, with the acceptable limit ranging from 15 to 50 liters per capita per day. Further, the definitions of "access" reported to the JMP in 1996 were more stringent in many countries than what those countries had previously submitted. Definitions of what is "safe" may also vary from country to country and are not included in JMP reports. The WHO considers safe drinking water to be treated surface water or untreated water from protected springs, boreholes, and wells. Thus, direct comparisons between countries, and across time within the same country, are difficult. Direct comparisons are additionally complicated by the fact that these data hide disparities between regions and socioeconomic classes.

Sources

a: United Nations Environment Programme, 1989, *Environmental Data Report*, GEMS Monitoring and Assessment Research Centre, Basil Blackwell, Oxford.

b: World Health Organization data, cited by the World Resources Institute, 1988, *World Resources 1988–89*, World Resources Institute and the International Institute for Environment and Development in collaboration with the United Nations Environment Programme, Basic Books, New York.

c: United Nations Environment Programme, 1993–94, *Environmental Data Report*, GEMS Monitoring and Assessment Research Centre in cooperation with the World Resources Institute and the UK Department of the Environment, Basil Blackwell, Oxford.

d: World Health Organization, 1996, *Water Supply and Sanitation Sector Monitoring Report: 1996 (Sector Status as of 1994),* in collaboration with the Water Supply and Sanitation Collaborative Council and the United Nations Children's Fund, UNICEF, New York.

TABLE 5. PERCENTAGE OF POPULATION WITH ACCESS TO SAFE DRINKING WATER BY COUNTRY, 1970 TO 1994

Region and Country	Urban						Rural						Total					
	1970	1975	1980	1985	1990	1994	1970	1975	1980	1985	1990	1994	1970	1975	1980	1985	1990	1994
Africa	66	68	69	77														
Algeria	84	100		85				61		55				77		68		
Angola			85	87	73	69			10	15	20	15			26	33	35	32
Benin	83	100	26	80	73	41	20	20	15	34	43	53	29	34	18	50	54	50
Botswana	1	95		84	100		26	39		46	88		29	45		53	91	
Burkina Faso	35	50	27	43			10	23	31	69	70		12	25	31	67		78
Burundi	77		90	98	92	92			20	21	43	49		23	25	45	52	
Cameroon	77			43	42		21			24	45		32		32	44		
Cape Verde			100	83		70			21	50		34			25	52		51
Central African Republic				13	19	18				26	18						23	18
Chad	47	43				48	24	23				17	27	26				24
Comoros																		
Congo	63	81	42				6	9	7				27	38	20			
Congo, Democratic Republic	33	38		52	68	37	4	12		21	24	23	11	19		32	36	27
Côte d'Ivoire	98				57	59	29				80	81	44				71	72
Djibouti			50	50		77			20	20		100			43	45		90
Egypt	94		88		95	82	93		64		86	50	93		84		90	64
Equatorial Guinea			47		65	88				18	100				32			95
Eritrea																		
Ethiopia	61	58		69				1		9			6	8		16		
Gabon																		
Gambia	97		85	97	100		3			50	48		12			59	60	76
Ghana	86	86	72	93	63	70	14	14	33	39		49	35	35	45	56	21	56
Guinea	68	69	69	41	100	61			2	12	37	62		14	15	18	53	62
Guinea-Bissau			18	17		38			8	22		57			10	21		53
Kenya	100	100	85			67	2	4	15			49	15	17	26			53
Lesotho	100	65	37	65		14	1	14	11	30		64	3	17	15	36		52
Liberia	100			100		58	6			23		8	15			53		30
Libya	100	100	100				42	82	90				58	87	96			
Madagascar	67	76	80	81		83	1	14	7	17		10	11	25	21	31		29
Malawi			77	97		52			37	50		44			41	56		45
Mali	29		37	46	41	36				10	4	38				16	13	37
Mauritania	98		80	73		84	10		85			69	17		84			76
Mauritius	100	100	100	100	100	95	29	22	98	100	100	100	61	60	99	100	100	98
Morocco	92		100	100	100	98	28			25	18	14	51			59	56	52
Mozambique				38		17				9		40				15		32
Namibia					90	87					37	42					52	57
Niger	37	36	41	35	98	46	19	26	32	49	45	55	20	27	33	47	56	53
Nigeria				100	100	63				20	22	26				38	49	39
Reunion																		
Rwanda	81	84	48	79	84		66	68	55	48	67		67	68	55	50	68	
Sao Tome and Principe									45							45		
Senegal	87	56	77	79	65	82			25	38	26	28			43	53	42	50
Seychelles									95							95		
Sierra Leone	75		50	68	80	58	1		2	7	20	21	12		14	24	39	34
Somalia	17	77		58			14	22		22			15	38		34		
South Africa																		70
Sudan	61	96	100			66	13	43	31			45	19	50	51			50
Swaziland		83		100		41		29		7		44		37		31		43
Tanzania	61	88		90			9	36		42			13	39		53		
Togo	100	49	70	100		74	5	10	31	41		58	17	16	38	54		63
Tunisia	92	93	100	100		100	17		17	31		89	49		60	70		99
Uganda	88	100		37	60	47	17	29		18	30	32	22	35		20	33	34
Zambia	70	86		76		64	22	16		41		27	37	42		58		43
Zimbabwe				95						32	80							84

continued

TABLE 5. CONTINUED

Region and Country	Urban 1970	1975	1980	1985	1990	1994	Rural 1970	1975	1980	1985	1990	1994	Total 1970	1975	1980	1985	1990	1994
North and Central America and Caribbean																		
Anguilla																		
Antigua and Barbuda																		
Aruba																		
Bahamas	100	100	100	100	98		12	13			75		65	65	100	100	90	
Barbados	95	100	99	100	100		100	100	98	99	100		98	100	99	99	100	
Belize			99	100	95	96			36	26	53	82			68	64	74	89
British Virgin Islands					100						100							
Canada																		
Cayman Islands			100	98														
Costa Rica	98	100	100	100		85	59	56	82	83		99	74	72	90	91		92
Cuba	82	96			100	96	15				91	85	56				98	93
Dominican Republic	72	88	85	85	82	74	14	27	34	33	45	67	37	55	60	62	67	71
Dominica																		
El Salvador	71	89	67	68	87	78	20	28	40	40	15	37	40	53	50	51	47	55
Grenada	100	100					47	77										
Guadeloupe																		
Guatemala	88	85	90	72	92		12	14	18	14	43		38	39	46	37	62	
Haiti		46	51	59	56	37		3	8	30	35	23		12	19	38	41	28
Honduras	99	99	93	56	85	81	10	13	40	45	48	53	34	41	59	49	64	65
Jamaica	100	100	55	99			48	79	46	93			62	86	51	96		
Martinique																		
Mexico	71	70	90	99	94	91	29	49	40	47		62	54	62	73	83	69	83
Montserrat																		
Netherlands Antilles																		
Nicaragua	58	100	67	76		81	16	14	6	11		27	35	56	39	48		61
Panama	100	100	100	100			41	54	62	64			69	77	81	82		83
Puerto Rico																		
St. Kitts																		
St. Lucia																		
St. Vincent																		
Trinidad/Tobago	100	79	100	100	100		95	100	93	95	88		96	93	97	98	96	
Turks/Caicos Islands				87						68						77		
United States of America																		
United States Virgin Islands																		
South America																		
Argentina	69	76	61	63			12	26	17	17			56	66	54	56		
Bolivia	92	81	69	75	76	78	2	6	10	13	30	22	33	34	36	43	53	55
Brazil	78	87	83	85	95	85	28		51	56	61	31	55		72	77	86	72
Chile	67	78	100	98		94	13	28	17	29		37	56	70	84	87		85
Colombia	88	86	93	100	87	88	28	33	73	76	82	48	63	64	86		86	76
Ecuador	76	67	79	81	63	82	7	8	20	31	44	55	34	36	50	57	55	70
Falkland Islands (Malvinas)																		
French Guiana																		
Guyana	100	100	100	100	100	90	63	75	60	65	71	45	75	84	72	76	81	61
Paraguay	22	25	39	53	61		5	5	9	8	9		11	13	21	28	34	
Peru	58	72	68	73	68	74	8	15	18	17	24	24	35	47	50	55	55	60
Suriname			100	71					79	94					88	83		
Uruguay	100	100	96	95	100		59	87	2	27			92	98	81	85	89	
Venezuela	92		93	93		80	38		53	65	36	75	75		86	89		79
Asia																		
Afghanistan	18	40	28	38	40	39	1	5	8	17	19	5	3	9	8	17	23	12
Armenia																		
Azerbaijan																		
Bahrain	100	100		100	100		94	100		100	0		99	100		100		
Bangladesh	13	22	26	24	39	100	47	61	40	49	89	97	45	56	39	46	81	97
Bhutan			50		60	75			5	19	30	54			7		32	64
Brunei Darus			100						95									
Cambodia																		
China					87	93					68	89					73	90

continued

TABLE 5. CONTINUED

Region and Country	Urban						Rural						Total					
	1970	1975	1980	1985	1990	1994	1970	1975	1980	1985	1990	1994	1970	1975	1980	1985	1990	1994
Cyprus	100	94		100	100		92	96		100	100		95	95		100	100	
East Timor																		
Gaza Strip																		
Georgia																		
Hong Kong			100	100					95		96						100	
India	60	80	77	76	86	85	6	18	31	50	69	79	17	31	42	56	73	81
Indonesia	10	41	35	43	35	78	1	4	19	36	33	54	3	11	23	38	34	62
Iran	68	76	82		100	89	11	30	50		75	77	35	51	66		89	83
Iraq	83	100		100	93		7	11	54	54	41		51	66		86	78	44
Israel																		
Japan																		
Jordan	98		100	100	100		59		65	88	97		77		86	96	99	89
Kazakhstan																		
Korea, Democratic People's Republic																		
Korea, Republic of	84	95	86	90	100		38	33	61	48	76		58	66	75	75	93	
Kuwait	60	100	86	97	100		51	89	87									
Kyrgyzstan																		
Laos	97	100	28		47	40	39	32	20		25	39	48	41	21		29	39
Lebanon						100						100						100
Macau																		
Malaysia	100	100	90	96	96		1	6	49	76	66		29	34	63	84	79	
Maldives			11	58	77	98			3	12	68	86			2	21		89
Mongolia				100						58							82	
Myanmar (Burma)	35	31	38	36	79	36	13	14	15	24	72	39	18	17	21	27	74	38
Nepal	53	85	83	70	66	66		5	7	25	34	41	2	8	11	28	38	44
Oman		100		90				48		49				52		53		63
Pakistan	77	75	72	83	82	77	4	5	20	27	42	52	21	25	35	44	55	60
Philippines	67	82	49	49	93	93	20	31	43	54	72	77	36	50	45	52	81	85
Qatar	100	100	76	100			75	83	43				95	97	71			
Saudi Arabia	100	97	92	100			37	56	87	88			49	64	90	94		
Singapore			100	100	100										100	100	100	
Sri Lanka	46	36	65	82	80	43	14	13	18	29	55	47	21	19	28	40	60	46
Syria	98		98			92	50		54			78	71		74			85
Tajikistan																		
Thailand	60	69	65	56			10	16	63	66	85		17	25	63	64		
Turkey		95						62						76				
Turkmenistan																		
United Arab Emirates			95						81						92			
Uzbekistan																		
Vietnam				70	47	53			32	39	33	32				45	36	36
Yemen Arab Republic	45		100	100			2		18	25			4		31	40		
Yemen, People's Democratic Republic	88		85				43		25				57		52			
Oceania																		
American Samoa																		
Australia																		
Cook Islands			100	99	100					88	100					92		
Fiji	78	89	94		96	100	15	56	66		69	100	37	69	77	80	100	
French Polynesia				100							18							
Guam																		
Kiribati			93		91					25	63							
Marshall Islands				100							45							
Micronesia					100						38	100					100	
Nauru																		
New Caledonia																		
New Zealand																		
Niue				0	100					100	100	100					100	
Northern Mariana Islands				100							0							
Palau				100							97							

continued

TABLE 5. CONTINUED

Region and Country	Urban						Rural						Total					
	1970	1975	1980	1985	1990	1994	1970	1975	1980	1985	1990	1994	1970	1975	1980	1985	1990	1994
Papua New Guinea	44	30	55	95	94	84	72	19	10	15	20	17	70	20	16	26	32	28
Pitcairn																		
Samoa	86	100	97		100			23	94		77		17	43				
Solomon Islands			96		82				45		58						62	
Tokelau						100				100		100						100
Tonga	100	100	86	99	92	100	53	71	70	99	98	100	63	83	17	99	96	100
Tuvalu				100		100				100		95						98
Vanuatu			65	95					53	54						64		
Wallis and Futuna Islands																		
Western Samoa			97	75					94	67						69		
Sources	a, b	a, b	a, b	a, b	c	d	a, b	a, b	a, b	a, b	c	d	a, b	a, b	a, b	a, b	c	d

Table 6. Percentage of Population with Access to Sanitation by Country, 1970 to 1994

Description

Adequate sanitation is also a fundamental requirement for basic human well-being. Data are given here for the percentage of urban, rural, and total populations with access to adequate sanitation for 1970, 1975, 1980, 1985, 1990, and 1994. Data for 1990 are based on original, unrevised World Health Organization (WHO) data from the end of the decade, and do not reflect recent revisions by the Joint Monitoring Programme (JMP) of WHO (see description of Tables 7 and 8). Access to sanitation was a focus of the WHO during the United Nations' International Drinking Water Supply and Sanitation Decade from 1981 to 1990. Most of the data presented here were assembled by the WHO during that decade and were drawn from responses by national governments to WHO question- naires. The JMP of WHO, the United Nations Children's Fund, and the Water Supply and Sanitation Collaborative Council have continued sector monitoring in the post-Decade years and aim to support and strengthen the monitoring efforts of individual countries.

Limitations

As is the case with drinking water data, definitions for access to sanitation vary from country to country and from year to year within the same country. Countries generally regard sanitation facilities that break the fecal–oral transmission route as adequate. In urban areas, adequate sanitation may be provided by connections to public sewers or by household systems such as pit privies, flush latrines, septic tanks, and communal toilets. In rural areas, pit privies, pour-flush latrines, septic tanks, and communal toilets are con- sidered adequate. Direct comparisons between countries, and across time within the same country, are difficult and are additionally complicated by the fact that these data hide disparities between regions and socioeconomic classes.

Sources

a: United Nations Environment Programme, 1989, *Environmental Data Report*, GEMS Monitoring and Assessment Research Centre, Basil Blackwell, Oxford.

b: World Health Organization data, cited by the World Resources Institute, 1988, *World Resources 1988–89*, World Resources Institute and the International Insti- tute for Environment and Development in collaboration with the United Nations Environment Programme, Basic Books, New York.

c: United Nations Environment Programme, 1993–94, *Environmental Data Report*, GEMS Monitoring and Assessment Research Centre in cooperation with the World Resources Institute and the UK Department of the Environment, Basil Blackwell, Oxford.

d: World Health Organization, 1996, *Water Supply and Sanitation Sector Monitor- ing Report: 1996 (Sector Status as of 1994),* in collaboration with the Water Supply and Sanitation Collaborative Council and the United Nations Children's Fund, UNICEF, New York.

TABLE 6. PERCENTAGE OF POPULATION WITH ACCESS TO SANITATION BY COUNTRY, 1970 TO 1994

Region and Country	Urban						Rural						Total					
	1970	1975	1980	1985	1990	1994	1970	1975	1980	1985	1990	1994	1970	1975	1980	1985	1990	1994
Africa	47	75	57	75														
Algeria	13	100		80			6	50		40			9	67		57		
Angola			40	29	25	34			15	16	20	8			20	19	21	16
Benin	83		48	58	60	54	1		4	20	35	6	14		16	33	44	20
Botswana				93	100					28	85					40	89	
Burkina Faso	49	47	38	44		42			5	6		11	4	4	7	9		18
Burundi	96		40	84	64	60			35	56	16	50			35	58	18	51
Cameroon				100						1						43		
Cape Verde			34	32		40			10	9		10			11	10		24
Central African Republic	64	100			45		96	100			46		72	100			46	46
Chad	7	9				73		1				7	1	1				21
Comoros																		
Congo	8	10					6	9					6	9				
Congo, Democratic Republic	5	65			46	23	5	6		9	11	4	5	22			21	9
Côte d'Ivoire	23				81	59					100	51	5				92	54
Djibouti			43	78		77			20	17		100			39	64		90
Egypt				80		20				10	26	5					50	11
Equatorial Guinea					54	61					24	48					33	54
Eritrea																		
Ethiopia	67	56		96			8	8		96			14	14				
Gabon																		
Gambia					100	83					27	23					44	37
Ghana	92	95	47	51	63	53	40	0	17	16	60	36	55	56	26	30	61	42
Guinea	70		54				2		1		0		13		11			70
Guinea-Bissau			21	29		32			13	18		17			15	21		20
Kenya	85	98	89			69	45	48	19			81	50	55	30			77
Lesotho	44	51	13	22		1	10	12	14	14		7	11	13	14	15		6
Liberia	100			6		38	9			2		2	19					18
Libya	100	100	100				54	69	72				67	79	88			
Madagascar	88		9	55		50		9				3						15
Malawi			100			70			81			51				83		53
Mali	63		79	90	81	58				3	10	21	8			19	27	31
Mauritania	100		5	8									7					
Mauritius	51	63	100	100	100	100	99	100	90	86	100	100	77	82	94	92	100	100
Morocco	75			62	100	69	4			16	18		29			20		40
Mozambique				53		70				12		70				20		
Namibia				24						11						15		
Niger	10	30	36		71	71		1	3		4	4	1	3	7		17	15
Nigeria					80	61				5	11	21					35	36
Reunion																		
Rwanda	83	87	60	77	88		52	56	50	55	17		53	57	51	56	21	
Sao Tome and Principe										15						15		
Senegal			100	87	57	83				2	38	40				36	46	58
Seychelles																		
Sierra Leone			31	60	55	17			6	10	31	8			12	24	39	11
Somalia		77		44				35		5				47		18		
South Africa						79						12						46
Sudan	100	100	73	73		79	4	10				4	16	22				22
Swaziland		99		100		36		25		25		37		36		45		36
Tanzania		88		93				14		58					17	66		
Togo	4	36	24	31		57		12	10	9		13	1	15	13	14		26
Tunisia	100		100	84		100	34			16		85	62			55		96
Uganda	84	82		32	32	75	76	95		30	60	55	76	94		30	57	57
Zambia	12	87		76		40	18	16		34		10	16	42		55		23
Zimbabwe				95					15	22							43	

continued

TABLE 6. CONTINUED

Region and Country	Urban 1970	1975	1980	1985	1990	1994	Rural 1970	1975	1980	1985	1990	1994	Total 1970	1975	1980	1985	1990	1994
North and Central America and Caribbean																		
Anguilla																		
Antigua and Barbuda																		
Aruba																		
Bahamas	100	100	88	100	98		13	13			2		66	65	88	100	63	
Barbados	100	100		100	100		100	100			100			100	100		100	
Belize			62	87	76	23			75	45	22	87			69	66	50	57
British Virgin Islands				100						100								
Canada																		
Cayman Islands			94	96						94				96				
Costa Rica	66	94	99	99		85	43	93	84	89		99	52	93	91	95		92
Cuba	57	100			100	71				68		51					92	66
Dominican Republic	63	74	25	41	95	76	54	16	4	10	75	83	58	42	15	23	87	78
Dominica																		
El Salvador	66	71	48	82	85	78	18	17	26	43	38	59	37	39	35	58	59	68
Grenada																		
Guadeloupe																		
Guatemala			45	41	72		11	16	20	12	52				30	24	60	
Haiti			42	42	44	42	43	1	10	13	17	16			19	21	25	24
Honduras	64	53	49	24	89	81	9	13	26	34	42	53	24	26	35	30	63	65
Jamaica	100	100	12	92			92	91	2	90			94	94	7	91		
Martinique																		
Mexico			77	77	85	81	13	14	12	13		26			55	58		66
Montserrat																		
Netherlands Antilles																		
Nicaragua			34	35		34	8	24		16		27				27		31
Panama	87	78	83	99			69	76	59	61				77	71	81		86
Puerto Rico																		
St. Kitts																		
St. Lucia																		
St. Vincent																		
Trinidad/Tobago				100							92						97	
Turks/Caicos Islands	51	83	96	100			96	97	88	95			81	92	93	98		
United States of America																		
United States Virgin Islands																		
South America																		
Argentina	87	100	80	75			79	83	35	35			85	97		69		
Bolivia	25		37	33	38	58	4	9	4	10	14	16	12		18	21	26	41
Brazil	85			86	84	55	24		1	1	32	3	58			63	71	44
Chile	33	36	100	100		82	10	11	10	4			29	32	83	84		
Colombia	75	73	93	96	84	76	8	13	4	13	18	33	47	48	61		64	63
Ecuador			73	98	56	87		7	17	29	38	34			43	65	48	64
Falkland Islands (Malvinas)																		
French Guiana																		
Guyana	95	99	73	100	97		92	94	80	79	81		93	96	78	86	86	
Paraguay	16	28	95	89	31				80	83	60		6	10	86	85	46	
Peru	52		57	67	76	62	16		0	12	20	10	36		36	49	59	44
Suriname			100	78					79	48					88	62		
Uruguay	97	97	59	59			13	17	6	59			82	83	51	59		
Venezuela			60	57		64	45		12	5	72	30			52	50		58
Asia																		
Afghanistan	69	63		5	13	38	16	15				1	21	21				8
Armenia																		
Azerbaijan																		
Bahrain			100	100						100	0					100		
Bangladesh	87	40	21	24	40	77			1	3	4	30	6	5	3	5	10	35
Bhutan					80	66					3	18					7	41
Brunei Darus																		
Cambodia			100					76										
China					100	58					81	7					86	21

continued

TABLE 6. CONTINUED

Region and Country	Urban 1970	1975	1980	1985	1990	1994	Rural 1970	1975	1980	1985	1990	1994	Total 1970	1975	1980	1985	1990	1994
Cyprus	100	94		100	96		92	95		100	100		95	95		100	98	
East Timor																		
Gaza Strip																		
Georgia																		
Hong Kong					90						50						88	
India	85	87	27	31	44	70	1	2	1	2	3	14	18	20	7	9	14	29
Indonesia	50	60	29	33	79	73	4	5	21	38	30	40	12	15	23	37	44	51
Iran	100	100	96		100	89	48	59	43		35	37	70	78	69		72	67
Iraq	82	75		100	96			1		11			47	47		74		36
Israel																		
Japan																		
Jordan			94	92	100				34		100					70	100	95
Kazakhstan																		
Korea, Democratic People's Republic																		
Korea, Republic of	59	80	100	100	67			50	100	100	12		25	64	100	100	52	
Kuwait			100	100					100						100			
Kyrgyzstan																		
Laos		10	13		30	70		2	4		8	13		3	5		12	24
Lebanon						100						100						100
Macau																		
Malaysia	100	100	100	100	94		43	43	55	60	94		59	60	70	75	94	
Maldives		21	60	100	95	95			1	2	4	26		3	13	22		44
Mongolia					100						47					78		
Myanmar (Burma)	45	38	38	33	50	42					13	40	35	33	20	24	22	41
Nepal	14	14	16	17	34	51			1	1	3	16	1	1	1		6	20
Oman	100	100		88				5		25					12	31		76
Pakistan	12	21	42	51	53	53			2	6	12	19	3	6	13	19	25	30
Philippines	90	76	81	83	79		40	44	67	56	63		57	56	75	67	70	
Qatar	100	100			100		16	100			85		83	100				
Saudi Arabia	67	91	81	100			11	35	50	33			21	47	70	82		
Singapore			80	99	99											80	99	
Sri Lanka	76	68	80	65	68	33	61	55	63	39	45	58	64	59	67	44	50	52
Syria			74			77			28			35			50			56
Tajikistan																		
Thailand	65	58	64	78			8	36	41	46	86		17	40	45	52		
Turkey			56															
Turkmenistan																		
United Arab Emirates			93						22							80		
Uzbekistan																		
Vietnam	100				23	43		2	55		10	15	26				13	21
Yemen Arab Republic			60	83														
Yemen, People's Democratic Republic			70						15							35		
Oceania																		
American Samoa																		
Australia																		
Cook Islands			100	100	100				76	99	100					99		
Fiji	100	100	85		91	100	87	93	60		65	85	91	96	70		75	92
French Polynesia				98							95							
Guam																		
Kiribati					91	100					49	100						100
Marshall Islands				100							45							
Micronesia					99	100					46	100						100
Nauru																		
New Caledonia																		
New Zealand																		
Niue					0	100				100	100	100						100
Northern Mariana Islands					100						71							
Palau					95						100							

continued

TABLE 6. CONTINUED

Region and Country	Urban						Rural						Total					
	1970	1975	1980	1985	1990	1994	1970	1975	1980	1985	1990	1994	1970	1975	1980	1985	1990	1994
Papua New Guinea	100	100	96	99	57	82	5	5	3	35		11	14	18	15	44		22
Pitcairn																		
Samoa	100	100	86		100		80	99	83		92	17	84	99				
Solomon Islands			80		73				21		2						13	
Tokelau						100			41			100						100
Tonga	100	100	97	99	88	100	100	100	94	40	78	100	100	100	19	52	82	100
Tuvalu			100	81		90			80	73		85						87
Vanuatu			95	86					68	25						40		
Wallis and Futuna Islands																		
Western Samoa			86	88					83	83						84		
Sources	a,b	a,b	a,b	a,b	c	d	a,b	a,b	a,b	a,b	c	d	a,b	a,b	a,b	a,b	c	d

Table 7. Access to Safe Drinking Water in Developing Countries by Region, 1980 to 1994

Description

The total population and the population without access to safe drinking water ("unserved") are shown here by World Health Organization (WHO) regions for 1980, 1990, 1990 (revised), and 1994—the most recent year for which data are available. The International Drinking Water Supply and Sanitation Decade was not successful in achieving its goal of access to safe drinking water for all by 1990, as this table shows. Included here are two sets of data for access in 1990. The first set is the original, unrevised data published by the WHO in 1990. The second set is revised based on current definitions of access to safe drinking water. Since 1980 countries have adopted increasingly stringent definitions of access, and WHO utilized the current and most stringent definition of access in each country to estimate what access *would have been* in 1990 if this *current* definition had been used. The most recent data available, from 1994, are also included. These data reflect the responses of approximately 80 developing countries, including five Middle Eastern countries (Iraq, Jordan, Lebanon, Oman, and Syria) included under the heading of Western Asia. The significant figures presented in the original source are retained for this table.

Limitations

These data give a good picture of current lack of access to safe drinking water, but comparisons between years should be done with extreme care, because definitions have changed over time. These data are confined to developing countries and do not include Eastern Europe, Central Asia, or countries classified as "developed" by the United Nations. The specific countries reporting in 1980 are in some cases different than those reporting in 1990 or 1994. Revised 1990 data are not available by country, and thus it is impossible to determine who was overlooked in the original 1990 data.

Sources

Christmas, J., and C. de Rooy, 1991, "The Decade and Beyond: At a Glance," *Water International*, 16: 127–134.

World Health Organization, 1996, *Water Supply and Sanitation Sector Monitoring Report: 1996 (Sector Status as of 1994),* in collaboration with the Water Supply and Sanitation Collaborative Council and the United Nations Children's Fund, UNICEF, New York.

TABLE 7. ACCESS TO SAFE DRINKING WATER IN DEVELOPING COUNTRIES BY REGION, 1980 TO 1994

Region and Country	1980			1990 (Unrevised)			1990 (Revised)			1994		
	Population (millions)	Percentage with Access	Number Unserved (millions)	Population (millions)	Percentage with Access	Number Unserved (millions)	Population (millions)	Percentage with Access	Number Unserved (millions)	Population (millions)	Percentage with Access	Number Unserved (millions)
Africa												
Urban	120	83	20.36	203	87	26.33	201	67	66	239	64	86
Rural	333	33	223.00	410	42	237.59	432	35	279	468	37	295
Total	453	46	243.36	612	57	263.92	633	45	345	707	46	381
Latin America and the Caribbean												
Urban	237	82	42.61	324	87	42.13	314	90	32	348	88	42
Rural	125	47	66.20	124	62	47.07	126	51	62	125	56	55
Total	362	70	108.81	448	80	89.20	440	79	94	473	80	97
Asia and the Pacific												
Urban	549	73	148.35	761	77	175.07	829	83	140	955	84	150
Rural	1,823	28	1,312.78	2,099	67	692.80	2,097	53	989	2,167	78	477
Total	2,373	38	1,461.12	2,861	70	867.87	2,926	61	1,129	3,122	80	627
Western Asia												
Urban	28	95	1.38	44	100	0.00	45	87	6	52	98	1
Rural	22	51	10.76	26	56	11.26	27	63	10	29	69	9
Total	49	75	12.13	70	84	11.26	72	78	16	81	88	10
Total												
Urban	933	77	214.70	1,332	82	239.80	1,389	82	244	1,594	82	279
Rural	2,303	30	1,612.09	2,659	63	983.65	2,682	50	1,341	2,789	70	836
Total	3,236	44	1,826.79	3,991	69	1,223.45	4,071	61	1,585	4,383	74	1,115

Table 8. Access to Sanitation in Developing Countries by Region, 1980 to 1994

Description

Total population and the population without access to sanitation services ("unserved") are shown here by World Health Organization (WHO) regions for 1980, 1990, 1990 (revised), and 1994—the most recent year for which data are available. Access to sanitation has historically lagged behind access to safe drinking water. As was the case with safe drinking water, the International Drinking Water Supply and Sanitation Decade was not successful in achieving its goal of access to sanitation for all by 1990. Included here are two sets of data for access in 1990. The first set is the original, unrevised data published by the WHO in 1990. The second set is revised 1990 data based on current definitions of access to sanitation. Since 1980 countries have adopted increasingly stringent definitions of access, and WHO utilized the current and most stringent definition of access in each country to estimate what access *would have been* in 1990 if this *current* definition had been used. The most recent data available, from 1994, are also included. These data reflect the responses of approximately 80 developing countries, including five Middle Eastern countries (Iraq, Jordan, Lebanon, Oman, and Syria) included under the heading of Western Asia. The significant figures presented in the original source are retained for this table.

Limitations

These data give a good picture of current lack of access to sanitation services, but comparisons between years should be done with extreme care, or not at all, because of definitions that have changed over time. These data are confined to developing countries and do not include Eastern Europe, Central Asia, or countries classified as "developed" by the United Nations. The specific countries reporting in 1980 are in some cases different than those reporting in 1990 or 1994. Revised 1990 data are not available by country, and thus it is impossible to determine who was overlooked in the original 1990 data.

Sources

Christmas, J., and C. de Rooy, 1991, "The Decade and Beyond: At a Glance," *Water International*, 16: 127–134.

World Health Organization, 1996, *Water Supply and Sanitation Sector Monitoring Report: 1996 (Sector Status as of 1994),* in collaboration with the Water Supply and Sanitation Collaborative Council and the United Nations Children's Fund, UNICEF, New York.

TABLE 8. ACCESS TO SANITATION IN DEVELOPING COUNTRIES BY REGION, 1980 TO 1994

Region and Country	1980			1990 (Unrevised)			1990 (Revised)			1994		
	Population (millions)	Percentage with Access	Number Unserved (millions)	Population (millions)	Percentage with Access	Number Unserved (millions)	Population (millions)	Percentage with Access	Number Unserved (millions)	Population (millions)	Percentage with Access	Number Unserved (millions)
Africa												
Urban	120	65	41.92	203	78	44.56	201	65	71	239	55	108
Rural	333	18	272.92	410	26	303.13	432	23	333	468	24	356
Total	453	30	314.84	612	43	347.69	633	36	404	707	34	464
Latin America and the Caribbean												
Urban	237	78	52.08	324	79	68.06	314	83	52	348	73	94
Rural	125	22	97.43	124	37	78.04	126	33	84	125	34	82
Total	362	59	149.51	448	67	146.09	440	69	136	473	63	176
Asia and the Pacific												
Urban	549	65	192.30	761	65	266.41	829	62	316	955	61	371
Rural	1,823	42	1,057.51	2,099	54	965.72	2,097	18	1,718	2,167	15	1,835
Total	2,373	47	1,249.82	2,861	57	1,232.14	2,926	30	2,034	3,122	29	2,206
Western Asia												
Urban	28	79	5.78	44	100	0.00	45	68	14	52	69	16
Rural	22	34	14.49	26	34	16.90	27	60	11	29	66	10
Total	49	59	20.27	70	76	16.90	72	65	25	81	68	26
Total												
Urban	933	69	289.38	1,332	72	373.02	1,389	67	453	1,594	63	589
Rural	2,303	37	1,450.88	2,659	49	1,355.84	2,682	20	2,146	2,789	18	2,284
Total	3,236	46	1,740.26	3,991	57	1,728.86	4,071	36	2,599	4,383	34	2,873

Table 9. Reported Cholera Cases and Deaths by Region, 1950 to 1997

Description

Annual cases and deaths from cholera are shown here for Africa, the Americas, Asia, Europe, and Oceania, along with the number of countries reporting. Cholera is one of many waterborne diseases that are a consequence of the lack of sanitation services and access to clean drinking water. Some data on cholera cases and deaths are available as early as 1922 from reports to the League of Nations, but consistent reports are available from the World Health Organization only since 1950. The data here are given for the years 1950 to 1997. These data reflect the spread of cholera during its seventh pandemic, which originated in 1961 in Indonesia and spilled into eastern Asia, Bangladesh, India, the USSR, Iraq, and Iran during the 1960s. In 1970 cholera entered West Africa, which had remained relatively free of the disease up to that time. The appearance of cholera in Peru in 1991 was the first manifestation of the seventh pandemic in the Americas (defined as North, Central, and South America and the Caribbean) and was the first time that the Americas had seen more than a handful of indigenous cases in over a century (see Chapter 2).

Limitations

These data are only the cases identified and reported to the World Health Organization and do not include undiagnosed or unreported cases. The number of countries reporting cholera cases and deaths increased in Asia in the 1960s and 1970s, thereby making it difficult to make comparisons over time within the region.

Source

Dr. M. Neira, and Dr. L. Kuppens, Global Task Force on Cholera, Division of Emerging and Other Communicable Diseases Surveillance and Control, World Health Organization, Geneva, personal communication, September 1997, and March 1998.

TABLE 9.   REPORTED CHOLERA CASES AND DEATHS BY REGION, 1950 TO 1997

	Africa			Americas			Asia		
Year	Cases	Deaths	Deaths as Fraction of Cases	Cases	Deaths	Deaths as Fraction of Cases	Cases	Deaths	Deaths as Fraction of Cases
1950							34,665 *5*	16,133 *5*	0.47
1951							114,518 *6*	58,661 *6*	0.51
1952							123,025 *5*	71,397 *5*	0.58
1953							240,927 *6*	141,173 *6*	0.59
1954							37,804 *6*	28,836 *6*	0.76
1955							40,218 *4*	17,476 *4*	0.43
1956							63,542 *4*	32,623 *4*	0.51
1957							63,031 *4*	54,853 *4*	0.87
1958							95,763 *7*	61,836 *7*	0.65
1959							38,138 *4*	19,387 *4*	0.51
1960							38,130 *6*	7,800 *5*	0.20
1961							62,111 *7*	30,649 *7*	0.49
1962							67,448 *8*	30,232 *8*	0.45
1963							69,690 *14*	19,147 *14*	0.27
1964							96,973 *14*	22,293 *14*	0.23
1965							56,609 *13*	13,996 *12*	0.25
1966							31,147 *13*	4,572 *12*	0.15
1967							24,748 *10*	637 *8*	0.03
1968							34,164 *12*	830 *11*	0.02
1969							36,196 *17*	2,209 *13*	0.06
1970	11,086 *16*	747 *15*	0.07				52,429 *22*	6,787 *19*	0.13
1971	72,654 *23*	11,427 *20*	0.16				103,308 *16*	14,701 *15*	0.14
1972	5,137 *18*	386 *12*	0.08				75,064 *17*	10,271 *16*	0.14
1973	6,337 *16*	636 *14*	0.10				102,148 *15*	9,422 *14*	0.09
1974	6,074 *17*	582 *15*	0.10				89,077 *14*	7,019 *13*	0.08
1975	6,650 *14*	504 *12*	0.08				85,136 *18*	6,567 *17*	0.08
1976	3,180 *14*	194 *10*	0.06				66,469 *20*	3,764 *18*	0.06
1977	9,502 *13*	462 *8*	0.05				52,460 *21*	1,694 *20*	0.03
1978	24,643 *20*	1,591 *16*	0.06				52,611 *21*	1,783 *20*	0.03
1979	21,586 *19*	1,869 *15*	0.09				36,732 *22*	1,602 *22*	0.04
1980	18,742 *16*	1,185 *15*	0.06				24,016 *17*	769 *16*	0.03
1981	19,415 *18*	1,681 *12*	0.09				32,343 *19*	860 *18*	0.03
1982	46,924 *17*	2,988 *14*	0.06				17,991 *14*	833 *15*	0.05
1983	37,383 *15*	1,903 *14*	0.05				27,877 *11*	765 *11*	0.03
1984	17,504 *19*	1,711 *16*	0.10				11,809 *9*	119 *9*	0.01
1985	31,884 *21*	3,837 *16*	0.12				13,389 *11*	276 *11*	0.02
1986	35,585 *18*	3,490 *15*	0.10				17,131 *11*	477 *11*	0.03
1987	31,324 *17*	2,658 *15*	0.08				17,668 *10*	238 *10*	0.01
1988	23,683 *13*	1,600 *10*	0.07				20,871 *12*	378 *12*	0.02
1989	35,951 *16*	1,445 *13*	0.04				18,007 *13*	224 *13*	0.01
1990	38,683 *11*	2,288 *10*	0.06	0 *0*	0		31,003 *13*	628 *12*	0.02
1991	153,367 *22*	13,998 *19*	0.09	391,220 *16*	4,002 *16*	0.01	49,791 *16*	1,286 *16*	0.03
1992	91,081 *20*	5,291 *18*	0.06	354,089 *21*	2,401 *21*	0.01	16,299 *18*	372 *17*	0.02
1993	76,713 *16*	2,532 *15*	0.03	209,192 *21*	2,438 *20*	0.01	80,862 *25*	1,809 *23*	0.02
1994	161,983 *28*	8,128 *24*	0.05	113,684 *17*	1,107 *17*	0.01	106,100 *27*	1,393 *26*	0.01
1995	71,081 *26*	3,024 *22*	0.04	85,809 *15*	846 *15*	0.01	63,169 *18*	1,158 *18*	0.02
1996	108,535 *29*	6,216 *24*	0.06	24,639 *17*	351 *17*	0.01	10,142 *13*	122 *11*	0.01
1997	118,349 *25*	5,853 *25*	0.05	17,760 *16*	226 *15*	0.01	10,321 *16*	191 *14*	0.02

Note: Number of countries reporting is shown in italics.

continued

TABLE 9. CONTINUED

	Europe			Oceania			Total		
Year	Cases	Deaths	Deaths as Fraction of Cases	Cases	Deaths	Deaths as Fraction of Cases	Cases	Deaths	Deaths as Fraction of Cases
1950							34,665	16,133	0.47
1951							114,518	58,661	0.51
1952							123,025	71,397	0.58
1953							240,927	141,173	0.59
1954							37,804	28,836	0.76
1955							40,218	17,476	0.43
1956							63,542	32,623	0.51
1957							63,031	54,853	0.87
1958							95,763	61,836	0.65
1959							38,138	19,387	0.51
1960				*0*	*0*		38,130	7,800	0.20
1961				*0*	*0*		62,111	30,649	0.49
1962				1,293 *1*	464 *1*	0.36	68,741	30,696	0.45
1963				*0*	*0*		69,690	19,147	0.27
1964				*0*	*0*		96,973	22,293	0.23
1965				*0*	*0*		56,609	13,996	0.25
1966				*0*	*0*		31,147	4,572	0.15
1967				*0*	*0*		24,748	637	0.03
1968				*0*	*0*		34,164	830	0.02
1969				1 *1*	0 *1*	0.00	36,197	2,209	0.06
1970	726 *4*	1 *1*	0.00	*0*	*0*		64,241	7,535	0.12
1971	97 *6*	4 *1*	0.04	*0*	*0*		176,059	26,132	0.15
1972	4 *2*	0	0.00	43 *2*	1 *2*	0.02	80,248	10,658	0.13
1973	303 *5*	23 *1*	0.08	*0*	*0*		108,788	10,081	0.09
1974	2,484 *6*	48 *1*	0.02	6 *1*	1 *1*	0.17	97,641	7,650	0.08
1975	1,089 *6*	8 *1*	0.01	*0*	*0*		92,875	7,079	0.08
1976	16 *8*	2 *2*	0.12	*0*	*0*		69,665	3,960	0.06
1977	8 *6*	0	0.00	1,310 *3*	21 *2*	0.02	63,280	2,177	0.03
1978	5 *2*	0	0.00	533 *3*	0 *3*	0.00	77,792	3,374	0.04
1979	289 *5*	8 *1*	0.03	64 *3*	0 *2*	0.00	58,671	3,479	0.06
1980	16 *5*	0	0.00	3 *2*	0 *2*	0.00	42,777	1,954	0.05
1981	46 *9*	0	0.00	6 *2*	0 *1*	0.00	51,810	2,541	0.05
1982	22 *4*	0	0.00	2,217 *3*	17 *3*	0.01	67,154	3,838	0.06
1983	12 *5*	0	0.00	326 *3*	1 *3*	0.00	65,598	2,669	0.04
1984	11 *4*	0	0.00	20 *2*	0 *2*	0.00	29,344	1,830	0.06
1985	6 *3*	1 *1*	0.17	7 *2*	0 *2*	0.00	45,286	4,114	0.09
1986	52 *5*	0	0.00	3 *1*	0 *1*	0.00	52,771	3,967	0.08
1987	14 *5*	0	0.00	2 *2*	0 *2*	0.00	49,008	2,896	0.06
1988	14 *4*	0	0.00	1 *1*	0 *1*	0.00	44,569	1,978	0.04
1989	11 *6*	0	0.00	*0*	*0*		53,969	1,669	0.03
1990	349 *10*	2 *2*	0.01	40 *5*	1 *5*	0.03	70,075	2,919	0.04
1991	320 *7*	9 *1*	0.03	*0*	*0*		594,698	19,295	0.03
1992	18 *7*	0	0.00	296 *2*	8 *2*	0.03	461,783	8,072	0.02
1993	73 *15*	0	0.00	5 *1*	0 *2*	0.00	366,845	6,779	0.02
1994	2,630 *20*	0	0.00	6 *3*	0 *3*	0.00	384,403	10,628	0.03
1995	937 *17*	0	0.00	7 *2*	0 *2*	0.00	221,003	5,028	0.02
1996	25 *7*	0	0.00	20 *3*	0 *1*	0.00	143,361	6,689	0.05
1997	18 *6*	1 *6*	0.06	5 *2*	0 *3*	0.00	146,453	6,271	0.04

Note: Number of countries reporting is shown in italics.

Table 10. Reported Cholera Cases and Deaths by Country, 1996 and 1997

Description

The regional data provided in Table 9 mask differences between countries. Country data are presented here, including the number of cases and deaths reported to the World Health Organization and deaths as a fraction of cases for 1996 and 1997. These data show that in Africa more than three-quarters of the cases were concentrated in only a few countries, while Brazil, Colombia, and Peru accounted for approximately half of all cases in the Americas.

Limitations

As was the case in Table 9, these data include only those cases identified and reported to the World Health Organization and therefore exclude undiagnosed and unreported cases. Using these data alone makes it difficult to make a true comparison between countries. Rather, the number of cases taken as a function of the population would be necessary to make such a comparison.

Source

Dr. M. Neira, and Dr. L. Kuppens, Global Task Force on Cholera, Division of Emerging and Other Communicable Diseases Surveillance and Control, World Health Organization, Geneva, personal communication, September 1997, and March 1998.

TABLE 10. REPORTED CHOLERA CASES AND DEATHS BY COUNTRY, 1996 AND 1997

Region and Country	1996 Cases	1996 Deaths	Deaths as Fraction of Cases	Region and Country	1997 Cases	1997 Deaths	Deaths as Fraction of Cases
Africa				**Africa**			
Angola	1,306	42	0.03	Angola	X	X	
Benin	6,190	203	0.03	Benin	778	16	0.02
Burkina Faso	425	58	0.14	Burkina Faso	X	X	
Burundi	418	X		Burundi	1,959	95	0.05
Cameroon	5,796	485	0.08	Cameroon	1,709	180	0.11
Cape Verde	426	3	0.01	Cape Verde	X	X	
Central African Republic	X	X		Central African Republic	443	75	0.17
Chad	7,830	448	0.06	Chad	8,801	443	0.05
Côte d'Ivoire	1,345	22	0.02	Côte d'Ivoire	X	X	
Congo	X	X		Congo	275	60	0.22
Congo, Democratic Republic	7,888	638	0.08	Congo, Democratic Republic	2,421	326	0.13
Djibouti	X	X		Djibouti	2,424	50	0.02
Gambia	7	X		Gambia	X	X	
Ghana	1,665	70	0.04	Ghana	379	12	0.03
Guinea	287	17	0.06	Guinea	X	X	
Guinea-Bissau	8,397	84	0.01	Guinea Bissau	20,555	905	0.04
Kenya	482	14	0.03	Kenya	17,200	555	0.03
Liberia	8,922	450	0.05	Liberia	91	0	0.00
Malawi	1	X		Malawi	130	15	0.12
Mali	5,723	761	0.13	Mali	6	3	0.50
Mauritania	4,534	148	0.03	Mauritania	462	7	0.02
Mozambique	X	X		Mozambique	8,739	259	0.03
Niger	3,957	206	0.05	Niger	259	13	0.05
Nigeria	12,374	1,193	0.10	Nigeria	1,322	134	0.10
Rwanda	106	10	0.09	Rwanda	274	16	0.06
Senegal	16,107	765	0.05	Senegal	371	11	0.03
Somalia	10,274	464	0.05	Somalia	6,814	252	0.04
Swaziland	2	X		Swaziland	X	X	
Tanzania	1,464	36	0.02	Tanzania	40,249	2,231	0.06
Togo	146	17	0.12	Togo	42	7	0.17
Uganda	291	40	0.14	Uganda	2,610	188	0.07
Zambia	2,172	42	0.02	Zambia	36	0	0.00
Total	**108,535**	**6,216**	**0.06**	**Total**	**118,349**	**5,853**	**0.05**
Asia				**Asia**			
				Afghanistan	4,170	125	
Cambodia	740	X		Cambodia	155	0	
China	312	4	0.01	China	1,163	29	0.02
Hong Kong	4	X		Hong Kong	14	0	
India	4,396	34	0.01	India	2,753	16	0.01
Iran	X	X		Iran	757	4	
Indonesia	X	X		Indonesia	68	0	
Iraq	X	X		Iraq	X	X	
Japan	39	0	0.00	Japan	89	1	0.01
Kazakstan	X	X		Kazakstan	4	0	
Laos	720	X		Laos	X	X	
Lebanon	X	33		Lebanon	X	X	

continued

TABLE 10.   CONTINUED

Region and Country	1996 Cases	Deaths	Deaths as Fraction of Cases	Region and Country	1997 Cases	Deaths	Deaths as Fraction of Cases
Malaysia	1,486	2	0.00	Malaysia	389	4	0.01
Mongolia	177	12	0.07	Mongolia	X	X	
Nepal	274	1	0.00	Nepal	245	X	
Philippines	1,402	14	0.01	Philippines	X	X	
Republic of Korea	7	X		Republic of Korea	10	0	
Singapore	19	0	0.00	Singapore	19	0	0.00
Sri Lanka	X	X		Sri Lanka	430	12	
Turkmenistan	X	X		Turkmenistan	55	0	
Vietnam	566	2	0.00	Vietnam	2	X	
Yemen	X	20		Yemen	X	X	
Total	**10,142**	**122**	**0.01**	**Total**	**10,323**	**191**	**0.02**
Americas				**Americas**			
Argentina	474	5	0.01	Argentina	637	12	0.02
Belize	28	0	0.00	Belize	2	0	0.00
Bolivia	2,847	68	0.02	Bolivia	1,632	16	0.01
Brazil	4,634	23	0.00	Brazil	2,881	37	0.01
Canada	2	0	0.00	Canada	X	X	
Chile	1	0	0.00	Chile	4	0	0.00
Colombia	4,428	70	0.02	Colombia	1,508	32	0.02
Costa Rica	19	1	0.05	Costa Rica	1	0	0.00
Ecuador	1,059	12	0.01	Ecuador	65	3	0.05
El Salvador	182	2	0.01	El Salvador	0	0	
Guatemala	1,568	14	0.01	Guatemala	1,263	0	0.00
Honduras	708	14	0.02	Honduras	90	1	0.01
Mexico	1,086	5	0.00	Mexico	2,356	X	
Nicaragua	2,813	107	0.04	Nicaragua	1,283	36	0.03
Peru	4,518	21	0.00	Peru	3,483	29	0.01
United States	3	0	0.00	United States	4	0	0.00
Venezuela	269	9	0.03	Venezuela	2,551	59	0.02
Total	**24,639**	**351**	**0.01**	**Total**	**17,760**	**225**	**0.01**
Europe				**Europe**			
France	6	X		France	3	0	0.00
Germany	0	0		Germany	2	0	0.00
Hungary	0	0		Hungary	1	0	0.00
Netherlands	3	X		Netherlands	2	0	0.00
Russian Federation	1	X		Russian Federation	4	1	0.25
Spain	1	X		Spain	0	0	
Sweden	1	X		Sweden	0	0	
United Kingdom	13	X		United Kingdom	6	0	0.00
Total	**25**	**0**	**0.00**	**Total**	**18**	**1**	**0.06**
Oceania				**Oceania**			
Australia	2	0	0.00	Australia	2	0	0.00
Guam	1	X		Guam	X	X	
Mariana Islands	17	X		Mariana Islands	3	0	0.00
Total	**20**	**0**	**0.00**	**Total**	**5**	**0**	**0.00**

Note: "X" indicates that there are no data available.

Table 11. Reported Cholera Cases and Deaths in the Americas, 1991 to 1997

Description

The cases and deaths reported to the World Health Organization (WHO) since the appearance of cholera in the Americas in 1991 are shown. Absolute numbers for each year and cumulative numbers since the beginning of the epidemic as reported to the WHO, are shown. In the seven years of data available, nearly 1.2 million people have been infected with cholera in the Americas, primarily in South America. The first cases were diagnosed on January 31, 1991, north of Lima, Peru, and nearly 33,000 cases were reported in the subsequent three weeks. During the first three years of the epidemic, the preponderance of cases were in Peru, although large numbers of cases were also reported in Bolivia, Brazil, Colombia, Ecuador, and Guatemala.

Limitations

Only those cases officially reported to the World Health Organization are shown here; undiagnosed and unreported cases are excluded.

Source

Dr. M. Neira, and Dr. L. Kuppens, Global Task Force on Cholera, Division of Emerging and Other Communicable Diseases Surveillance and Control, World Health Organization, Geneva, personal communication, September 1997, and March 1998.

TABLE 11.   REPORTED CHOLERA CASES AND DEATHS IN THE AMERICAS, 1991 TO 1997

Year	Actual Cases	Cumulative Cases	Actual Deaths	Cumulative Deaths
1991	391,220	391,220	4,002	4,002
1992	354,089	745,309	2,401	6,403
1993	209,192	954,501	2,438	8,841
1994	113,684	1,068,185	1,107	9,948
1995	85,809	1,153,994	846	10,794
1996	24,639	1,178,633	351	11,145
1997	17,760	1,196,393	226	11,371

Table 12. Reported Cases of Dracunculiasis by Country, 1972 to 1996

Description

Dracunculiasis, or guinea worm disease, is a disease directly related to drinking unclean water. A global campaign is under way to eradicate dracunculiasis by 2000. Dracunculiasis cases reported to the World Health Organization are shown here by country. The number of cases worldwide is declining rapidly. Pakistan and Iran (not included in the table) were declared "free of dracunculiasis transmission" by the World Health Organization in early 1997 after not having any cases for three consecutive years. National eradication programs, funded by international agencies, have contributed to the overall decline of dracunculiasis (see Chapter 2).

TABLE 12.   REPORTED DRACUNCULIASIS CASES BY COUNTRY, 1972 TO 1996

Region and Country	1972	1973	1974	1975	1976	1977	1978	1979	1980	1981	1982	1983
Africa												
Benin	1,480		820									
Burkina Faso	5,822	4,404	4,008	6,277	1,557		2,885	2,694	2,565			4,362
Cameroon				251								
Chad							172					
Côte d'Ivoire	4,891	4,654	6,283	4,971	4,656	5,207	6,993		6,712	7,978		2,259
Ethiopia												
Ghana	693	1,606	1,226	4,052	1,421	1,617	1,676		2,703	853	3,413	3,040
Kenya												
Mali	668	786	737	542	760	1,084			816	777	401	428
Mauritania						127			651	663	903	1,612
Niger					2,600	3,000	5,560		1,906	2,113	1,530	
Nigeria	98			1,007					1,693			
Senegal		334	208	65	137				161			
Sudan												
Togo				3,261	1,648		2,617	2,673	1,748	951	2,592	
Uganda												
Asia and the Middle East												
India						7,052	6,827	2,846	2,729	5,406	42,926	44,818
Pakistan							250		14,155			
Yemen		25										
Number of countries reporting	6	6	6	8	7	6	8	3	11	7	6	6

Limitations

Dracunculiasis cases are prevalent in remote, rural areas and sufferers often do not believe hospital visits can do much to improve the course of disease. Neither mortality nor morbidity is regularly reported. Care should be taken in looking at the time-series data. National surveillance became more thorough in many countries in the 1980s (see Chapter 2 for details), thus causing significant (but not real) increases in reported cases (for example, from 2,821 in 1986 to 216,484 in 1987 in Nigeria).

Sources

World Health Organization, "Dracunculiasis Surveillance," *Weekly Epidemiological Record*, 57(9): 65–7, 5 March 1982; 60(9): 61–3, 1 March 1985; 61(5): 31–6, 31 January 1986; 66(31): 225–30, 2 August 1991; 68(18): 125–6, 30 April 1993; 69(17): 121–8, 29 April 1994; 70(18): 125–6, 5 May 1995; 71(19): 141–7, 10 May 1996; 72(19): 133–9, 9 May 1997.

1984	1985	1986	1987	1988	1989	1990	1991	1992	1993	1994	1995	1996
			400	33,962	7,172	37,414	4,006	4,315	16,334	4,302	2,273	1,427
1,739	458	2,558	1,957	1,266	45,004	42,187		11,784	8,281	6,861	6,281	3,241
0	168	86		752	871	742		127	72	30	15	17
1,472	9	314						156	1,231	640	149	127
2,573	1,889	1,177	1,272	1,370	1,555	1,360	12,690		8,034	5,061	3,801	2,794
2,882	1,467	3,385	2,302	1,487	3,565	2,333		303	1,120	1,252	514	371
4,244	4,501	4,717	18,398	71,767	179,556	123,793	66,697	33,464	17,918	8,432	8,894	4,877
					5	6			35	53	23	0
5,008	4,072	5,640	435	564	1,111	884	16,024		12,011	5,581	4,218	2,402
1,241	1,291		227	608	447	8,036		1,557	5,882	5,029	1,762	562
	1,373		699		288		32,829	500	25,346	18,562	13,821	2,956
8,777	5,234	2,821	216,484	653,492	640,008	394,082	281,937	183,169	75,752	39,774	16,374	12,282
	62	128	132	138		38	1,341	728	815	195	76	19
		822	399	542				2,447	2,984	53,271	64,608	118,578
1,839	1,456	1,325		178	2,749	3,042	5,118	8,179	10,349	5,044	2,073	1,626
6,230	4,070			1,960	1,309	4,704		126,369	42,852	10,425	4,810	1,455
39,792	30,950	23,070	17,031	12,023	7,881	4,798	2,185	1,081	755	371	60	9
			2,400	1,110	534	160	106	23	2	0	0	0
										94	82	62
12	14	12	13	15	15	15	10	15	18	19	19	19

Table 13. Waterborne Disease Outbreaks in the United States by Type of Water Supply System, 1971 to 1994

Description

The 1993 cryptosporidium outbreak in Milwaukee, Wisconsin, in which an estimated 404,000 people fell ill, was the largest recorded waterborne disease outbreak in the United States since surveillance began in 1920. Although this outbreak may have altered the perception that the United States is free of water-related illnesses, it is still not subject to the high level of waterborne disease found in developing countries.

This table shows the number of waterborne disease outbreaks reported in the United States since 1971 and whether they originated in community (municipal), noncommunity (semi-public), or individual water supply systems. In additional to *Cryptosporidium parvum*, the most common etiologic agents are *Giardia lamblia*, *Salmonella*, *Campylobacter*, *Shigella*, *Norwalk* virus, chemicals, and metals. These agents are introduced into water supplies because surface or groundwater sources go untreated, or because of problems in the distribution or treatment system.

Limitations

These data represent only waterborne disease outbreaks that are reported. No breakdown is given on the type of disease involved, though the vast majority is associated with *Giardia*, *Cryptosporidium*, and the Norwalk virus. No data are given on the severity of the outbreaks. These data exclude waterborne diseases imported from other countries.

Sources

Centers for Disease Control and Prevention, *Waterborne Disease Outbreaks, 1989–1990, CDC Surveillance Summaries: Morbidity and Mortality Weekly Report* 40(SS-3): 1–21, U.S. Department of Health and Human Services, Public Health Service Centers for Disease Control, Massachusetts Medical Society, Waltham, Massachusetts.

Centers for Disease Control and Prevention, *Waterborne Disease Outbreaks, 1991–1992, CDC Surveillance Summaries: Morbidity and Mortality Weekly Report* 42(SS-3): 1–30, U.S. Department of Health and Human Services, Public Health Service Centers for Disease Control, Massachusetts Medical Society, Waltham, Massachusetts.

Centers for Disease Control and Prevention, *Waterborne Disease Outbreaks, 1993–1994, CDC Surveillance Summaries: Morbidity and Mortality Weekly Report* 45(SS-1): 1–33, U.S. Department of Health and Human Services, Public Health Service Centers for Disease Control, Massachusetts Medical Society, Waltham, Massachusetts.

Levine, W., and C. Craun, 1990, *Waterborne Disease Outbreaks, 1986–1988, CDC Surveillance Summaries: Morbidity and Mortality Report* 39(SS-1): 1–13, U.S. Department of Health and Human Services, Public Health Service Centers for Disease Control, Massachusetts Medical Society, Waltham, Massachusetts.

TABLE 13. WATERBORNE DISEASE OUTBREAKS IN THE UNITED STATES BY TYPE OF WATER SUPPLY SYSTEM, 1971 TO 1994

Year	Community	Noncommunity	Individual	Total Outbreaks	Total Cases
1971	8	8	4	20	5,184
1972	9	19	2	30	1,650
1973	6	16	3	25	1,762
1974	11	9	5	25	8,356
1975	6	16	2	24	10,879
1976	9	23	3	35	5,068
1977	14	18	2	34	3,860
1978	10	19	3	32	11,435
1979	24	13	8	45	9,841
1980	26	20	7	53	20,045
1981	14	18	4	36	4,537
1982	26	15	3	44	3,588
1983	30	9	4	43	21,036
1984	12	5	10	27	1,800
1985	7	14	1	22	1,946
1986	10	10	2	22	1,569
1987	8	6	1	15	22,149
1988	4	8	1	13	2,128
1989	6	5	1	12	2,540
1990	5	7	2	14	1,758
1991	2	13	0	15	12,960
1992	9	14	4	27	4,724
1993	9	4	5	18	404,190
1994	5	5	2	12	1,176

Table 14. Hydroelectric Capacity and Production by Country, 1996

Description

Hydroelectric generation is an important "instream" use of water and continues to play a significant role in electricity generation worldwide. Hydropower supplies approximately 18 percent of the world's electricity and over 90 percent of electricity in 18 countries. Despite environmental opposition and complex financial management arising from increased private investment, plans are being pursued for large-scale pumped storage and small-scale hydropower projects alike. Given here are 1996 data on installed hydroelectric capacity in megawatts (MW), hydroelectric production in gigawatt-hours per year (GWh/yr), and the fraction of total national electricity generated by hydropower for 170 countries. These data were obtained from national and regional power authorities, government departments, the International Commission on Large Dams, and other organizations associated with hydropower or water resources development. The number of significant figures in the original source was retained here. Data on capacity and production are also shown for the major continental regions.

Limitations

Data are not available for all countries or for the type of hydroelectric facilities. Annual hydroelectric generation depends on the annual flow of water, which varies from year to year, as well as on the way a dam is operated. Some countries are missing from the tabulation, and therefore true regional totals cannot be calculated.

Source

International Journal on Hydropower and Dams, 1997, *World Atlas and Industry Guide,* Aqua-Media International, Surrey, England.

TABLE 14.   HYDROELECTRIC CAPACITY AND PRODUCTION BY COUNTRY, 1996

Country and Region	Installed Hydroelectric Capacity (MW)	Hydroelectric Production (GWh/yr)	Fraction of Electricity Generated with Hydropower
Africa			
Algeria	275	353	0.9
Angola	554	1,800	75
Benin	0	0	0
Botswana	0	0	0
Burkina Faso	32	95	39
Burundi	41	147	100
Cameroon	723	2,778	98.5
Central African Republic	22	78	81
Chad	0	0	0
Comoros	1	2	13
Congo	89	352	99.5
Congo, Democratic Republic (formerly Zaire)	2,523	5,550	99.9
Côte d'Ivoire	895	1,098	68
Egypt	2,825	8,500	20
Equatorial Guinea	1	2	11
Ethiopia	540	2,000	87
Gabon	166	710	80
Gambia	0	0	0
Ghana	1,072	6,100	97
Guinea	52	228	32
Guinea-Bissau	0	0	0
Kenya	598	2,920	80
Lesotho	3	0	0
Liberia	81	175	43
Libya	0	0	0
Madagascar	225	411	75
Malawi	218	850	99.5
Mali	50	200	69.6
Mauritania	61	26	18
Mauritius	59	140	12.8
Morocco	693	443	25
Mozambique	2,184	336	92.2
Namibia	240	1,134	90.1
Niger	0	0	0
Nigeria	1,938	13,365	37
Reunion	125	500	61
Rwanda	56	230	98
Sao Tome and Principe	2	8	53
Senegal	(200)	0	0
Seychelles	0		
Sierra Leone	4	1	
Somalia	5	6	
South Africa	661	2,090	1
Sudan	240	950	71
Swaziland	43	135	15
Tanzania	380	1,539	85
Togo	69	154	12

continued

TABLE 14.   CONTINUED

Country and Region	Installed Hydroelectric Capacity (MW)	Hydroelectric Production (GWh/yr)	Fraction of Electricity Generated with Hydropower
Tunisia	64	40	0.3
Uganda	186	940	98
Zambia	1,648	8,102	99.8
Zimbabwe	666	2,500	20
Total	**20,112**	**66,981**	
North and Central America and Caribbean			
Belize	0	0	0
Canada	64,770	330,690	62
Costa Rica	924	3,960	75
Cuba	60		
Dominica	8	38	77.9
Dominican Republic	375	1,294	15
El Salvador	388	1,505	55
Grenada	0	0	0
Guadeloupe	5		
Guatemala	450	2,100	94
Haiti	70	280	50
Honduras	482	2,278	90.8
Jamaica	23	98	2
Mexico	10,034	26,235	28.8
Nicaragua	106	340	25
Panama	551	2,610	77.6
Puerto Rico	85	260	2
St. Vincent	6	40	75
Trinidad/Tobago	0	0	0
United States	74,856	296,378	9.9
Total	**153,193**	**668,106**	
South America			
Argentina	8,123	30,000	44.7
Bolivia	315	1,582	59
Brazil	51,100	250,000	97.4
Chile	3,515	17,300	86.9
Colombia	7,838	31,771	76.1
Ecuador	1,480	6,566	80
French Guiana	0	0	0
Guyana	0	0	0
Paraguay	7,320	29,000	100
Peru	2,500	15,000	85
Suriname	290		81
Uruguay	1,356	6,190	88
Venezuela	13,216	63,000	72
Total	**97,053**	**450,409**	
Asia			
Afghanistan	292	478	68
Armenia	750	475	90
Azerbaijan	1,700	2,426	
Bangladesh	230	796	10

continued

TABLE 14. CONTINUED

Country and Region	Installed Hydroelectric Capacity (MW)	Hydroelectric Production (GWh/yr)	Fraction of Electricity Generated with Hydropower
Bhutan	342	100	
Cambodia	1	<1	
China	52,180	166,800	18
Georgia	2,728	7,000	89
India	20,576	72,283	25
Indonesia	3,046	12,390	13.7
Iran	1,952	7,520	12.9
Iraq	910		
Israel	5	23	>0
Japan	21,171	91,301	9.2
Jordan	6	15	1
Kazakhstan	2,129	7,000	10
Korea, Republic of	1,493	2,760	3
Korea, Democratic People's Republic	5,000	24,000	60
Kyrgyzstan	2,950	11,113	90
Laos	202	1,000	90
Lebanon	200	731	13
Malaysia	1,800	>5,800	30
Mongolia	1	>0	
Myanmar (Burma)	290	1,200	48
Nepal	254	1,116	85.7
Pakistan	4,825	22,858	45.5
Philippines	2,278	6,142	26
Russian Federation	39,986	162,800	26.8
Saudi Arabia	0	0	0
Sri Lanka	1,135	4,514	94
Syria	900	6,709	45
Taiwan	1,688	4,090	4
Tajikistan	4,036	16,600	97.8
Thailand	2,858	6,685	20
Turkey	9,993	36,104	41
Turkmenistan	10	27	
Uzbekistan	1,704	6,750	12
Vietnam	2,836	10,000	72.1
Total	**192,457**	**693,706**	
EUROPE			
Albania	1,437	5,200	87.5
Austria	11,400	39,000	68
Belgium	102	337	0.4
Bosnia and Herzegovina	1,624	8,900	35
Bulgaria	2,407	2,507	6
Byelarus	6	0	>0
Croatia	1,932	5,193	42
Cyprus	1		0.1
Czech Republic	908	2,002	4
Denmark	10	4	>0
Estonia	1	2	0.02

continued

TABLE 14. CONTINUED

Country and Region	Installed Hydroelectric Capacity (MW)	Hydroelectric Production (GWh/yr)	Fraction of Electricity Generated with Hydropower
Faeroe Islands	30	72	40
Finland	2,827	12,788	19
France	23,100	65,500	15.4
Germany	4,389	17,500	4.2
Greece	2,524	4,487	9.3
Greenland	30	165	59
Hungary	50	214	0.1
Iceland	880	4,950	95
Ireland	220	750	5
Italy	12,925	41,425	18.6
Latvia	1,512	3,000	74
Lithuania	107	373	5.5
Luxembourg	33	89	7.2
Macedonia, TFYR	422	1,283	13.8
Moldova, Republic of	64		
Netherlands	30	60	0.2
Norway	26,000	112,676	99.6
Poland	535	1,700	1
Portugal	4,125	12,415	25.6
Romania	5,871	16,700	28.7
Slovakia	1,943	4,116	18
Slovenia	847	3,190	33.3
Spain	14,803	32,000	20
Sweden	16,450	63,500	52
Switzerland	10,118	35,597	59
Ukraine	4,465	10,800	8.7
United Kingdom	1,343	3,973	0.5
Yugoslavia	2,711	12,204	35.4
Total	**158,181**	**524,672**	
Oceania			
Australia	7,400	15,940	11
Fiji	84	383	89.3
French Polynesia	10	35	
New Caledonia	78	250	
New Zealand	4,980	24,400	66
Pacific Islands Trust	10	30	
Papua New Guinea	180	460	75
Solomon Islands	<1	<1	1
Western Samoa	9	22	
Vanuatu	0	0	0
Total	**12,751**	**41,520**	

Table 15. Populations Displaced as a Consequence of Dam Construction, 1930 to 1996

Description

One of the more devastating social impacts of dam construction is the displacement of people caused by the filling of reservoirs and the subsequent flooding of land. Included here are 181 completed and 81 planned or proposed projects that each has displaced, or will displace, more than 100 people. These data include the name of the dam project, country or countries involved, the number of people displaced, and, when available, the reservoir area (km^2) and installed hydroelectric capacity (MW).

Limitations

These data are culled from a wide variety of sources, since no one institution is responsible for keeping consistent estimates of population displacements from dam construction. Estimates for the total number of people ever displaced by dam construction range from 20 to 60 million (see McCully 1996, and the discussion in Chapter 3). The number of displaced people from projects listed in this table amount to 6.5 million. Clearly, not all dam projects are included, and estimates from those that are included may be too low. For political reasons, data for population displacement are not maintained or gathered.

Sources

Biswas: Biswas, A.K., 1982, "Environment and Sustainable Water Development." In *Water for Human Consumption*. IRWA Water Resources Series, Vol. 2.

Cernea: Cernea, M.C. (ed.) 1991, "Involuntary Resettlement; Social Research, Policy, and Planning," in *Putting People First*. Oxford University Press, New York, pp. 188–215.

IDN: International Rivers Network, 1987, *International Dams Newsletter*, Vols. 1(1)–2(5), Berkeley and San Francisco, California.

Maloney: Maloney, C., 1990, *"Environmental and Project Displacement of Population in India: Part 1. Development and Deracination."* Field Staff Reports, Universities Field Staff International and the Natural Heritage Institute, San Francisco, California.

McCully: McCully, P., 1996, *Silenced Rivers: The Ecology and Politics of Large Dams*. Zed Books, London.

Nachowitz: Nachowitz, T., 1992, Department of Anthropology, Syracuse University, personal communication.

WRD: International Commission on Large Dams, 1988, *World Registry of Dams*. Paris.

WRR: International Rivers Network, 1987–1991, *World Rivers Review*, Vols. 2–6. San Francisco and Berkeley, California.

TABLE 15.   POPULATIONS DISPLACED AS A CONSEQUENCE OF DAM CONSTRUCTION, 1930 TO 1996

Dam	Country	Installed Capacity (MW)	Area of Reservoir (km²)	Number of People Displaced	Date	Source
Completed						
Sanmenxia	China			410,000–870,000	1960	Biswas, McCully
Danjiang Kou	China			383,000	1974	Cernea
Xinanjiang	China	663		306,000	1960	McCully
Dongpinghu	China			278,000	1958	McCully
Maduru Oya	Sri Lanka		64	200,000	1983	WRR 6(1)6, WRD
Pong	India			150,000	1974	Cernea, McCully
Zhaxi	China			141,000	1961	McCully
Bargi	India	105	8,090	113,600	1990	McCully
Hirakud	India	270	743	110,000	1957	McCully
Srisailam	India	440	247	100,000–330,000	1984	Cernea
Aswan	Egypt, Sudan	1,815	6,500	100,000–120,000	1970	WRD, Biswas
Kaptai	Bangladesh		777	100,000	1962	WRR 5(3)10, WRD
Damodar (4 projects)	India			93,000	1959	Biswas
Mangla	Pakistan	600	253	90,000–110,000	1967	Nachowitz, McCully
Nanela	Pakistan			90,000	1967	Biswas
Xijin	China	234		89,300	1964	McCully
Tarbela	Pakistan	1,750	243	86,000	1976	WRD, Biswas
Wuqiangxi	China	1,200	170	84,800	1995	McCully
Ukai	India	300	600	80,000–88,000	1972	Maloney, McCully
Akasombo	Ghana	882	9,000	80,000–84,000	1965	WRR 6(5)8, Nachowitz
Narayanpur	India		132	80,000	1990s	Nachowitz
Rengali	India	60	414	80,000	1980s	McCully
Lower Manair (Pochampad Project)	India		81	78,000	1980s	McCully
Kossou	Ivory Coast		1,700	75,000–85,000	1972	WRD, Biswas
Sobradinho	Brazil			65,000	1978	Cernea
Shuikou	China	1,400	930	62,000–67,000	1996	Nachowitz, McCully
Saguling	Indonesia	700	53	60,000–65,000	1984	Cernea, McCully
Rihand (Singrauli)	India	300	469	60,000	1962	McCully
Tabqua (Thawra/Assad)	Syria	800	600	60,000	1976	McCully
TVA (about 20 projects)	United States			60,000	1930s–present	Biswas
Hoa Binh	Vietnam	1,920	200	58,000	1993	McCully
Bhima (Ujjani)	India		34	57,000	1963	McCully
Cirata	Indonesia	500	62	56,000	1987	McCully
Dongjiang	China	500	160	53,000	1989	McCully
Gandhi Sagar	India			52,000	1960	Biswas
Kariba	Zambia, Zimbabwe	1,266	5,100	50,000–57,000	1959	WRD, Biswas
Chung Ju	South Korea	460	97	46,500	1985	McCully
Kadana	India			45,200	1978	McCully
Victoria	Sri Lanka	210	23	45,000	1984	McCully
Itaipu	Brazil, Paraguay	12,600	1,350	42,400	1982	McCully
Kainji	Nigeria	760	1,250	42,000–50,000	1968	WRD, Biswas, McCully
Ataturk	Turkey	2,400	817	40,000–60,000	1991	IDN 2(2)1, McCully
Itaparica	Brazil	1,500		40,000–50,000	1988	Nachowitz
Yantan	China			40,000	1994	McCully
Dhom (Dhon)	India		25	39,000	1978	McCully
Chandil (Subarenarekha Multipurpose Project)	India		166	38,000	1990s	McCully
Bhakra	India	450		36,000	1963	Biswas
Keban	Turkey	1,360	675	30,000	1974	WRD, Biswas
Lam Pao	Thailand		400	30,000	1970	Biswas, WRD
Ubolratana	Thailand	25	410	30,000	1965	McCully
Kedung Ombo	Indonesia	29	46	29,000	1993	McCully

continued

TABLE 15. CONTINUED

Dam	Country	Installed Capacity (MW)	Area of Reservoir (km²)	Number of People Displaced	Date	Source
Mython (Jharkh)	India	200		28,030	1955	Nachowitz
Kedong Ombo	Indonesia			27,000	1992	WRR 6(4)4
Geheyan	China	1,200	72	26,700	1995	McCully
Dadin Kowa	Nigeria		530	26,000	1980s	McCully
Gezhouba	China	2,715		26,000	1988	McCully
Culiacan	Mexico			25,200	1967	McCully
Nam Pong	Thailand		20	25,000–30,000	1965	Biswas, WRD
Nagarjunasagar	India	810	285	25,000–28,000	1974	Maloney, McCully
Cabora Bassa	Mozambique	2,250	3,800	25,000	1974	McCully
Tucurui	Brazil	4,000	2,430	23,871	1984	WRR 6(4)8
Pujal-Coy I	Mexico			23,400	1982	McCully
Portile de Fier (Iron Gates)	Romania, Yugoslavia	2,100	52	23,000	1972	Cernea, McCully
Katse (Lesotho Highlands Water Project 1A)	Lesotho		36	21,700[a]	1996	McCully
Presidente M. Aleman	Mexico	154		21,000	1955	McCully
Awash Project (3 dams)	Ethiopia			20,000	1960s	McCully
Bhumibol	Thailand	535	300	20,000	1964	McCully
Karakaya	Turkey	1,800	298	20,000	1987	McCully
Kiri	Nigeria		130	19,000	1982	McCully
Rajghat	India		224	19,000	1980s	McCully
Cerro de Oro	Mexico		170	18,000	1989	McCully
Dez	Iran	840	63	17,000	1978	McCully
Sri Ramu Sagar	India	36	434	16,000	1983	McCully
Watrak	India			16,000	1990s	McCully
Kabini	India	32	61	15,000	1974	McCully
Sao Simao	Brazil	2,680	680	14,000	1978	McCully
Upper Pampanga	Philippines			14,000	1973	Biswas
Colorado (El Tapiro)	Mexico		3	13,300	1982	McCully
Bakolori	Sokoto		120	13,000	1978	McCully
Kotmale	Sri Lanka	200	9.5	13,000	1988	McCully
Pantabangan	Philippines	100	89	13,000	1973	McCully
Ruzizi II	Rwanda, Democratic Republic of Congo	40		12,600–15,000	1980s	WRR 5(2)8, McCully
Saravathi (Sharavathy)	India	510	59	12,500	1964	McCully
Selingue	Mali	44	409	12,500	1980	McCully
Nangbeto	Togo, Benin	63	180	12,000	1987	Cernea, McCully
Sukhi	India		29	11,200	1980s	McCully
Khao Laem	Thailand	300	388	10,800	1986	McCully
Pujal-Coy II	Mexico			10,800	1982	McCully
Sipu	India		29	10,400	1968	McCully
Manantali	Mali	200	480	10,000–11,000	1988	WRR 5(2)5, McCully
Cerron Grande (Silencio)	El Salvador			10,000	1973	McCully
Roseires	Sudan	130		10,000	1965	McCully
Salvajina	Colombia		22	10,000	1985	IDN 2(1)4, WRD
Lang Suan	Thailand	135		9,800	1980s	McCully
Srinakharin	Thailand	720	419	9,400	1981	McCully
Piedra del Aguila	Argentina	1,400	292	9,000	1991	McCully
Furnas	Brazil	1,216	1,440	8,500	1963	McCully
Ravishankar	India			8,400	1980s	McCully
Salto Grande	Argentina, Paraguay	3,100	172,000	8,000	1979	McCully
Kiambere	Kenya	142	25	7,000	1988	McCully
Kpong	Ghana	160	35	7,000	1982	McCully
Sir	Turkey	284	48	7,000	1991	McCully

continued

TABLE 15. CONTINUED

Dam	Country	Installed Capacity (MW)	Area of Reservoir (km²)	Number of People Displaced	Date	Source
Und	India			6,500	1990s	McCully
Bajo Candelaria	Mexico			5,800	1995	McCully
Al Massira (Sidi Cheho)	Morocco		137	5,500	1979	McCully
Guavio	Colombia	1,600	14.4	5,500	1990	McCully
La Angostura	Mexico	1,100	644	5,500	1974	McCully
Brokopondo	Surinam	30	1,500	5,000	1971	Biswas, McCully
Caracol (Carlos Ramirez Ulloa)	Mexico			5,000	1986	WRR 6(2)11, WRD
Lubuge I	China	450	4	5,000	1992	McCully
Nova Ponte	Brazil	510	443	5,000	1994	McCully
Chandoli (Warna)	India			4,900	1990s	McCully
Pak Mun	Thailand	136	60	4,900	1994	McCully
Pak Mun	Thailand	136	60	4,900	1994	McCully
El Cajon	Honduras	292	94	4,700	1985	McCully
Bayano	Panama	300	300	4,500	1976	McCully
Pedra de Cavalo	Brazil	600		4,400	1985	McCully
Tuttle Creek	United States		64	4,000	1962	McCully
Itumbiara	Brazil	2,080	760	3,700	1980	McCully
Aji III	India			3,500	1980s	McCully
Sidi Salem	Tunisia	36	550	3,500	1984	McCully
Chixoy (Pueblo Viejo)	Guatemala	300	14	3,400	1985	McCully
Diama	Senegal, Mauritania			3,400	1986	McCully
Bang Lang (Pattani)	Thailand	72	51	3,300	1981	McCully
Venu II	India			3,300	1990s	McCully
Selaulim	India			3,200	1990s	McCully
Riano	Spain	680	20	3,100	1987	McCully, WRR 2(3)3
Batang Ai	Malaysia	92	85	3,000	1985	WRR 5(3)12, McCully
Foum-Gleita	Mauritania			3,000	1980s	McCully
Marsayangdi	Nepal	69	0.6	3,000	1989	McCully
Mtera	Tanzania	280	650	3,000	1988	McCully
Nam Ngum (1)	Laos	150	370	3,000	1971	Biswas, McCully
Santa Rita	Colombia			3,000	1978	McCully
Gustavo Diaz Ordaz Presidente (Bacurato)	Mexico		56	2,900	1982	McCully
Zankhari	India			2,900	1990s	McCully
Segredo	Brazil	1,260	82	2,700	1993	McCully
Arenal	Costa Rica	157	83	2,500	1980	McCully
Conemaugh	United States		121.4	2,500	1952	McCully
Kulekhani	Nepal	92	2.2	2,500	1982	McCully
Zimapan	Mexico	292	23	2,500	1994	McCully
Sukhbhadar	India			2,400	1980s	McCully
Amli (Ver II)	India			2,300	1984	McCully
Machhanala	India			2,100	1982	McCully
Weija	Ghana			2,000	1978	McCully
La Grande Project	Canada	15,719	15,900	1,900	1996	McCully
Garrison	United States		1,490	1,800	1953	McCully
Samuel	Brazil	216	579	1,800	1989	McCully
Chiew Larn	Thailand	240	165	1,600	1987	McCully
Thac Mo	Vietnam	150		1,600	1995	McCully
Tres Irmaos	Brazil	1,292	820	1,600	1990	McCully
Guri	Venezuela	10,300	4,260	1,500	1986	McCully
Netzahualcoyotl	Mexico	1,080	292	1,500–3,000	1964	Biswas, McCully
Magat	Philippines	360	45	1,500	1983	McCully
Salto Santiago	Brazil	2,000	225	1,500	1980	McCully

continued

TABLE 15. CONTINUED

Dam	Country	Installed Capacity (MW)	Area of Reservoir (km²)	Number of People Displaced	Date	Source
Temengor	Malaysia	348	150	1,500	1977	McCully
Chivor (La Esmeralda)	Colombia	1,000		1,400	1982	McCully
Playas	Colombia	200	10	1,400	1985	McCully
Uben	India			1,400	1982	McCully
Navajo	United States		30	1,250	1963	McCully
Rio Grande	Colombia	324	10	1,200	1985	McCully
Balbina	Brazil	250	3,147	1,100	1989	McCully
Jhuj	India		2.72	1,100	1980s	McCully
Aguamilpa	Mexico	960	130	1,000	1995	McCully
Moxoto	Brazil	2,440	88	1,000	1977	McCully
Oahe	United States		1,453	900	1958	McCully
Kinzua	United States		49	700	1996	McCully
M'bali	Central African Republic			700	1991	McCully
Fort Randall	United States		385	680	1952	McCully
Fortuna	Panama	255	10	600	1982	McCully
Cana Brava	Brazil	480		500	1983	McCully
Dau Tieng	Vietnam			500	1980s	McCully
Thika	Kenya	2	5	500	1990s	McCully
Big Bend	United States		59	445	1950s	McCully
Apatzingan (Chilatan)	Mexico	28	24	400	1995	McCully
San Carlos	Colombia	1,550		350	1987	McCully
Lupohlo	Swaziland	20	1.2	300	1984	McCully
Youghiogheny	United States	1,948	11.5	300	1948	McCully
Clyde	New Zealand	430	200	280	1979	McCully
Taquaracu	Brazil	500		200	1985	McCully
Xingo	Brazil	3,000	60	150	1996	McCully
Avila	Brazil	28		100	1990	McCully
Planned, Postponed, or Under Construction						
Three Gorges	China	30,000	1,100	750,000–1,300,000	uc:2009	Nachowitz, Cernea, McCully
Pa Mong	Thailand, Laos			310,000–480,000	planned	Biswas
Xiaolangdi	China	1,800	247	170,000	uc:2001	Cernea, McCully
Almatti	India			160,000–240,000	postponed	Nachowitz, Cernea, McCully
Ta Bu	Vietnam	3,600		112,400	planned	McCully
Chico	Philippines	1,100		90,000	postponed	Nachowitz
Tehri	India	1,000	42.5	85,600	postponed	WRR 3(2)8, McCully
Kalabagh	Pakistan	2,400	550	80,000–250,000	planned	WRR 5(3)4, 4(6)3, Cernea, McCully
Longtan	China	4,200	370	73,000	planned	McCully
Sardar Sarovar	India	1,450	40,000	70,000–90,000	postponed	WRR 6(4)1, Nachowitz, Cernea, McCully
Mahaweli (15 dams)	Sri Lanka	500	360	65,000	1996	WRR 5(3)10, Cernea
Subarnarekha (Bihar) (5 dams)	India	10		64,500–80,000	uc	Nachowitz, Cernea
Pancheswar	Nepal	7,200	121	60,000	planned	McCully
Saguling	India			55,000	uc	Nachowitz
Low Pa Mong	Laos, Thailand	2,670	560	52,000	planned	McCully
Karnali (Chisapani)	Nepal	3,000	341	50,000–55,000	planned	Nachowitz, Cernea, McCully
Tillari	India			50,000	postponed	McCully
Tianshengqiao 1	China	1,200		48,800	uc:1999	McCully

continued

TABLE 15.   CONTINUED

Dam	Country	Installed Capacity (MW)	Area of Reservoir (km²)	Number of People Displaced	Date	Source
Yacyreta	Argentina, Paraguay	2,700	1,700	40,000–45,000	uc:1998	WRR 6(1)12, 5(4)5, Cernea, McCully
Soubre	Ivory Coast			40,000	planned	Nachowitz
Ertan	China	3,300	101	35,000	uc:1999	McCully
Rogun	Tadjikistan	3,600		28,200	postponed	McCully
Daguangba	China	240	99	23,800	uc	McCully
Pa Mong A	Laos, Thailand	2,030	120	23,300	planned	McCully
Pasak	Thailand			23,000	uc	McCully
San Juan Tetelcingo	Mexico	280	140	22,000–30,000	planned	WRR 6(1)4, Cernea, McCully
Dai Thi	Vietnam			20,600	planned	McCully
Kayraktepe	Turkey			20,000	uc	Nachowitz
Ita	Brazil	1,620	103	19,200	uc:1999	McCully
Upper Indravati	India	1,000	128.65	16,100	postponed	McCully
Machadinho	Brazil	1,200	252	15,700	postponed	McCully
Ban Mai	Vietnam	450		15,000	planned	McCully
Garabi	Brazil, Argentina	1,800	810	15,000	planned	McCully
Dai Ninh	Vietnam	300		14,100	planned	McCully
Maheshwar	India	200		14,000	planned	Nachowitz
Riau	Indonesia			14,000	planned	WRR 6(1)4
Chiang Khan	Laos, Thailand	570	90	13,000	planned	McCully
Omkareshwar	India	390		13,000	planned	Nachowitz
Castanhao	Brazil	75	229	12,000	planned	McCully
Jiangya	China			12,000	planned	McCully
Pak Lay	Laos	1,320	110	11,800	planned	McCully
An Khe	Vietnam	116		10,800	planned	McCully
Bhopalpatnam	India	1,000		10,000	postponed	Nachowitz
Inchampalli	India	393		10,000	postponed	Nachowitz
Stung Treng	Cambodia	980	640	9,200	planned	McCully
Bodhghat	India	107		9,000	postponed	WRR 3(2)8
Mohale (Lesotho Highlands Water Project 1B)	Lesotho			8,400[a]	uc:2004	McCully
Cachoeira Porteira	Brazil	1,400	911	8,000	planned	McCully
Jafuri	Indonesia	30	100	8,000	planned	IDN 1(1)9
Wadaslingtan Project (2 dams)	Indonesia			7,500–9,500	planned	WRR 3(2)5
Yali Falls	Vietnam	720		7,400	uc	McCully
Porto Primavera	Brazil	1,800		7,000	postponed	McCully
Serra de Mesa	Brazil	1,200		6,800	uc:1998	McCully
Luang Prabang	Laos	1,410	110	6,600	planned	McCully
Jingping 1-Y	China			5,800	planned	McCully
Nam Tha 1	Laos	230	265	5,700	planned	McCully
Sambor	Cambodia	3,300	880	5,100	planned	McCully
Upper Mazaruni	Guyana		2,590	5,000	planned	Nachowitz
Nam Theun 2	Laos	681	450	4,500	planned	McCully
Nam Ngum 3	Laos	400	58.7	4,400	planned	McCully
Casecna(n)	Philippines	156	36	4,000–20,000	uc	Cernea, McCully
Nam Ngum 2	Laos	320	110	4,000	planned	McCully
Buan Kuop	Vietnam	81		3,600	planned	McCully
Itati-Itacora	Argentina	1,000		3,000	planned	McCully
Ji-Parana	Brazil	612		2,700	planned	McCully
Muela (Lesotho Highlands Water Project)	Lesotho	72		2,700[a]	uc:2003	McCully

continued

TABLE 15.   CONTINUED

Dam	Country	Installed Capacity (MW)	Area of Reservoir (km^2)	Number of People Displaced	Date	Source
Ban Koum	Laos, Thailand	2,330	130	2,600	planned	McCully
Nam Choan	Thailand	580	147	2,000	postponed	McCully
Urra I	Colombia	340	70	2,000	uc	McCully
Pak Beng	Laos	1,230	110	1,700	planned	McCully
Sayaburi	Laos	1,260	30	1,700	planned	McCully
Garafiri	Guinea	75	88	1,500	planned	McCully
Nam Ngiep 2	Laos	440	160	1,400	planned	McCully
Caruachi	Venezuela	2,076	238	1,000	uc	McCully
Pak Mun	Thailand	136		1,000	uc	WRR 4(6)4
Ghazi Barotha	Pakistan	1,450	26.4	900	uc:2,000	McCully
Ralco	Chile	570	34	700	planned	McCully
Nam Mang 3	Laos	30	14	675	planned	McCully
Rio 12 de Outubro	Brazil	12		347	planned	WRR 6(2)4
Lam Takhong	Thailand	250	45	225	uc	McCully
Berke	Turkey	510	7.8	140	uc	McCully

Notes: (1) uc = under construction. (2) A range of data is included when different sources offered different numbers. (3) [a] These numbers are people who will have their land or livelihood disrupted, not people who will be displaced.

Table 16. Desalination Capacity by Country (January 1, 1996)

Description

Desalination is considered an alternative source for fresh water in wealthy coastal regions. The data shown here on total installed capacity by country include land-based desalting plants rated at more than 500 cubic meters per day and delivered or under construction as of January 1, 1996. Six of the ten countries with the greatest desalination capacity are in the Middle East, as is 60 percent of total global capacity.

Limitations

These data were collected from a wide range of sources, from desalting plant suppliers to plant operators, and therefore depend on the accuracy of the information supplied. Plants with capacities less than 500 cubic meters per day are not included.

Source

Wangnick, K., 1996, *1996 IDA Worldwide Desalting Plants Inventory, Report No. 14*, Wangnick Consulting, Gnarrenburg, Germany. With permission.

TABLE **16**. DESALINATION CAPACITY BY COUNTRY (JANUARY 1, 1996)

Country	Total Capacity (m³/day)	Country	Total Capacity (m³/day)
Saudi Arabia	5,006,194	Venezuela	19,629
United States	2,799,000	Cuba	18,926
United Arab Emirates	2,134,233	Lebanon	17,083
Kuwait	1,284,327	Cayman Islands	16,986
Libya	638,377	Maldives	16,940
Japan	637,900	Argentina	15,960
Qatar	560,764	Austria	14,540
Spain	492,824	Malaysia	13,699
Italy	483,668	Bermudas	13,171
Iran	423,427	Azerbaijan	12,680
Iraq	324,476	Belarus	12,640
Bahrain	282,955	Czech Republic	11,085
Korea	265,957	Cabo Verde	10,500
Netherland Antilles	210,905	French Antigua	10,400
Algeria	190,837	Colombia	7,165
Hong Kong	183,079	Jordan	7,131
Oman	180,621	Sahara	7,002
Kazakhstan	167,379	Cyprus	6,275
Malta	145,031	Jamaica	6,094
Singapore	133,695	Nigeria	6,000
Russia	116,140	Denmark	5,960
India	115,509	Portugal	5,920
Holland	110,438	Philippines	5,648
Mexico	105,146	Syria	5,488
Indonesia	103,244	Pakistan	4,560
Egypt	102,051	Mauritania	4,440
Great Britain	101,397	Ecuador	4,433
Taiwan	101,180	Belgium	3,900
Israel	90,378	Ireland	2,725
Chile	83,509	Marshall Island	2,650
Australia	82,129	Switzerland	2,506
South Africa	79,531	Yugoslavia	2,204
Virgin Islands St. Croix	71,940	Sudan	1,450
Tunisia	47,402	Acsension	1,362
Virgin Islands St. Thomas	46,807	Bulgaria	1,320
Turmenistan	43,707	Sweden	1,300
Bahamas	37,474	Norway	1,200
Yemen	36,996	Nauru Pacific	1,136
Canada	35,629	Dominican Republic	1,135
Greece	35,620	Namibia	1,090
Virgin Islands Tortola	31,702	Brazil	1,079
Uzbekistan	31,200	Paraguay	1,000
France	29,112	Belize	757
Antigua	28,533	Virgin Islands Handsome Bay	681
Peru	24,538	Virgin Islands Road Town	681
Thailand	24,075	Honduras	651
Ukraine	21,000	Turks and Caicos	640
Poland	20,564	Turkey	600
Gibralter	20,079	Virgin Islands St. John	568
Morocco	19,700	Hungary	500

Note: Only desalting plants rated at 500 cubic meters per day or more are included.

Table 17. Desalination Capacity by Process (January 1, 1996)

Description

Several processes are used for desalinating water. These processes, and their installed capacity, are listed here. Both multistage flash and reverse osmosis continue to dominate the desalination field. Total desalting capacity is given here and is broken down by type of process. The data include land-based desalting plants rated at more than 500 cubic meters per day and in operation as of January 1, 1996.

Limitations

These data were collected from a wide range of sources, from desalting plant suppliers to plant operators, and therefore depend on the accuracy of the information supplied. Plants with capacities less than 500 cubic meters per day are not included.

Source

Wangnick, K., 1996, *1996 IDA Worldwide Desalting Plants Inventory, Report No. 14*, Wangnick Consulting, Gnarrenburg, Germany. With permission.

TABLE 17.   DESALINATION CAPACITY BY PROCESS (JANUARY 1, 1996)

Process	Total Capacity (m³/day)
Multistage flash	8,615,950
Reverse osmosis	6,612,381
Electrodialysis	1,019,521
Multi-effect evaporation	762,948
Vapor compression	684,494
Membrane softening	103,249
Other	46,353
Hybrid	36,991
TOTAL	17,881,888

Table 18. Threatened Reptiles, Amphibians, and Freshwater Fish, 1997

Description

Human manipulation of aquatic systems can have adverse effects on aquatic species. Dams, irrigation systems, and other major engineering projects disrupt habitat and impinge on feeding and breeding habits. Changes in water quality, including pH and temperature, also impact species viability, as do fishing and the introduction of exotic species. Reptile, amphibian, and freshwater fish species known to be threatened are shown here by country. Species included are those that are critically endangered, endangered, or vulnerable according to the World Conservation Union's definition. Critically endangered species face an "extremely high risk of extinction in the wild in the immediate future"; endangered species face a "very high risk of extinction in the near future"; and vulnerable species face a "high risk of extinction in the wild in the medium-term future." Information on total number of species and endemic species is included when available.

Limitations

Reporting of threatened species is not consistent because of problems with definitions, lack of adequate research, and inadequate reporting. No data are available for many countries. These are marked with an "X" in the table. Differences in the number of species listed from country to country often reflect differences in funding and research rather than the true state of aquatic species' well-being. Aquatic invertebrates, waterfowl, and aquatic plant species are not included.

Sources

World Conservation Monitoring Centre, 1997, World Conservation Monitoring Centre searchable databases at *http://www.wcmc.org.uk/*.

World Resources Institute, 1996, *World Resources 1996–97,* A Joint Publication of the World Resources Institute, United Nations Environment Programme, United Nations Development Programme, and the World Bank, Oxford University Press, New York.

TABLE 18.   THREATENED REPTILES, AMPHIBIANS, AND FRESHWATER FISH BY COUNTRY, 1997

	Reptiles			Amphibians			Freshwater Fish	
	All Species	Endemic Species	Threatened/ Endangered Species	All Species	Endemic Species	Threatened/ Endangered Species	All Species	Threatened/ Endangered Species
Africa								
Algeria	X	3	1	X	0	0	X	1
Angola	X	18	5	X	22	0	X	0
Benin	X	1	2	X	0	0	X	0
Botswana	157	2	0	38	0	0	92	0
Burkina Faso	X	3	1	X	0	0	X	0
Burundi	X	0	0	X	2	0	X	0
Cameroon	X	19	3	X	66	0	X	24
Cape Verde	X	X	3	X	X	0	X	1
Central African Republic	X	0	1	X	0	0	X	0
Chad	X	1	1	X	0	0	X	0
Comoros	X	X	2	X	X	0	X	1
Congo	X	1	2	X	1	0	X	0
Congo, Democratic Republic of (formerly Zaire)	X	33	3	X	53	0	X	1
Côte d'Ivoire	X	2	4	X	3	1	X	0
Djibouti	X	X	2	X	X	0	X	0
Egypt	83	0	6	6	0	0	70	0
Equatorial Guinea	X	3	2	X	2	1	X	0
Eritrea	X	0	3	X	0	0	X	0
Ethiopia	X	6	1	X	32	0	X	0
Gabon	X	3	3	X	4	0	X	0
Gambia	X	1	1	X	0	0	79	1
Ghana	X	1	4	X	4	0	X	0
Guinea	X	3	18	X	3	2	X	14
Guinea-Bissau	X	2	3	X	1	0	X	1
Kenya	187	15	5	88	11	0	X	18
Lesotho	X	2	0	X	1	0	8	1
Liberia	62	2	3	38	4	1	X	0
Libya	X	1	3	X	0	0	X	0
Madagascar	252	197	12	144	143	2	40	13
Malawi	124	6	0	69	3	0	X	0
Mali	16	2	1	X	1	0	X	0
Mauritania	X	1	3	X	0	0	X	0
Mauritius	11	8	6	0	0	0	X	0
Morocco	X	8	2	X	1	0	X	2
Mozambique	X	5	5	62	1	0	X	2
Namibia	X	26	3	32	1	1	102	3
Niger	X	0	5	X	0	0	X	0
Nigeria	>135	7	4	>109	1	0	260	0
Reunion	X	X	4	X	X	0	X	0
Rwanda	X	1	0	X	0	0	X	0
Sao Tome and Principe	X	X	2	X	X	0	X	0
Senegal	X	1	7	X	1	0	83	1
Seychelles	X	X	4	X	X	4	X	0
Sierra Leone	X	1	3	X	2	0	X	0
Somalia	193	48	2	27	3	0	X	3
South Africa	299	81	19	95	45	9	94	27
Sudan	X	6	3	X	1	0	X	0
Swaziland	102	1	0	40	0	0	40	0
Tanzania	245	56	4	121	43	0	X	18
Togo	X	1	3	X	3	0	X	0
Tunisia	X	1	2	X	0	0	X	0
Uganda	149	2	2	50	1	0	291	29
Western Sahara	X	X	2	X	X	0	X	0

continued

TABLE 18.   CONTINUED

	Reptiles			Amphibians			Freshwater Fish	
	All Species	Endemic Species	Threatened/ Endangered Species	All Species	Endemic Species	Threatened/ Endangered Species	All Species	Threatened/ Endangered Species
Zambia	X	2	0	93	1	0	X	0
Zimbabwe	153	2	0	120	3	0	112	0
North and Central America and Caribbean								
Anguilla	X	X	5	X	X	0	X	0
Antigua and Barbuda	X	X	5	X	X	0	X	0
Aruba	X	X	2	X	X	0	X	0
Bahamas	X	X	14	X	X	0	X	1
Barbados	X	X	2	X	X	0	X	0
Belize	107	2	5	32	1	0	63	4
British Virgin Islands	X	X	0	X	X	0	X	0
Canada	41	0	3	41	0	1	177	13
Cayman Islands	X	X	4	X	X	0	X	0
Costa Rica	214	36	7	162	34	1	130	0
Cuba	102	80	8	41	42	0	28	4
Dominican Republic	105	22	11	35	15	1	16	0
Dominica	X	X	15	X	X	1	X	0
El Salvador	73	4	6	23	0	0	16	1
Grenada	X	X	4	X	X	0	X	0
Guadeloupe	X	X	6	X	X	0	X	0
Guatemala	231	18	9	99	28	0	220	1
Haiti	102	29	7	46	23	1	16	0
Honduras	152	12	7	56	16	0	46	1
Jamaica	36	26	8	21	21	4	6	0
Martinique	X	X	5	X	X	0	X	0
Mexico	687	368	18	285	179	3	384	88
Montserrat	X	X	5	X	X	0	X	0
Netherlands Antilles	X	X	6	X	X	0	X	0
Nicaragua	161	5	7	59	2	0	50	1
Panama	226	25	7	164	21	0	84	2
Puerto Rico	X	X	12	X	X	0	X	3
St. Kitts	X	X	1	X	X	0	X	7
St. Lucia	X	X	6	X	X	0	X	0
St. Vincent	X	X	4	X	X	0	X	0
Trinidad/Tobago	70	2	5	26	2	0	76	0
Turks/Caicos Islands	X	X	5	X	X	0	X	0
United States of America	280	71	28	233	146	25	822	124
United States Virgin Islands	X	X	0	X	X	0	X	0
South America								
Argentina	220	64	0	145	45	2	410	1
Bolivia	208	17	3	112	26	0	389	0
Brazil	468	177	15	502	349	5	X	13
Chile	72	33	1	4i	30	3	44	4
Colombia	584	106	15	585	208	0	X	6
Ecuador	374	114	21	402	160	0	706	0
Falkland Islands (Malvinas)	X	X	0	X	X	0	X	0
French Guiana	X	X	8	X	X	0	X	0
Guyana	X	2	8	X	13	0	X	0
Paraguay	120	3	3	85	3	0	X	0
Peru	298	95	9	315	122	1	X	0
Suriname	151	0	6	95	8	0	300	0
Uruguay	X	1	0	X	4	0	X	0
Venezuela	259	57	14	199	116	0	X	5
Asia								
Afghanistan	103	4	1	6	1	1	84	0
Armenia	46	1	1	6	0	0	X	0

continued

TABLE 18.   CONTINUED

	Reptiles			Amphibians			Freshwater Fish	
	All Species	Endemic Species	Threatened/ Endangered Species	All Species	Endemic Species	Threatened/ Endangered Species	All Species	Threatened/ Endangered Species
Azerbaijan	52	0	0	8	0	0	X	1
Bahrain	X	X	14	X	X	0	X	1
Bangladesh	119	1	13	19	0	0	X	0
Bhutan	19	2	1	24	0	0	X	0
Brunei Darus	X	X	4	X	X	0	X	2
Cambodia	82	1	9	28	0	0	>215	5
China	340	74	15	263	154	1	686	29
Cyprus	X	X	4	X	X	0	X	0
East Timor	X	X	0	X	X	0	X	0
Georgia	46	0	8	11	0	0	X	4
Hong Kong	X	X	1	X	X	0	X	0
India	389	185	18	197	120	3	X	4
Indonesia	511	298	19	270	109	0	X	61
Iran	164	26	8	11	5	2	269	12
Iraq	81	1	2	6	0	0	X	2
Israel	X	1	5	X	0	0	26	0
Japan	66	27	7	52	41	8	186	9
Jordan	X	0	5	X	0	0	26	0
Kazakhstan	37	0	1	10	0	1	X	11
Korea, Democratic People's Republic	19	1	X	14	0	X	X	X
Korea, Republic of	25	3	X	14	2	X	130	X
Kuwait	29	0	2	2	0	0	X	0
Kyrgyzstan	X	X	1	X	X	0	X	0
Laos	66	1	7	37	1	0	244	4
Lebanon	X	2	2	X	0	0	X	0
Malaysia	268	68	14	158	57	0	449	14
Maldives	X	X	2	X	X	0	X	0
Mongolia	21	0	0	8	0	0	70	0
Myanmar (Burma)	203	38	20	75	10	0	X	1
Nepal	80	3	5	36	9	0	120	0
Oman	64	9	4	X	0	0	3	3
Pakistan	172	21	6	17	3	0	156	1
Philippines	190	153	7	63	55	2	X	22
Qatar	X	X	2	X	X	0	X	0
Saudi Arabia	84	4	2	X	0	0	8	0
Singapore	X	0	1	X	0	0	73	1
Sri Lanka	144	75	8	39	21	0	65	9
Syria	X	X	3	X	X	0	X	0
Taiwan	X	X	3	X	X	0	X	6
Tajikistan	38	0	1	2	0	0	X	1
Thailand	298	35	16	107	17	0	>600	15
Turkey	102	4	13	18	2	2	>152	21
Turkmenistan	80	2	2	2	0	0	X	8
United Arab Emirates	37	1	2	X	0	0	73	1
Uzbekistan	51	0	0	2	0	0	X	3
Vietnam	180	39	12	80	27	1	X	3
Yemen	77	31	2	X	1	0	5	0
Europe								
Albania	31	0	1	13	0	0	39	7
Andorra	X	X	X	X	X	X	X	X
Austria	14	0	2	20	0	0	60	3
Belarus	8	0	0	10	0	0	X	0
Belgium	8	0	0	17	0	0	X	1
Bosnia and Herzegovina	X	0	1	X	0	0	X	6
Bulgaria	33	0	1	17	0	0	X	9
Channel Islands	X	X	X	X	X	X	X	X
Croatia	X	0	0	X	0	1	X	21

continued

TABLE 18. CONTINUED

	Reptiles			Amphibians			Freshwater Fish	
	All Species	Endemic Species	Threatened/ Endangered Species	All Species	Endemic Species	Threatened/ Endangered Species	All Species	Threatened/ Endangered Species
Czech Republic	X	0	0	X	X	0	X	6
Denmark	5	0	0	14	0	0	41	2
Estonia	5	0	0	11	0	0	30	1
Faeroe Islands	X	X	0	X	X	0	X	0
Finland	5	0	0	5	0	0	66	1
France	32	0	3	32	3	2	53	3
Germany	12	0	0	20	0	0	71	7
Gibraltar	X	X	0	X	X	0	X	0
Greece	51	3	7	15	0	1	98	16
Holy See	X	X	0	X	X	0	X	0
Hungary	15	0	3	17	0	0	X	17
Iceland	0	0	0	0	0	0	7	0
Ireland	1	0	0	3	0	6	25	1
Isle of Man	X	X	X	X	X	X	X	X
Italy	40	0	5	34	7	6	45	9
Latvia	7	0	0	13	0	0	109	1
Liechtenstein	X	X	0	X	X	0	X	0
Lithuania	7	0	0	13	0	0	X	1
Luxembourg	X	X	0	X	X	0	X	0
Macedonia, TFYR	X	0	1	X	0	0	X	4
Malta	X	X	0	X	X	0	X	0
Moldova, Republic of	9	0	1	13	0	0	82	9
Netherlands	7	0	6	16	0	0	X	1
Norway	5	0	0	5	0	0	X	1
Poland	9	0	0	18	0	0	X	2
Portugal	29	2	0	17	0	1	28	9
Romania	25	0	3	19	0	0	87	17
Russian Federation	X	X	6	X	X	0	X	29
San Marino	X	X	0	X	X	0	X	0
Slovakia	X	X	0	X	X	0	X	9
Slovenia	21	0	0	X	0	1	X	6
Spain	53	9	6	X	0	3	98	10
Sweden	6	0	0	25	2	0	50	1
Switzerland	14	0	0	13	0	1	X	4
Ukraine	19	0	2	18	0	0	48	17
United Kingdom	8	0	0	7	0	0	36	1
Yugoslavia	X	X	1	X	X	0	X	18
Oceania								
American Samoa	X	X	2	X	X	0	X	0
Australia	748	596	16	205	188	18	216	16
Cook Islands	X	X	2	X	X	0	X	0
Fiji	25	11	6	2	2	1	X	3
French Polynesia	X	X	2	X	X	0	X	3
Guam	X	X	2	X	X	0	X	1
Kiribati	X	X	2	X	X	0	X	0
Marshall Islands	X	X	2	X	X	0	X	0
Micronesia	X	X	2	X	X	0	X	0
Nauru	X	X	0	X	X	0	X	0
New Caledonia	X	X	3	X	X	0	X	0
New Zealand	40	36	11	3	3	1	29	8
Niue	X	X	1	X	X	0	X	0
Northern Mariana Islands	X	X	2	X	X	0	X	0
Palau	X	X	2	X	X	0	X	0
Papua New Guinea	280	77	10	197	115	0	282	13
Pitcairn	X	X	0	X	X	0	X	0
Samoa	X	X	X	X	X	X	X	X

continued

TABLE 18.   CONTINUED

	Reptiles			Amphibians			Freshwater Fish	
	All Species	Endemic Species	Threatened/ Endangered Species	All Species	Endemic Species	Threatened/ Endangered Species	All Species	Threatened/ Endangered Species
Solomon Islands	61	10	4	17	9	X	X	X
Tokelau	X	X	2	X	X	0	X	0
Tonga	X	X	3	X	X	0	X	0
Tuvalu	X	X	2	X	X	0	X	0
Vanuatu	X	X	3	X	X	0	X	0
Wallis and Futuna Islands	X	X	0	X	X	0	X	0
Western Samoa	X	X	2	X	X	0	X	0

Note: An "X" indicates that no data are available.

Table 19. Irrigated Area by Country and Region, 1961 to 1994

Description

Total irrigated areas by country and continental region are listed here for 1961, 1965, 1970, 1975, 1980, 1985, 1990, and 1994—the latest year for which reliable data are available. Units are thousands of hectares. The rate of change over each time period, and an average annual rate of change, are also given.

Limitations

These data depend on in-country surveys, national reports, and estimates by the Food and Agriculture Organization. In some regions, multiple cropping may increase the apparent area in production. These data are not reported here. No differentiation is made about the quality of the land in production.

Source

Food and Agriculture Organization, 1997, web site at *www.fao.org*.

TABLE 19.   IRRIGATED AREA BY COUNTRY, 1961 TO 1994 (THOUSAND HECTARES)

Country and Region	1961	1965	1970	1975	1980	1985	1990	1994
Africa	7,364	7,747	8,426	8,937	9,407	10,229	11,121	12,202
Algeria	229	233	238	244	253	338	384	555
Angola	75	75	75	75	75	75	75	75
Benin	0	2	2	4	5	6	6	10
Botswana	1	2	1	1	2	2	2	1
Burkina Faso	2	2	4	8	10	12	20	24
Burundi	3	5	5	5	10	14	14	14
Cameroon	2	4	7	10	14	21	21	21
Cape Verde	2	2	2	2	2	2	3	3
Chad	5	5	5	6	6	10	14	14
Congo	0	0	1	2	1	1	1	1
Congo, Democratic Republic of (formerly Zaire)				0	7	9	10	11
Côte d'Ivoire	4	6	20	34	44	54	66	73
Djibouti	1	1	1	1	1	1	1	1
Egypt	2,568	2,672	2,843	2,825	2,445	2,497	2,648	3,265
Eritrea								28
Ethiopia	150	150	155	158	160	162	162	190
Gabon	4	4	4	4	4	4	4	4
Gambia	1	1	1	1	1	1	1	2
Ghana	0	0	7	7	7	7	6	6
Guinea	20	20	50	50	90	90	90	93
Buinea-Bissau	17	17	17	17	17	17	17	17
Kenya	14	14	29	40	40	42	54	67
Lesotho	3	3	3	3	3	3	3	3
Liberia	0	0	2	2	2	2	2	2
Libya	121	130	175	200	225	300	470	470
Madagascar	300	330	330	465	645	826	1,000	1,087
Malawi	1	1	4	13	18	18	20	28
Mali	60	60	61	60	60	60	78	82
Mauritania	20	20	30	30	49	49	49	49
Mauritius	8	12	15	15	16	17	17	18
Morocco	875	895	920	1,060	1,217	1,245	1,258	1,258
Mozambique	8	16	26	40	65	93	105	107
Namibia	4	4	4	4	4	4	4	6
Niger	16	16	18	18	23	30	66	66
Nigeria	200	200	200	200	200	200	230	233
Reunion	3	5	5	5	5	8	11	12
Rwanda	4	4	4	4	4	4	4	4
Sao Tome and Principe	10	10	10	10	10	10	10	10
Senegal	70	85	78	78	62	90	94	71
Sierra Leone	1	2	6	13	20	28	28	29
Somalia	90	90	95	100	125	180	180	200
South Africa	808	890	1,000	1,017	1,128	1,128	1,290	1,270
Sudan	1,480	1,550	1,625	1,700	1,800	1,946	1,946	1,946
Swaziland	36	40	47	56	58	62	67	67
Tanzania	20	28	38	52	120	127	144	170

continued

TABLE 19. CONTINUED

Country and Region	1961	1965	1970	1975	1980	1985	1990	1994
Togo	2	2	4	6	6	7	7	7
Tunisia	100	100	200	200	243	300	300	352
Uganda	2	3	4	4	6	9	9	9
Zambia	2	2	9	18	19	28	30	46
Zimbabwe	22	34	46	70	80	90	100	125
North and Central America	17,959	19,535	20,952	22,856	27,697	27,605	29,069	30,126
Barbados	1	1	1	1	1	1	1	1
Belize	0	0	1	1	1	2	2	3
Canada	350	380	421	500	596	748	718	710
Costa Rica	26	26	26	36	61	110	118	126
Cuba	230	330	450	580	762	861	900	910
Dominican Republic	110	115	125	140	165	198	225	250
El Salvador	18	20	20	33	110	110	120	120
Guadeloupe	1	1	2	1	2	2	2	2
Guatamala	32	43	56	72	87	102	117	125
Haiti	35	40	60	70	70	70	75	85
Honduras	50	66	66	70	72	72	74	74
Jamaica	22	24	24	32	33	33	33	33
Martinique	1	1	1	2	5	4	4	3
Mexico	3,000	3,200	3,583	4,479	4,980	5,285	5,600	6,100
Nicaragua	18	18	40	67	80	83	85	88
Panama	14	18	20	23	28	30	31	32
Puerto Rico	39	39	39	39	39	39	39	39
St. Lucia	1	1	1	1	1	1	2	2
St. Vincent	0	1	1	1	1	1	1	1
Trindad and Tobago	11	11	15	18	21	22	22	22
United States	14,000	15,200	16,000	16,690	20,582	19,831	20,900	21,400
South America	4,521	4,892	5,468	6,194	7,079	7,594	8,733	9,638
Argentina	980	1,110	1,280	1,440	1,580	1,620	1,680	1,700
Bolivia	72	75	80	120	140	125	110	100
Brazil	490	610	796	1,100	1,600	2,100	2,700	3,000
Chile	1,075	1,100	1,180	1,242	1,255	1,257	1,265	1,265
Colombia	226	235	250	300	400	465	680	1,037
Ecuador	440	450	470	506	500	300	290	250
French Guiana	1	1	1	1	1	2	2	2
Guyana	90	109	115	120	125	127	130	130
Paraguay	30	30	40	55	60	65	67	67
Peru	1,016	1,060	1,106	1,130	1,160	1,210	1,450	1,700
Suriname	14	15	28	33	42	55	59	60
Uruguay	27	35	52	57	79	97	120	140
Venezuela	60	62	70	90	137	171	180	187
Asia	90,166	97,073	109,446	121,165	132,199	140,792	153,623	174,298
Afghanistan	2,160	2,260	2,340	2,430	2,505	2,586	3,000	2,800

continued

TABLE 19. CONTINUED

Country and Region	1961	1965	1970	1975	1980	1985	1990	1994
Armenia								287
Azerbaijan								1,000
Bahrain	1	1	1	1	1	1	2	3
Bangladesh	426	572	1,058	1,441	1,569	2,073	2,936	3,288
Bhutan	8	10	18	22	26	30	39	39
Brunei Darsm				0	1	1	1	1
Cambodia	62	100	89	89	100	130	160	173
China	30,402	33,579	38,113	42,776	45,467	44,581	47,965	49,368
Cyprus	30	30	30	30	30	30	36	40
Gaza Strip	8	8	9	10	10	11	11	12
Georgia								469
Hong Kong	9	8	8	6	3	3	2	2
India	24,685	26,510	30,440	33,730	38,478	41,779	45,144	50,100
Indonesia	3,900	3,900	3,900	3,900	4,301	4,300	4,410	4,581
Iran	4,700	4,900	5,200	5,900	4,948	6,800	7,000	7,264
Iraq	1,250	1,350	1,480	1,567	1,750	1,750	3,525	3,525
Israel	136	151	172	180	203	233	206	193
Japan	2,940	2,943	3,415	3,171	3,055	2,952	2,846	2,737
Jordan	31	32	34	36	37	48	63	73
Kazakstan								2,300
Korea, Democratic People's Republic	500	500	500	900	1,120	1,270	1,420	1,460
Korea, Republic of	1,150	1,199	1,184	1,277	1,307	1,325	1,345	1,335
Kuwait	0	0	1	1	1	2	3	5
Kyrgyzstan								1,050
Laos	12	13	17	40	115	119	130	155
Lebanon	41	61	68	86	86	86	86	88
Malaysia	228	236	262	308	320	334	335	340
Mongolia	5	5	10	23	35	60	77	80
Myanmar (Burma)	536	753	839	976	999	1,085	1,005	1,336
Nepal	70	86	117	230	520	760	900	885
Oman	20	23	29	34	38	41	58	62
Pakistan	10,751	11,472	12,950	13,630	14,680	15,760	16,940	17,200
Philippines	690	730	826	1,040	1,219	1,440	1,560	1,580
Qatar	1	1	1	1	3	5	6	13
Saudi Arabia	343	353	365	375	600	800	900	1,473
Sri Lanka	335	341	465	480	525	583	520	550
Syria	558	522	451	516	539	652	693	1,082
Tajikistan								718
Thailand	1,621	1,768	1,960	2,419	3,015	3,822	4,238	4,590
Turkey	1,310	1,400	1,800	2,200	2,700	3,200	3,800	4,186
Turkmenistan								1,300
United Arab Emirates	30	35	45	50	53	58	63	67
Uzbekistan								4,000
Vietnam	1,000	980	980	1,000	1,542	1,770	1,840	1,998
West Bank	10	10	9	8	9	10	10	9
Yemen	207	231	260	282	289	302	348	481

continued

TABLE 19. CONTINUED

Country and Region	1961	1965	1970	1975	1980	1985	1990	1994
Europe	8,324	9,225	10,351	12,287	13,967	15,438	16,726	25,158
Albania	156	205	284	331	371	399	423	340
Austria	4	4	4	4	4	4	4	4
Bel-Lux	1	1	1	1	1	1	1	1
Belarus								124
Bulgaria	720	945	1,001	1,128	1,197	1,229	1,263	800
Bosnia and Herzegovinia								2
Croatia								2
Czechoslovakia	108	116	126	136	123	187	282	24
Denmark	40	65	90	180	391	410	430	465
Finland	2	7	16	40	60	62	64	64
France	360	440	539	680	870	1,050	1,300	1,500
Germany	321	390	419	448	460	470	482	475
Greece	430	576	730	875	961	1,099	1,195	1,327
Hungary	133	100	109	156	134	138	204	210
Italy	2,400	2,400	2,400	2,400	2,400	2,425	2,711	2,710
Macedonia								70
Malta	1	1	1	1	1	1	1	1
Moldova, Republic of								310
Netherlands	290	330	380	430	480	530	555	565
Norway	18	25	30	40	74	90	97	97
Poland	295	275	213	231	100	100	100	100
Portugal	620	621	622	625	630	630	630	631
Russia								5,367
Romania	206	230	731	1,474	2,301	2,956	3,109	3,104
Slovakia								299
Slovenia								2
Spain	1,950	2,226	2,379	2,818	3,029	3,217	3,402	3,657
Sweden	20	22	33	45	70	99	114	115
Switzerland	20	23	25	25	25	25	25	25
United Kingdom	108	105	88	86	140	152	164	108
Ukraine								2,591
Yugoslav SFR	121	118	130	133	145	164	170	68[a]
Former Soviet Union	9,400	9,900	11,100	14,500	17,200	19,689	20,800	—[b]
Oceania	1,079	1,368	1,588	1,620	1,684	1,957	2,113	2,605
Australia	1,001	1,274	1,476	1,469	1,500	1,700	1,832	2,317
Fiji	1	1	1	1	1	1	1	3
New Zealand	77	93	111	150	183	256	280	285
TOTAL	138,813	149,740	167,331	187,559	209,233	223,304	242,185	254,027
RATE OF CHANGE		0.08	0.12	0.12	0.12	0.07	0.08	0.05
ANNUAL CHANGE (%)		1.97	2.35	2.42	2.31	1.35	1.69	1.22

[a]Data from Yugoslavia are now split among several independent states.

[b]Data for the former Soviet Union in 1994 are split among the separate independent states, now included in Asia and Europe.

Index